AF517334

TECHNOLOGICAL INNOVATIONS IN INTEGRATED PEST MANAGEMENT

Biorational and Ecological Perspective

By

Dharam P Abrol

Published by
SCIENTIFIC PUBLISHERS (INDIA)

Jodhpur –
5 A, New Pali Road
P.O. Box 91
Jodhpur - 342 001 INDIA

Delhi –
4806/24, Ansari Road
Daryaganj
New Delhi - 110 002 INDIA

ISBN: 978-93-86102-82-9

Visit the Scientific Publishers (India) website at
http://www.scientificpub.com

Digitally Printed at Replika Press Pvt. Ltd.

Printed in India

FOREWORD

M.S. SWAMINATHAN RESEARCH FOUNDATION

M.S. Swaminathan
Founder Chairman
Ex-Member of Parliament (Rajya Sabha)

The world is facing food deficit as the growing human population disproportionately to available resources is a major challenge facing agriculture in the 21st century. The human population is expected to exceed 9 billion in mid-century doubling demand for food, feed and fodder. The excessive use of pesticides and intensification in crop production practices has encouraged the pest incidence, development of insecticide resistant populations, resurgence of pest populations, elevation of secondary pests to a status of primary importance, deleterious effects on populations of non-target organisms, pollution of the environment, air water, soil and food. These limitations necessitated a change in the concept and practice of pest management strategies relying on pesticide management to a system approach based primarily on biological knowledge of pests and their interaction with crops.

Evidently, the need was felt to develop new pest management systems that are long-term, cost-effective, solve unmet needs, and protect human and environmental health. Conventional, chemically based pest management strategies encourage short-term solutions that can be harmful to the environment and to human health. Broad-spectrum chemicals also are ineffective against some pest problems. Agricultural practices that augment natural processes that suppress pests, where available, will replace existing practices that disrupt natural processes; and these practices will be supplemented with the judicious use of biological control organisms and products, target-specific chemical pesticides, and pest-resistant plants. There should be a minimum disturbance to the existing agri-horticultural ecosystem through the use of chemical pesticides. The present book **'Technological Innovations In Integrated Pest Management- Biorational and Ecological Perspective'** by **Professor Dharam P. Abrol** is an attempt to cover integrated pest management from multidisciplinary, and multi-faceted components providing a unique overview of the efforts made to develop IPM strategies focusing on the conservation of natural enemies in various ecologically undisturbed habitats wherein the traditional agriculture is still being practiced. The purpose of this book is to present a balanced overview of environmentally safe and ecologically sound approaches applied to commonly used practices in insect pest management. I congratulate the contributors and editor for this excellent book. I hope the book will be widely read and utilized by policy makers, scientists and research scholars and farmers.

M S Swaminathan

3rd Cross Road, Taramani Institutional Area, Chennai (Madras) - 600 113, India
Phone: +91-44-2254 2790, 2254 1698 Fax: +91-44-2254 1319
E-mail: founder@mssrf.res.in, swami@mssrf.res.in

PREFACE

Problems ranging from pesticide resistance in target species to environmental degradation and contamination of food products by pesticide residues have proven that reliance on single tactics seriously detracts from sustainability. Reducing the risks associated with pest management tactics like pesticides in plant protection is the challenge of pest control for the twenty-first century in view of the fast growing human population. Therefore, one of the goals would be in the deployment of ecologically based integrated pest management practice. The idea is to shift paradigm from focusing on pest management strategies relying on pesticide management to a system approach relying primarily on biological knowledge of pests and their interaction with crops. Evidently, the need was felt to develop new pest-management systems that are long-term, cost-effective, solve unmet needs, and protect human and environmental health. Conventional, chemically based pest-management strategies encourage short-term solutions that can be harmful to the environment and to human health. Broad-spectrum chemicals also are ineffective against some pest problems. Agricultural practices that augment natural processes that suppress pests, where available, will replace existing practices that disrupt natural processes; and these practices will be supplemented with the judicious use of biological-control organisms and products, target-specific chemical pesticides, and pest-resistant plants. There should be a minimum disturbance to the existing agri-horticultural ecosystem through the use of chemical pesticides. Therefore, the present book focuses on the conservation of natural enemies in various ecologically undisturbed habitats wherein the traditional agriculture is still being practiced.

This book provides multidisciplinary perspective about the different facets of IPM and shall serve as reference book for students, teachers, researchers, extension functionaries and policy planners. This book is the outcome of my personal experiences and the contributions of several workers which have been incorporated. All the contributors deserve special appreciation for writing chapters in their respective fields in great depth with dedication. I thank university authorities for the excellent working atmosphere and needed encouragement for compiling such a voluminous book. *Dr. Uma Shankar* and *Dr. Debjyoti Chaterjee* needs special mention for their help in various ways. My sincere thanks are due to my wife Professor *Dr. Asha Abrol*, daughter *Er. Vitasta* and son *Er. Rajat* for their endurance and help while writing this book.

Place: Jammu **Dharam P. Abrol**

Date : 15 March, 2017

CONTRIBUTORS

D. P. Abrol
Dean Faculty of Agriculture,
Sher-e-Kashmir University of
Agricultural Sciences & Technology,
Faculty of Agriculture
Chatha, Jammu – 180009, J&K, India.
E-mail: dharam_abrol@rediffmail.com

Hafeez Ahmad
Division of Entomology,
Sher-e-Kashmir University of
Agricultural Sciences & Technology,
Faculty of Agriculture,
Chatha –180009, Jammu, J&K, India.
E-mail: hafeezskuastj786@gmail.com

Kamlesh Bali
Division of Sericulture,
Sher-e- Kashmir University of
Agricultural Sciences & Technology,
Faculty of Agriculture,
Udheywalla, Jammu.
E-mail: balikamlesh76@yahoo.com

A. K. Bhowmick
Department of Entomology,
JNKVV, Jabalpur, Madhya Pradesh.
E-mail: dr.bhowmick@gmail.com

Sudhakar Dwivedi
Division of Agricultural Economics,
Agri-Business Management,
Sher-e-Kashmir University of
Agricultural Sciences and Technology
of Jammu,
Chatha, Jammu-180009.
E-mail: dwivedi.sudhakar@gmail.com

Suheel Ahmad Ganie
Division of Entomology,
Sher-e-Kashmir University of
Agricultural Sciences & Technology,
Faculty of Agriculture,
Chatha –180009, Jammu, J&K, India.
E-mail: suhail_jalib@yahoo.com

R. K. Gupta
Division of Entomology,
Sher-e-Kashmir University of
Agricultural Sciences & Technology,
Faculty of Agriculture,
Chatha Jammu – 180009, J&K, India.
E-mail: rkguptaentoskuast@gmail.com

S. K. Gupta
Division of Plant Breeding & Genetics,
Sher-e-Kashmir University of
Agricultural Sciences & Technology,
Faculty of Agriculture,
Chatha –180009, Jammu, J&K, India.
E-mail: guptaskpbg@rediffmail.com

Vishal Gupta
Division of Plant Pathology,
Faculty of Agriculture,
Sher-e-Kashmir University of
Agriculture Sciences and Technology
of Jammu, Jammu.
E-mail: vishalgupta@rediffmail.com

C. S. Kalha
Division of Plant Pathology,
Faculty of Agriculture,
Sher-e-Kashmir University of
Agriculture Sciences and Technology
of Jammu, Jammu.
E-mail: kalhacs@gmail.com

Virender Kaul
Division of Entomology,
Sher-e-Kashmir University of
Agricultural Sciences & Technology,
Faculty of Agriculture,
Chatha –180009, Jammu, J&K, India.
E-mail: kaulvirendra@gmail.com

S. K. Kher
Division of Agricultural Extension
Education,
Faculty of Agriculture,
Sher-e-Kashmir University of
Agricultural Sciences and Technology
of Jammu,
Chatha, Jammu-180009.
E-mail: shivan16kher@gmail.com

Rakesh Kumar
Division of Agricultural Extension Education,
Faculty of Agriculture,
Sher-e-Kashmir University of Agricultural Sciences and Technology of Jammu,
Chatha-Jammu-180009.
E-mail: sharmar1975@gmail.com

R. K. Patel
ICAR-National Research Cente on Litchi,
Musahari, Muzaffarpur (842002)
E-mail: patelrk5@yahoo.com

Vijay K. Razdan
Division of Plant Pathology,
Faculty of Agriculture,
Sher-e-Kashmir University of Agriculture Sciences and Technology of Jammu, Jammu.
E-mail: vijayrazdan@rediffmail.com

Karandeep Singh Sasan
Division of Entomology,
Sher-e-Kashmir University of Agricultural Sciences & Technology,
Faculty of Agriculture,
Chatha –180009, Jammu, J&K, India.
E-mail: karandeep@yahoo.com

Uma Shankar
Division of Entomology,
Sher-e-Kashmir University of Agricultural Sciences & Technology,
Faculty of Agriculture,
Chatha –180009, Jammu, J&K, India.
E-mail: umashankar.ento@gmail.com

Devinder Sharma
Division of Entomology,
Sher-e-Kashmir University of Agricultural Sciences & Technology,
Faculty of Agriculture,
Chatha –180009, Jammu, J&K, India.
E-mail: devskuast@gmail.com

Magdeshwar Sharma
Mega Seed Project,
Sher-e- Kashmir University of Agricultural Sciences & Technology,
Faculty of Agriculture,
Chatha, Jammu–180009, J&K, India.
E-mail: magdeshwars@gmail.com

Amit K. Singh
Division of Entomology,
Sher-e-Kashmir University of Agricultural Sciences & Technology,
Faculty of Agriculture
Chatha –180009, Jammu, J&K, India.
E-mail: amitskuast@gmail.com

P. S. Slathia
Division of Agricultural Extension Education,
Faculty of Agriculture,
Sher-e-Kashmir University of Agricultural Sciences and Technology of Jammu
Chatha-Jammu-180009.
E-mail: slathia2007@gmail.com

Ajay K. Sood
Department of Entomology, CSK Himachal Pradesh Krishi Vishvavidyalaya, Palampur (H.P.)
E-mail: sood_pau@yahoo.co.in

Kuldeep Srivastava
ICAR-National Research Cente on Litchi,
Musahari, Muzaffarpur (842002)
E-mail: kuldeep.ipm@gmail.com

Deepika Sud
CSK HPKV,
Krishi Vigyan Kendra, Kangra,
Himachal Pradesh.
E-mail: deepika_agri@rediffmail.com

CONTENTS

ECOLOGICAL BASED PEST MANAGEMENT

1

D. P. Abrol

1.0 INTRODUCTION

Plant pests have been a part of the man's environment, since ages. Traditional agriculture which included diverse crops and indigenous crop management practices provided little opportunity for the pests to build up and spread rapidly (Smith and Allen, 1954). Crop production is constantly threatened by increasing difficulties in controlling insect pests, which are a major impediment to improved livelihoods of the poor farmers. However, in the quest for increasing food production for the growing population, use of pesticides became a norm in crop management. Pesticides became an integral part of the process owing to losses from weeds, diseases and pests reducing the harvestable produce. Use of pesticides along with high yielding varieties, irrigation and improved seed did result in increased productivity. But the rampant use of these chemicals, under the adage, "if little is good, a lot more will be better" has played havoc with human and other life forms.

During 80's, the negative effects of pesticides were already being seen - many pest species were becoming resistant (thus needing heavier dosages and more frequent applications), secondary pests were emerging, the number of acute poisonings was very high, and the overall impact of pesticide use in the ecosystem was becoming apparent. This clearly showed the need for "alternative methods" and for a "different pest management system" for worldwide agriculture. Farmers needed to look at pests and diseases as part of the ecosystem, and focus on prevention through the diversification of production systems. A major component of managing pests and diseases was the relationship between pest incidence, soil fertility and soil organic matter content. It was increasingly realized that poor soils not only result in poor plant growth but also increase plant vulnerability to pest attacks.

During late 90s, Integrated Pest Management (IPM) developed into a comprehensive methodology, based on farmers' better understanding of their own agro-ecosystems. IPM first considered the integration of control methods and target pests as a combination of biological control methods, host plant resistance, cultural control and selective chemical control (Knipling, 1979, Reuveni, 1995). When it was realized that many agricultural practices influence pest incidence, farm management was also added to the equation as an important part of an IPM approach, together with management of natural resources (Altieri, 1994, Bartlett, 1956, 1964, van den Bosch and Stern, 1962, Metcalf and Luckmann, 1994, Bottrell, 1979). Gradually, social aspects were also included which included pest and disease management, the role of local organizations and to the importance of indigenous knowledge in the process of developing the necessary skills and confidence to make ecologically sound and cost-effective decisions on crop health.

In the last 20 years, farmers have been able to stabilize and increase their yields and incomes. Many examples have shown a reduced reliance on pesticides and their use. Lobbying and advocacy actions have led to new rules and regulations in many countries, forbidding the commercialization and use of the most toxic products, and at the same time greatly increasing the awareness of farmers and consumers. However, the problems still persist. Also, these are not limited to large-scale farmers or to intensive agriculture situations. Majority of the farming communities are still grappling with chemical options.

Although reliable statistics are difficult to find, the impact of pests and diseases in worldwide agriculture is considerable, lowering yields and overall production, resulting in losses that are equivalent to millions of dollars (Dahlberg, 1979). Agricultural production in India is limited by many factors like soil, water, genetic potential of the crop and the organisms that feed on or compete with food plants. Food crops the world over are damaged by more than 10,000 species of insects and the overall estimated yield losses from different pests are reported to be US$

500 billion globally and Rs 100,000 crores in India alone. About 42.1% of attainable production is lost due to pest attack; however, if no control measures are adopted, the figure would be 69.8%. These losses are not only unpredictable but are also greater in fragile ecosystems on which majority of the small and marginal farmers are dependant.

Furthermore, the most common and easy access to chemical pesticides has only made problems worse. Farmer suicides driven by indiscriminate investments and use of chemical pesticides have become common in countries as different as India and Ecuador. Pesticide poisoning is a common story, having reached endemic proportions in many countries. The first report of poisoning owing to pesticides in India came from Kerala in 1958, where, over 100 people died after consuming wheat flour contaminated with parathion. Farmers and labourers who are regularly in contact with pesticides suffer from severe health problems, seriously affecting them and their families. Indiscriminate use of pesticides has also resulted in the contamination of soils and groundwater, leading to the disappearance of fish and birds. Because of the lack of regulations, or of the difficulties in effectively enforcing them, these problems are greater in the rural areas of the non industrialized world.

1.2 Ecological Pest Management

IPM has had many successes, but the magnitude of the problem which farmers currently face forces us to look once again at pests and diseases and at their management. Pests and diseases are not an isolated part of agriculture, but rather a symptom of a broader problem, and need thus to be seen –and managed– accordingly (Levins and Wilson, 1980). There is a need for an ecological based pest management when addressing the presence of pests and diseases in agriculture (Turner, 1992, Higley and Pedigo, 1996). Building on the wide range of experiences gained with IPM during the last two decades, an Ecological Pest Management (EPM) approach is one which focuses on managing pests as one component of a larger ecosystem (National Research Council 1996). As such, EPM is based on:

1.2.1 Minimising the disturbances which are caused by agriculture

A successful pest management strategy is based on mimicking nature, redesigning a farm so that it resembles a complex ecosystem. This will mean maximizing a farm's positive ecological processes (such as nitrogen fixation, nutrient mineralization); while at the same time, minimizing undesirable processes such as nutrient loss or erosion. In many cases, it may be convenient to reduce tillage and thus achieve minimal soil disturbance; in many others, it will be necessary to include perennial species and enhance a farm's overall diversity. The interaction of different species can have interesting results, contributing to the system's overall resilience. Undisturbed ecosystems include a host of plants, birds, animals – ensuring a balance between the crop-pest-predator relationships. Abundance of natural predators minimizes the need for specific plant protection measures.

1.2.2 Decreasing plants' vulnerability

Field observations show that pests prefer to attack plants under stress. Current studies, such as those falling under the theory known as trophobiosis, talk of a plant's "internal balance", directly related to its nutritional state. The best way to prevent the attack of pests and diseases is thus by providing a healthy and balanced environment and food supply. There are many factors, which can affect a plant's internal balance and thus lessen or increase its susceptibility to pest and disease attacks. These are related to the plant itself (such as adaptation to the local climate or its age) or to the environment (climate, light, temperature, humidity, wind). Plant vulnerability is also related to the different management practices, which regularly take place in a farm, such as spacing, tilling, pruning, or the time of planting.

1.2.3 Understanding pests and acting accordingly

In pest and disease management, one of the main considerations is the way the pest species behave - some show abilities to reproduce often and disperse widely, others are able to withstand competition or adverse conditions. To "know your enemy" is thus a key strategy in every pest management approach. This knowledge needs then to be translated into action, considering, for example, the release of beneficial insects at a particular moment, adding bird nesting sites to a farm, or changing the sowing time of certain crops.

These principles are clearly visible in many traditional low input agricultural systems, where ecological principles form the basis for all pest management strategies. Some of the examples to keep pests away are: raising companion crops emitting strong odour and use of botanical insect repellants. In brief, these refer to working with nature, and not against it. In agricultural terms, this means growing plants in the right soil, at the right time; nourishing the soil and relying on a system's biodiversity as a natural means to safeguard the whole system's health. Traditional wisdom is being maintained by many societies, while it is also being recreated in many "modern" farms. Managing a farm, however, and relying on its ecological processes, requires a thorough understanding of how these work. Therefore EPM is, above all, based on farmers' skills, abilities and knowledge.

1.3 Building knowledge

If EPM is based on farmers' understanding of their ecosystem and of the processes taking place in it, then training, education and knowledge building processes are essential. Many different participatory learning approaches for promoting sustainable agriculture have been developed, most of which work towards improving farmer decision making capacities and stimulating local innovations. Through Farmer Field Schools, farmers are trained to make an analysis of their agro ecosystem. In this way, they become aware of the pest predator balance and of the damaging effect of pesticides on such balance. FFSs are being popularized by NGOs and governments, on a small as well as on a large scale. They are learner-centred, field-based, experiential learning methodologies.

They involve observation, analysis, assessment and experimentation over a time period sufficient to understand the key agro-ecological and socio-ecological relationships. These group learning processes empower farmers in decision-making and taking up leadership roles, ultimately resulting in building social capital. FFS methodology is being used extensively by incorporating changes as necessary – to suit the local conditions and needs; enlarging its scope from IPM to holistic farm management; from training tool to a scaling up tool.

1.4 Challenges

Ecological pest management is about bringing the balance back to disturbed ecosystems; it is also about learning to observe such balances (Geier and Clark, 1961). The enormous impact which pests and diseases have on today's agriculture forces us to pay special attention to this issue in terms of resource management, cost-benefit analysis, impact on health and the environment. Our main challenge is therefore, to keep the collective learning spirit that has been built by approaches like the FFS alive. There are no standard recipes or solutions available. We need to continue to strive to enrich this learning methodology building on the wide range of experiences and wisdom. The challenge is to apply holistically the agro ecological and socio-economic relationships through knowledge empowering methodologies.

Keeping in view the guiding princlples in the selection of pest-control methods, it is generally realized that any single approach to the problem of insect control is not feasible (Stern, 1966, 1973). Each method has its own advantages as well as its disadvantages and also has limitations under a particular set of conditions. The current thinking all over the world is to suitably synthesize the use of as many methods as possible to control the insect pests effectively, economically and without any adverse after-effect. Progress in pest control technologies has contributed to the improved yield and quality of food, fiber, and ornamental crops that have occurred during the twentieth century. However, the development and widespread adoption of some pest-control technologies did not occur without environmental impacts and societal concerns about food safety. Integrated pest management (IPM) arose in the second half of the century as the paradigm of choice for pest control, and stressed the need to incorporate basic ecological concepts in the design and implementation of pest control systems. Integrated pest management requires detailed understanding of pest biology and ecology, including interactions at the community and ecosystems levels. The ultimate goal of IPM is to achieve complete integration of management tactics, breaking the traditional barriers often imposed by the pest disciplines. In reality, the application and realization of IPM varies among cropping systems, with some agroecosystems being managed in a more integrated manner than are others. The future of IPM is promising, with opportunities and challenges ahead.

Crop production is constantly threatened by increasing difficulties in controlling insect pests, which are a major impediment to improved livelihoods of the poor farmers. Of the several methods available to control these pests, conventional insecticides are still relied on by a vast majority of the farmers (Kogan, 1986, Kogan and McGrath, 1993). However, an 'insecticide treadmill' situation is developing due to the use of poor application equipment, poor choice of insecticides, and the development of resistance to insecticides (Carson, 1962). In addition to huge direct economic costs, the indirect costs such as deleterious effects of pesticides on the environment and human health are becoming increasingly severe all over the world. Therefore, there is a need to have a critical look at the adverse effects of pesticides in the environment, and develop appropriate technologies and guidelines for judicious application of pesticides for crop production. Several natural plant products, biopesticides (such as entomopathogenic bacteria, fungi, protozoa, NPVs, and nematodes), parasites, and predators have been proposed as a safer alternative to the synthetic pesticides. However, there is little information on the bio-safety of these products to human beings and the environment. These products have not been subjected to the same rigorous bio-safety testing as the synthetic chemicals. Therefore, there is a need to look at the bio-safety of these biological methods in the context of environmental safety and sustainable crop production. Significant progress has been made over the past three decades in the introduction of exotic genes into genetically modified organisms for pest management. The recombinant DNA technology has helped to extend the range of conventional genetic manipulation to meet the increasing need for food and fiber. Genetically engineered organisms have been deregulated for use in pest management for increasing crop production in several countries. Genetic engineering for improving the effectiveness of crop protection technologies offers an environmental friendly method of crop protection, and has resulted in a drastic reduction in pesticide sprays, and in increased crop production. The products of genetic engineering are not conceptually different than those derived through conventional technologies. Crop cultivars derived through conventional breeding have not been subjected to bio-safety assessment. The potential of recombinant technologies to allow a greater modification than is possible with the conventional technologies has raised some concerns, and therefore, genetically modified organisms have been subjected to rigorous bio-safety assessment. Procedures employed for assessment of bio-safety of transgenic crops to the non-target organisms, coupled with the interpretation and utilization of such information are important components for sustainable deployment of transgenic crops for food and environmental safety in the future.

Therefore, there is a need to have a critical look at the procedures employed for assessing the bio-safety of synthetic pesticides, biopesticides, and the transgenic crops, their impact on the environment, and frame appropriate procedures and guidelines for testing and deployment of these technologies for sustainable crop production and food and environmental safety in future

There is a need to develop ecologically based new pest-management systems (EBPM) that are long-term, cost-effective, solve unmet needs, and protect human and environmental health (Kadir and Barlow, 1992). Future pest-management systems will be based on a broad knowledge of the agroecosystem and will seek to manage rather than eliminate pests. Agricultural practices that augment natural processes that suppress pests, where available, will replace existing practices that disrupt natural processes; and these practices will be supplemented with the judicious use of biological-control organisms and products, target-specific chemical pesticides, and pest-resistant plants. It will also be necessary to reopen and develop channels of communication at all levels that increase the flow of information and cooperative action, subsequently lowering risk to users, fostering interdisciplinary interaction, and improving the profitability of alternative pest-control methods.

The fundamental goals of EBPM are (1) safety, (2) profitability, and (3) durability. Pest-management systems must be safe for growers and workers who use them, and for consumers of the food produced. Minimizing health risks must be a primary criterion of acceptability of novel management systems. Pest control strategies must be cost-effective as well as effective, easy to implement, and readily integrated with other crop-production practices. Economic factors involving cost-effectiveness include crop prices, costs and availability of labor, land, equipment, and other production inputs. Pest-management programs must ensure that pests in the agroecosystem can be managed over the long-term without adverse environmental, economic, or safety consequences. Current pest-management strategies that rely on repeated applications of conventional broad-spectrum pesticides encourage the development of resistant species. In new ecologically based approaches, addressing potential development of pest resistance will be important. EBPM promotes the economic and environmental viability of agriculture by using knowledge of interactions between crops, pests, and naturally occurring pest-control organisms to modify cropping systems in ways that reduce damage associated with pests. Ecologically based management relies on a comprehensive knowledge of the ecosystem, including the natural biological interactions that suppress pest populations. It is based on the recognition that many conventional agricultural practices disrupt natural processes that suppress pests. Agricultural practices recommended by EBPM will augment natural processes, supplemented by biological-control organisms and products, resistant plants, and targeted pesticides.

An ecosystem is dynamic with interacting physical, chemical, and biological processes. The coexisting crops, herbivores, predators, pathogens, weeds, and other organisms interact with one another and respond to their environment. Each organism has developed a repertoire of offensive and defensive maneuvers in response to changes in the behavior of other organisms in the cropping system. This web of interrelated interactions also confers stability on the system; while a population increases and decreases, it is subject to the checks and balances imposed by populations of the other organisms. The focus of this book is on the processes which are ecologically based, economically feasible and environmental friendly.

References

Altieri, M.A. 1994. *Biodiversity and Pest Management in Agroecosystems.* New York: Food Products. 185 pp.

Bartlett, B.R. 1956. Natural predators— can selective insecticides help preserve biotic control? *Agric. Chem.* 11:42–44, 107

Bartlett, B.R. 1964. Integration of chemical and biological control. In *Biological Control of Insect Pests and Weeds*, ed. P Debach, pp. 489–511. London: Chapman & Hall. 844 pp.

Bottrell, D.G. 1979. *Integrated Pest Management.* Washington, DC: Council Environ. Qual. US Gov. Print. Off. 120 pp.

Carson, R. 1962. *Silent Spring.* Boston, MA: Houghton Mifflin. 368 p

Dahlberg, K.A. 1979. Beyond the Green Revolution: The Ecology and Politics of Global *Agricultural Development.* New York: Plenum. 256 pp.

Higley, L.G and Pedigo, L.P, eds. 1996. Economic Thresholds for Integrated Pest Management. Lincoln, NE: Univ. Nebraska Press. 328 pp.

Kadir AASA, Barlow HS, eds. 1992. *Pest Management and the Environment in 2000.* Wallingford, UK: CAB Int. 401 pp.

Knipling, E.F. 1979. *The Basic Principles of Insect Population Suppression and Management.* Washington, DC: USDA, Agric. Handb. 512. 659 pp.

Kogan, M, ed. 1986. *Ecological Theory and Integrated Pest Management Practice.* New York: Wiley. 362 pp.

Kogan, M and McGrath , D. 1993. Integrated pest management: present dilemmas and future challenges. In *Anais 14_Congr. Bras. Entomol.*, pp. 1–16. Piracicaba, SP, Brazil: Soc. Entomol. Bras. 124 pp.

Levins, R. Land Wilson, M. 1980. Ecological theory and pest management. *Annu. Rev. Entomol.* 25:287–308

Metcalf, R. Land Luckmann,W.H, eds. 1994. *Introduction to Insect Pest Management.* NewYork: Wiley. 650 pp. 3rd ed. (1st ed., 1975)

National Research Council. 1996. Ecologically Based Pest Management: New Solutions for a New Century. Washington, DC. National Academy Press.

Reuveni R, ed. 1995. *Novel Approaches to Integrated Pest Management.* Boca Raton, FL: CRC Press. 369 pp.

Smith, R.F and Allen, W.W. 1954. Insect control and the balance of nature. *Sci. Am.* 190(6):38–92

Stern, V.M., Smith, R.F, van den Bosch, R and Hagen, K.S. 1959. The integrated control concept. *Hilgardia* 29:81–101

Stern VM. 1966. Significance of the economic threshold in integrated pest control. Proc. FAO Symposium on Integrated Control 2: 41-56

Stern VM. 1973. Economic thresholds. *Annu. Rev. Entomol.* 18:259–80

Turner, C. E. 1992. Biological approaches to weed management. Pp. 32-67 in Beyond Pesticides: Biological Approaches to Pest Management in California, T. Beall, ed. Oakland, Calif.: University of California.

van den Bosch R and Stern VM. 1962. The integration of chemical and biological control of arthropod pests. *Annu. Rev. Entomol.* 7:367–86

2

ROLE OF BIOTIC STRESS IN INTEGRATED PEST MANAGEMENT

Magdeshwar Sharma

2.0 INTRODUCTION

Agricultural sector in India has long been recognized for its dependence on chemical control for the management of biotic stresses (insect, diseases and weeds). The increasing population often demands more and more food grain production. The crop yields in farms are generally low and there are wide gaps between the farmer's yield and the potential yields of several crops.

Biotic stress, a loss of fitness caused to an individual by other organisms acting as parasites, places a major constraint on plant growth. Parasitism is a strategy adopted by a large number of species – perhaps the majority of species on Earth (Windsor, 1998) and plants have co-evolved with their parasites for several hundred million years. It affects metabolism of crops and livestock to reduce the expression of genes that are responsible for high yield of crop commodities and productivity traits in animals influencing the national agricultural productivity. Various processes of the invasion of biotic stresses in crops are genetically controlled by both pests (insect, mites, diseases and nematodes) and their hosts. Such pests along with weeds, vertebrate pests including rodents are major causes for crop yield reduction. Oerke *et al.* (1995) estimated about 42% loss in global output due to insect pests, diseases and weeds despite the use of plant protection options. The loss could have been up to 70% in the absence of plant protection. In India, the pre-harvest loss was up to 30% in cereals and pulses and it can be up to 50% in cotton and oil seeds crops (Dhaliwal and Arora, 1993). Annual Economic loss due to *Helicoverpa* alone was estimated at Rs. 2,000 crores despite the use of pesticides worth Rs. 500 crores (Pawar, 1998) (Table 2.1 & Table 2.2).

Table 2.1. Economically important pests of major crops in India (Ranga Rao *et al.*, 2009)

Crop	Common Name	Scientific name	ETLs	Existing control Methods
Cereals				
Rice	Stem borer	*Scirpophaga incertulus* Walker	5% white ears/ One egg mass sqm-1	IPM
	Brown plant hopper	*Nilaparvata lugens* Stal.	10 hoppers per clump.	IPM
	Gall Midge	*Orseolia oryzae* wood-mason	5-10% silver Shoots	Host plant resistance (HPR)
	Leaf folder	*Cnaphalocrocis medinalis* Guen	10-15% webbed Foliage	HPR
Wheat	Aphid	*Schizaphis graminum* (Rondani)	5-10% of plants with infestation	HPR
Termite				
Maize	Stem borer	*Chilo partellus* (Swinhoe)	5-10% infestation	Chemical
	Shoot fly	*Atherigona* spp.	5-10% dead Hearts	Chemical
	Earworm	*Helicoverpa armigera* Hubner	25-30% damage to cobs	Chemical
Legumes				
Pigeonpea	Pod borer	*Helicoverpa armigera* (Hubner)	5 eggs or 3 small larvae per plant	IPM
	Pod fly	*Melanagromyza obtusa* (malloach)	In all endemic Locations	Chemical
	Leaf webber	*Maruca vitrata* (Geyer)	5 webs per plant	Chemical
	Pod sucking Bugs	*Clavigralla gibbosa* Spinola	One egg mass per plant	Chemical
Chickpea	Pod borer	*Helicoverpa armigera* (Hubner)	3 eggs or 2 small larvae per plant	IPM
	Cutworm	*Agrotis ipsilon* (Hufnagel)	5% plant Mortality	Chemical
Soybean	Stem fly	*Ophiomyia phasioli* (Tryon)	5% plant Infestation	Chemical
	Girdle beetle	*Obereopsis brevis* (Swed)	5% incidence	Chemical
	Hairy caterpillar	*Spilosoma obliqua* (Walker)	5 larvae meter row	Chemical

Crop	Common Name	Scientific name	ETLs	Existing control Methods
Oil Seeds				
Groundnut	Leaf miner	*Aproaerema midicella* Deventer	5 mines per plant at 30 days of crop age	IPM
	Tobacco Caterpillar	*Spodoptera litura* (Fab)	20-25% defoliation at 40 days	IPM
	Thrips	*Scirtothrips dorsalis* (Hood)	5 thrips/ terminal at seedling stage	Chemical
	Aphids	*Aphis craccivora* (Kouch)	5-10 aphids per terminal at seedling stage stage in dry spells only in rainy season	IPM
Sunflower	Gram pod Borer	*Helicoverpa armigera* Hubner	One larva per Head	Chemical
Sesame	Leaf webber	*Antigastra catalaunalis* Dub	2-5 webbs per Plant	Chemical
Rapeseed	Aphids	*Lipaphis erysimi* (Kalt)	5-10 aphids per Plant	Chemical
Vegetables				
Brinjal	Fruit and stem Borer	*Leucinodes orbanalis*	1-5% shoot/ fruit Infestation	IPM
Cabbage & Cauli-flower	Dimond back Moth	*Plutella xylostella* (Linn)	1-5% incidence	IPM
	Tobacco caterpillar	*Spodoptera litura* (Fab)	1-5% incidence	IPM
	Cabbage butterfly			
Tomato	Fruit worm	*Helicoverpa armigera* Hubner	1-5% fruit Damage	IPM
Fruits				
Apple	San Jose Scale	*Quadraspidiotus perniciosus* (Comstock)	Appearance of pest in 5% trees	Chemical & miscible oils
	Wooly aphis			
	Codling moth	*Cydia pomonella* (L.)	1-2% incidence	IPM
	Phytophag ous Mites	*Panonychus ulmi* (Koch)	5-10% foliage Infestation	Miscible oil & IPM

Crop	Common Name	Scientific name	ETLs	Existing control Methods
Grapes	Flea beetle	*Scelodonta stricollis* (Mots.)	20% foliar Damage	Chemical
	Thrips	*Retithrips syriacus* (Mayet)	5 thrips/young Leaf	Chemical
	Mealy bugs	*Maconellicoccus hirstutus* Green	1% bunch Infestation	Chemical
Oranges	Fruit flies	*Carpomyia vesuviana* costa.	1-2% incidence	Chemical
	Defoliators	*Papilio demoleus L.*	20-30 % foliar Damage	Chemical
Mango	Hopper	*Amritodes atkinsoni* (Leth).	2-5 hoppers per Inflorescence	Chemical
	Stem borer	*Batocera rufomaculatus* Deg	Appearance of the pest	Chemical
Cash Crops				
Cotton	American Bollworm	*Helicoverpa armigera* Hub.	5-10 % boll Infestation	IPM
	Pink bollworm	*Pectinophora gossipiella* Saund	5-10% boll infestation	IPM
	Whitefly	*Bemisia tabaci* (Genn).	8-10 adults/leaf	IPM
	Spoted Bollworm	*Earias insulana* (Boisd).	5-10% boll infestation	IPM
Sugarcane	Stem borer	*Chilo sacchariphagus indicus* (Kapur).	10% shoot damage	IPM
	Scale insect	*Melanapsis glomerata* (Green)	20-30% canes with scale incidence	IPM
Tobacco	Tobacco Caterpillar	*Spodoptera litura* Fab.	5-10% leaves with damage	IPM
Storage pests				
Cereals	Rice weevil	*Sytophilus oryzae*	Appearance of live insects	Chemical
	Paddy moth	*Sitotroga cerealella*	Appearance of adult moths	Chemical
	Rice moth	*Corcyra cephalonica*	Appearance of adult moths	Chemical
	Red flour Beetle	*Tribolium castaneum*	Appearance of adult beetles	Chemical
Pulses	Bruchids	*Bruchus sp. Laria* spp.	Appearance of adult insects	Chemical
Oil Seeds				
Groundnut	Groundnut Bruchid	*Caryedon serratus*	Appearance of adult beetles	Chemical

Table 2.2 . Economically important diseases of major crops in India (Ranga Rao *et. al.*, 2009)

Crop	Disease name	Causal organism	Existing control methods
Cereals			
Rice	Blast	*Pyricularia oryzae*	IDM
	Sheath blight	*Rhizoctonia solani*	IDM
	Bacterial leaf blight	*Xanthomonas oryzae*	IDM
Wheat	Leaf or brown rust	*Puccinia recondite* f.sp. tritici	HPR & IDM
	Stem or black rust	*Puccinia graminis* f.sp. tritici	HPR & IDM
	Karnal bunt	*Neovossia indica*	HPR & IDM
	Loose smut	*Ustilago segetum*	IDM
	Yellow rust		
Maize	Maydis leaf blight	*Cochliobolus heterostrophus*	HPR & Chemical
	Common rust	*Puccinia sorghi*	HPR & Chemical
	Downy mildew	*Peronosclerospora* sp	Chemical
	Fusarium wilt & stalk rot	*Fusarium moniliforme*	HPR
	Charcoal rot	*Macrophomina phaseolina*	HPR
Legumes			
Pigeon pea	Wilt	*Fusarium udum*	HPR
	Phytophthora blight	*Phytophthora drechsleri* f.sp. cajani	IDM
	Sterility mosaic	*Virus transmitted by Aceria cajani*	HPR
Crop	Disease name	*Causal organism*	Existing control methods
Chick pea	Wilt	*Fusarium oxysporum* f.sp. ciceri	HPR
	Dry root rot	*Rhizoctinia bataticola*	HPR
	Collar rot	Sclerotium rolfsii	HPR
	Ascochyta blight	*Ascochyta rabiei*	IDM
	Botrytis gray mold	*Botrytis cinerea*	IDM
	Stunt	Bean leaf roll virus	HPR
Soybean	Pod blight	*Colletotrichum dematium* f.sp. Truncate	Chemical & HPR
	Bacterial pustule	*Xanthomonas campestris*	HPR
	Bacterial Blight	*Pseudomonas* sp.	Cultural & HPR
	Charcoal rot	*Macrophomina phaseolina*	Cultural & HPR
	Collar rot	*Sclerotium rolfsii*	HPR

Crop	Disease name	Causal organism	Existing control methods
Oil Seeds			
Groundnut	Crown rot	*Aspergillus niger*	Chemical
	Stem & pod rots	*Sclerotium rolfsii*	HPR & cultural
	Aflatoxin	*Aspergillus flavus*	Integrated management
	Early leaf spot	*Cercospora arachidicola*	IDM
	Late leaf spot	*Phaeoisariopsis personata*	IDM
Sun-flowers	Gray mold	*Botrytis cinerea*	Chemical
	Alternaria blight	*Alternaria helianthi*	Chemical
	Wilt	*Verticillum dahliae*	HPR
	Scorch	*Maacrophomina phaseoli*	HPR
Sesame	Phytophthora blight	*Phytophthora parasitica*	Chemical
	Phyllody		
	Charcoal rot	*Macrophomina phaseolina*	HPR
	Wilt	*Fusarium oxysporum* f.sp. sesami	HPR
	Cercospora leaf spot	*Cercospora sesami*	HPR
	Alternaria leaf spot	*Alternaria sesami*	HPR
	Bacterial blight	*Xanthomonas campestris*	HPR
Rape-seed	Alternaria blight	*Alternaria brassicae*	HPR
	White rust	*Albugo candida*	
	Downy mildew	*Peronospora parasitica*	HPR
	Powdery mildew	*Erysiphe cruciferarum*	HPR
Vegetables			
Brinjal	Damping-off	*Phytophthora/Pythium* sp	Chemical
	Wilt	*Fusarium ozonium*	HPR
	Phomopsis blight	*Phomopsis vexans*	HPR
Cabbage	Downy mildew	*Perenospora parasitica*	Chemical
	Alternaria blight	*Alternaria solani*	Chemical
	Black rot	*Xanthomonas campestris*	Chemical
Cauli-flower	Stalk rot	*Sclerotinia sclerotiorum*	Chemical
Tomato	Late blight	*Phytophthora infestans*	Chemical
	Leaf blight	*Septoria lycopersici*	Chemical
Tomato	Spotted wilt	*Vial disease*	HPR + cultural
	Wilt	*Psuedomonas solanacearum*	HPR
Fruits			
Apple	Scab	*Venturia inaequalis*	HPR + Chemical

Crop	Disease name	Causal organism	Existing control methods
Grapes	Anthracnose	*Gloeosporium ampelophagum*	Chemical
	Downy mildew	*Plasmopara viticola*	Chemical
	Powdery mildew	*Uncinula necator*	Chemical
Oranges	Canker	*Xanthomonas campestris* pr. citri	Chemical
	Gummosis	*Diaporthe citri*	Chemical
Mango	Powdery mildew	*Oidium mangiferae*	Chemical
	Anthracnose	*Colletitrichum gloeosporiodes*	Chemical
Cash Crops			
Cotton	Verticillium wilt	*Verticillium dahliae*	HPR
	Root rot	*Rhizoctonia* sp	HPR
	Alternaria leaf spot	*Alternaria macrospora*	IDM
	Anthracnose	*Colletotrichum gossypii*	Chemical
Sugarcane	Red rot	*Colletotrichum falcatum*	HPR
	Smut	*Ustilago scitaminea*	HPR
	Wilt	*Fusarium sacchari*	HPR
Tobacco	Damping-off	*Pythium aphanidermatum*	Chemical
	Frog-eye spot	*Cercospora nicotianae*	Chemical

Losses to pests and disease in crop plants continue to pose a significant threat to agriculture and food security. In the past five decades there was a steady increase in the chemical utilization from 2.2 gm ha^{-1} of active ingredient (a.i.) in 1950 to the current level of 650 gm ha^{-1} which is a 300 fold increase (David, 1995). In recent years farmers' income is declining particularly due to increased cost of plant protection inputs. Among various pesticides, the use of insecticides was much in India compared to the global scenario (Verma, 1998). During the latter half of the 20th century, agriculture became increasingly reliant on synthetic chemical pesticides to provide control of pests and diseases, especially within the intensive farming systems common in the developed world. However, as we move into the 21st century, we recognise that this reliance on chemical control is becoming unsustainable. Pesticides tend to have a limited lifespan due to the emergence of resistance in the target pests, and are increasingly recognised in many cases to have unacceptable impacts on biodiversity, and on the health of agricultural workers and even consumers. As a consequence, there is a much greater focus being placed on using alternative approaches to crop protection, both by individual growers and national policy makers. The research in stress tolerance response in crops and animals has increasing importance, not only because agricultural production need to keep pace with increasing demand for agricultural produce from dwindling resources, but also due to possible changes in pest scenario with climate change that may make the environment much more hostile for agricultural production than what is today.

Since alternatives to excessive dependence on chemical pesticides led to the development of resistance in pests to pesticides, out breaks of secondary pests and pathogens/ biotypes, and occurrence of residues in food chain. To overcome such situations and minimize damage to human-and animal-health, several organizations have started advocating the concept of IPM with better profits. Besides damage to human health, total dependence on chemical pesticides has eliminated bio-diversity, resulting in the reduction of natural enemies. Though Indian plant protection in the modern age is making larger strides of progress, it is necessary to consider the treasure of ancient knowledge, particularly the use of safer pesticides/chemical pesticides rarely provide the same level of control, the use of a combination of approaches is required, tailored specifically for the particular crop and local conditions. IPM programmes are built on the concept of minimizing chemical use by maximizing natural controls, which might include a range of cultural practices, the use of genetic resources and the introduction or facilitation of biological control. Research has generated new technologies using naturally occurring enemies of insect pests (parasitoids, predators and pathogens) for use in IPM. Some important commercially available products include *Trichogramma, Bracons, Crysoperla carnea, Crytaemus montrouzieri, Bacillus thuringiensis, Bacillus sphaericus,* Nuclear polyhedrosis viruses (NPV) and *Trichoderma.* In addition, a number of plant products such as azadirachtin (neem), pyrethrum, nicotine, etc. are also valuable as biopesticides. In India, more than 160 natural enemies have been studied for their utilization against insect pests (Singh, 1997). Technologies have been standardized for multiplication of 26 egg parasitoids, 39 larval/nymphal parasitoids, 26 predators and 7 species of weed. Importantly, good IPM should not only be 'environmentally friendly', but should maximize protection with minimum economic input. IPM is grounded firmly in ecology, the original principle being that understanding agro-ecosystems provides the best way to control pest populations. It is clear that we need to be able to better exploit plants' wide range of natural resistance mechanisms to provide future solutions for use in IPM.

2.1 Types of biotic stress

The biotic stresses like weeds, insect, disease, nematodes, mites etc. are major /contributors to reduction of crop yield. Increasing industriali-zation could affect agricultural ecosystem by possible contamination of ground water and land with organic pollutant substances and heavy metals, in addition to possible bio-magnification through the vegetation of the region. Like-wise increasing ozone CO_2 and greenhouse gasses in the farming atmosphere are possible reasons for increasing pest pressure and change in pest species community patterns to agricultural and horticultural crops. These factors are changing the pest scenario and there are great shift in pest dynamics and cycles. Commercial agriculture has to sustain farming system approach to offer profitability. Ventures and investments in agriculture shall shoot up if mitigation of biotic stress is contained and reduce the loss of agricultural commodities to acceptable levels.

2.1.1 Biostress and Agriculture: Indian scenario

Indian agriculture went through the change-over during last century when the 'Green-revolution era was ushered with intensification of agricultural input use towards maximization of genetic yield of crops. The high yielding crop varieties / hybrids responded to increased agrochemical inputs such as chemical nutrients (fertilizers) along with the recommended farm yard manure quantities. However, this also triggered the upsurge in communities of herbivores from disease causing viruses, bacteria fungi, insects / mites and even big animals such as birds and other vertebrates (rodents, blue bull, wild boar etc.) that differentially exploited more nutritious crop tissues.

Metabolic stresses on crop plants as well as physical damage to commodities due to the invasion of insects, disease, nematodes, larger animals such as rodents, weeds, etc. are threat to both crop production and productivity as well as for securing the commodities in the country from any form of loss. Similar stresses are encountered in other sectors such as in taking up animal husbandry and fisheries. There is a perceived drag in productivity due to biotic stresses in agriculture all over the world, as threatful challenge to the sustainable production of agricultural commodities. Indian agriculture is spread across more than fourteen agro-climatic regions and under each there are several sub-regions that have typical weather patterns and edaphic factors that guide the agricultural activities. Farmers have to adjust their judgement, every season, to decide upon commencing sowing of crops that are permitted by the agro-ecology and edaphic situation of the parcel of land they have to till. Crop production of field and horticultural crops are ridden with a host of pests and diseases both in the field as well as in storage of pre-processed commodities. Such crop loss, amounting to almost 30% all over the country shall endanger both food and nutritional security that is often called upon as national requirements. Similarly, the loss due to biotic stresses to live stock and fisheries is also noticed to be over 15%.

Traditionally, focus was given in India, since colonial times, to undertake research on mitigation of arising biotic stresses in agriculture of every season. Although several research institutes of Indian Council of Agricultural research (ICAR), State Agricultural Universities (SAUs), and other agencies/organizations are working on biotic stresses, the efforts are far too inadequate considering the magnitude of the problem. Integrated Pest Management of crop pests was one of the toolbox that was developed to tackle pests and diseases in crops. The research of tackling the biotic stresses most of the times is on isolated lines, even though the multiple biotic stresses occur in conjunction. Moreover, during the last decade, several new tools in the areas of biotechnology, nanotechnology, remote sensing, information technology, and polymer sciences have become available for application in crop and animal health management. Therefore, it is of national importance to not only initiate high quality research programme with global standards in this important area, but also capture, synthesize, adapt, and apply the technological advancements taking place in scientific fields. Conjunctively, it is also indispensable to build up human resources through education and capacity building to address the all the emerging challenges of biotic stress management such as biosecurity and biosafety of agricultural systems.

2.2 Emerging issues in biotic stress *vis a vis* biosecurity/ biosafety

The emerging issues on biosafety and biosecurity in agriculture, particularly in the light of emerging new pests as well as possible introduction of new pests (organisms that cause diseases, insects, mites etc.) could bear economic burden on Indian agriculture. Futuristic research in the management on emerging new pests could address certain aspects of agricultural biosecurity of the country. As an umbrella body, this institution aspires to oversee both the funding and management of research in the areas of managing biotic stresses in agriculture.

Yet another aspect of increased consumer awareness on safety of agricultural commodities has been the perceived biosafety threat from the contamination of agro-chemicals, particularly those for pest management and nutrient supply to crops. Biotic stresses that emanated from intensive agriculture was primarily due to anomalous expansion of nutritious food source, and the crop husbandry practices with commercial scale production plan did upset the natural food chain and agri-biodiversity by selecting a few organisms in the agro-ecologies. In comparison to dryland and rainfed agro-ecosystems, irrigated crop production ecologies were more ravaged by various biotic stresses. Communities of life forms that got freedom natural pressures of multiplication in agri-land did make panic stricken farmers to resort to injudicious use of pesticides. The avoidable crop loss in India is estimated to be about 18-20% due to all the above types of pests. Upsurge in pestilence has been in recent times was due to weather aberrations too.

2.3 Biostress and IPM: Indian agriculture

Pesticides are chemical toxicants that are designed to check pest-build up when deployed at the identified dosage and time of application using the right applicator. However, in the excitement to enhance the production and commercial agriculture tended to ignore the natural balance of food chains and webs, the sudden spurt in pestilence became the order of the day. India adopted integrated pest management (IPM) as the policy drive to manage the various crop pests. The ideology of integration of all pest suppression tools in the IPM tool box brought in balanced approach to restore natural balance of food chains and webs in the agricultural farms. Today diseases such as potato late blight, rice blast and wheat rust are fairly contained due to intense focus on developing tolerant crop varieties and other management tools. Similarly, the extensive cultivation of GM cotton could reduce the pestilence due to pulse pod borer in pulses. Large-scale intensive break out of mites, aphids, jassids and other pests has been reduced due to combination of modern chemistry of pesticides that have environmental friendly formulation chemistry. Such pesticides can be therapeutic and could enable restoration of natural agro-ecological food webs. Hence the upsurge of diseases and other pests did get suppressed to farmers' advantage of reduced pestilence in farms. Metabolic stresses on crop plants as well as physical damage to commodities due to the invasion of insects, disease, nematodes, larger animals such as rodents, weeds, etc. are threat to both crop production and productivity as well as for securing the commodities in the country from any form of loss.

Indian agriculture is spread across more than fourteen agro-climatic regions and under each there are several sub-regions that have typical weather patterns and edaphic factors that guide the agricultural activities. Farmers have to adjust their judgement, every season, to decide upon commencing sowing of crops that are permitted by the agro-ecology and edaphic situation of the parcel of land they have to till. Crop production of field and horticultural crops are ridden with a host of pests and diseases both in the field as well as in storage of pre-processed commodities. Such crop loss, amounting to almost 30% all over the country shall endanger both food and nutritional security that is often called upon as national requirements. Similarly, the loss due to biotic stresses to live stock and fisheries is also noticed to be over 15%.

The biotic stresses that challenge agriculture enterprises are the major risks for sustaining agricultural productivity across all sectors. Before initiating biotic stress management at watershed level, one should take up in-depth farmer participatory appraisal (PRA) for diagnosis and categorize various biotic stresses to design appropriate management strategies. To achieve this, general PRA needs to be organized at each location and the results should be discussed with the group. The whole farming community needs to be involved at every level of decision making. The biotic stress at lases should be developed and updated at regular intervals. These atlases should be in a language that could be easily communicated to the farmers. Failure of the agricultural production process across seasons due to biotic stresses is addressed

A. to start with by garnering available knowledge in the field,

B. standardizing these for fitting under the given situation and

C. evolve suitable management strategy to be commercially viable.

Strategic priority areas for the 12th Plan Period of ICAR included

1. Durable resistance breeding research in major food crops, animals and plants.
2. Utilization of novel plant and animal defense mechanisms against biotic stresses as a part of IPM.
3. Diagnostics and vaccines for diseases of plants, animal and fish.
4. Use of plants, small animals and fish for production of high value biologicals.
5. Identification of genes and gene systems related to defense.
6. Isolation and characterization of biomolecules from biological sources from land and water for human, animal, plant and soil health applications.

References

David BV. 1995. *The pesticide industry.* Kothari's Desk book series. PP 464.

Dhaliwal GS and Arora R. 1993. Changing status of insect pests and their management strategies. *In:* Changing scenario of Indian agriculture (eds: Gill KS, Dhaliwal GS and Hansara BS) Commonwealth Publishers, New Delhi.

Oerke EC, Delhne HW, Schohnbeck F and Weber A. 1995. *Crop production and crop protection: Estimated losses in major food and cash crops.* Elsevier, Amsterdam.

Pawar C.S. 1998. Helicoverpa – *A national problem which needs a national policy and commitment for its management.* Pestology 22(7):51-57.

Ranga Rao GV, S Desai, OP Rupela, K Krishnappa and Suhas P Wani 2009. Integrated Pest Management Options for Better Crop Production, 155-75. *In* Proceedings of the Comprehensive Assessment of Watershed Programs in India, 25-27 July 2007. International Crops Research Institute for the Semi-Arid Tropics (ICRISAT) Patancheru 502 324, Andhra Pradesh, India.

Singh, S.P. 1997. Progress and problems in the use of biocontrol agents. *In:* The National Conference on Biopesticides. Ministry of Agriculture, Govt. of India, New Delhi, and All India Biotech Association, New Delhi. April 10, 1997.

Verma A. 1998. *Plant protection in India: Vision 2020.* Pesticide world, June-July 1998. pp. 28-33.

Windsor, D.A. 1998. Controversies in parasitology. Most of the species on Earth are parasites. *International Journal for Parasitology* 28: 1939-1941.

BIOLOGICAL CONTROL: CURRENT PERSPECTIVES AND LIMITATIONS IN HILL ECO-SYSTEM

3

Kamlesh Bali and R. K. Gupta

3.0 INTRODUCTION

Hill agriculture is largely traditional-characterized by a large number of smallholdings of not more than one ha per household. Crop production takes place under extremely variable agro-ecological conditions, with annual rainfall ranging from 300 to 850 mm. Farmers often select well-adapted, stable crop varieties, and cropping systems are such that two or more crops are grown in the same field at the same time. These diverse traditional systems enhance natural enemy abundance and generally keep pest numbers at low levels. Pest management practice in traditional agriculture is a built-in process in the overall crop production system rather than a separate well-defined activity. Increased population pressure and the resulting demand for increased crop production have necessitated agricultural expansion with the concomitant decline in the overall biodiversity. Increases in plant material movement in turn facilitated the accidental introduction of foreign pests. At present about two dozen arthropod pests, both introduced and native, are recognized as one of the major constraints to agricultural production and productivity.

Arthropod pests and weeds are major constraints to agriculture production throughout the world. Current management strategies based on Integrated Pest Management have three important goals *viz.*, safety, profitability and durability. Of the several tactics available in pest management, biological control holds greater promise for sustainable pest management. Our understanding of the role of natural enemies in the control of insect populations has evolved over time. Biological control forms part of the ecosystem services which operate on vast scales and have been recognised as irreplaceable.

What is biological control ?

It is the purposeful use of an organism or organisms to reduce a plant or animal population that is inimical to man. An undesirable organism may be eliminated locally or, more usually its numbers are suppressed to a level where it no longer causes economic damage (Samways, 1981). The concepts of population equilibrium and population regulation lie at the heart of biological control. The proponents of biological control considered it as a sub discipline of applied ecology. Ecologists consider biological control as a *Stabilizing process* of ecosystem services. For a crop protection specialist, biological control is a form of population management. To adequately practice it one should have a firm understanding of population and behavioural ecology. Biological control has traditionally often been seen as a vision. DeBach (1964) gave this wording of the vision:

'We would point out that people fortunate enough to have witnessed a striking example of biological control taking place usually become 'true believers', but some of those, who happen later to see only the final result can be unimpressed if not downright skeptical.'

Natural enemies are doing an important job for us, in both perennial habitats like forests and ephemeral habitats found in annual crops in agriculture. We must find ways to enhance their biological control potential so that they can hold pest populations at levels where no economic damage occurs. To do this we should seek ways to increase natural enemy numbers and enhance their predation rates. Some factors that have been shown to be important and warrant further investigation are the influence of habitat quality, landscape structure and management practices. We also need to know more about the composition of communities of natural enemies found in differently managed ecosystems and the interactions occurring between different species of enemies. Many natural occurring predators and parasitoids often play an important role in suppressing pest populations. Sometimes, however, this "natural" control does not limit pest numbers sufficiently to avoid damage of economic importance. To enhance "natural" control we need better knowledge of the factors that influence natural enemy efficacy. Pest dynamics in perennial habitats are often characterized by episodic outbreaks (Myers, 1987). The reason for these outbreaks is often unclear (Liebhold & Kamata, 2000) and two competing theories are invoked to explain insect outbreaks. One puts climatic factors at the forefront while the other suggests that biotic interactions such as natural enemy attack are most important. Recent work (Dwyer *et al.*, 2004) proposes that at low pest densities generalist enemies may hold pest populations at a low level and at high pest densities cycles may be driven by specialist parasitoids and pathogens. Complex dynamics are generated when effects of weather cause unpredictable shifts between low and high pest densities. Agricultural systems are often dominated by ephemeral habitats. Pests in agricultural fields are usually not permanent residents and will recolonize fields each season (Wissinger, 1997) making it difficult for pests and enemies to have a close relationship. Traits of natural enemies in disturbed habitats are not the same as those associated with natural enemies in perennial habitats (Wiedenmann & Smith, Jr., 1997). Annual crop systems are frequently disturbed, which can make it difficult for natural enemies to perform successful pest suppression. Therefore, for predators to occur in fields after disturbances, farmers are dependent on the predators' ability to recolonize. It is important that the predators maintain a high population in proximity to the field and have a high dispersal ability to reduce the time lag between recolonization of natural enemies and establishment of pest. If this is the case, a synchrony will be created between predator and prey, which will lead to more successful pest suppression (Wiedenmann & Smith, Jr., 1997).

3.1 Natural Control

In nature all populations of organisms (animal / plant) have restraints preventing their unlimited increase. Paul Debach (1964) distinguished natural control from biological control.

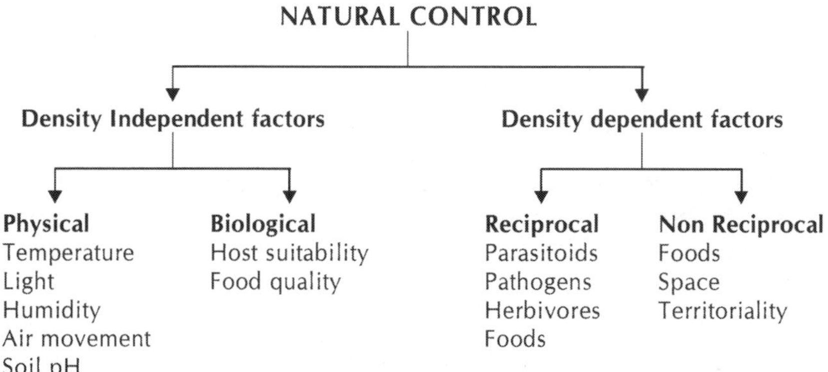

In this opinion, "natural control is the maintenance of a more or less fluctuating population density of an organism within certain definable upper and lower limits over a period of time by actions of abiotic and/ or biotic environmental factors."

Definitions on biological control

The term biological control has been used both in a fundamental ecological sense and utilitarian sense to designate a field of human endeavour. It can be considered as a phase of natural control or limitation imposed by man. It rests on the premise that the densities of many noxious species of plants and animals are amenable to control largely by their natural enemies. In 1919, A. Harrry Scott Smith used the term Biological Control to denote "the utilization of organisms for the control of population densities of animals and plants".

Jost M. Franz (1961) simplified it and defined it as "the active manipulation of antagonistic organisms by man to reduce pest population densities, both plant and animal to non-economically important levels". Paul DeBach (1964) defined it as "the action of parasitoids, predators and pathogens in maintaining other organism's density at a lower average than would occur in their absence". However, Johnson (2000) considers it as a science "that deals with the role natural enemies play in the regulation of the numbers of their hosts, especially as it applies to animals or plant pests".

Applied biological control

It is the deliberate pitting of beneficial organisms against harmful ones in required numbers so that the economic plant damage is prevented in crop pests management programmes. In applied biological control, physical factors may be of secondary concern (Paul Debach, 1964). In applied biological control a natural enemy (usually, parasitoid, predator, pathogen or weed killer) is introduced so that the population of the noxious pest is brought down to a desirable level (Fig 1).

The latest explanations provided by various authors in modern times consider biological control/suppression as prevention of a pest outbreak due to purposeful manipulation of natural enemies. It is very important to consider similar approaches to pest management are not biological control. Although natural control is very important in management of pests it is not biological control as there is no element of active manipulation by human. Plant products, toxins etc., derived from living autotrophic and heterotrophic organisms should not be considered as agents of biological control although they may complement biological control.

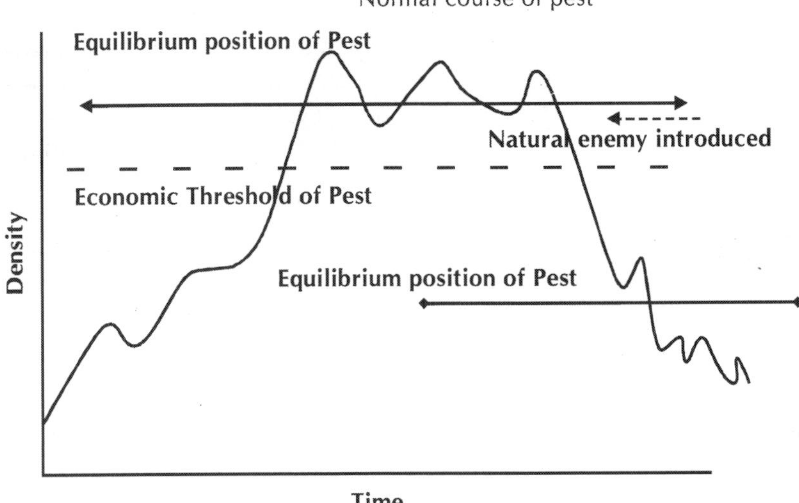

Fig 1. The concept of thresholds and applied biological control

Applicability of Biological control

It finds useful applications in several contexts. It extends from field, crops, horticultural and silvipastoral segments to post harvest stages. I summarise below the Wood-Mckenzie Report (1995) that recognizes three principal areas of applicability:

a. where the value of crop is too low
b. where over use of insecticides has created pesticide related problems so that other control methods become necessary
c. where biological control is not a part of already adequate pest management.

Objectives of biological control

In applied sense, it has three objectives *viz.*, reduction, prevention and delay of effects caused by the noxious pests.

i. *Reduction*

This approach is similar to the practice of chemical control. The use of biological control agent is done as and when the pest has crossed a certain threshold. The principal objective here is to reduce the level of pest population where it is no longer a pest problem at least for a long period of time in a cropping season (Fig. 1). Reduction is achieved by a single release/application of a biocontrol agent or periodic release or application of one or several categories of agents. Reduction is often a part of the biointensive pest management packages.

ii. *Prevention*

It requires early intervention before a potential pest builds up. Hence, the tactic here is to utilize natural enemies before the pest reaches a threshold that could be the damaging level (Fig. 2). This requires information of the pest is detail within the agroecosystem and the potential of the natural enemy that is to be exploited.

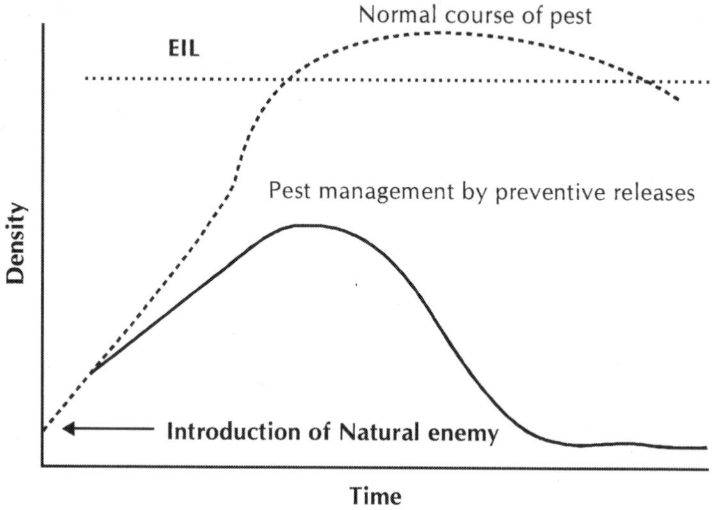

Fig 2. Hypothetical example of prevention

iii. *Delay*

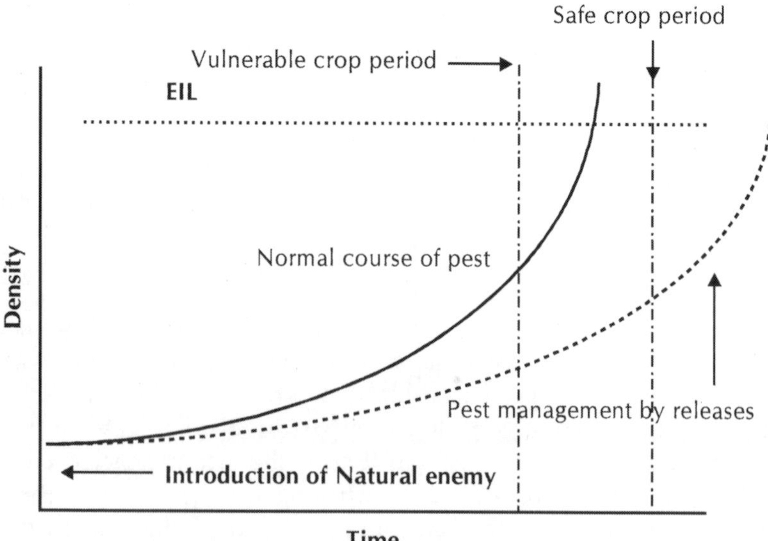

Fig 3. Hypothetical example of delay

This objective is similar to prevention. However, the pest populations will eventually buildup to a level that is very high, but it occurs at a time when the species is no longer considered a pest (Fig. 3).

Biological control can provide lasting highly selective and effective control. In practice, effective biological control may require several tactics to be combined to achieve effective pest management. Earlier concepts in biological control relied more on the classical approaches.

Van Driesche *et al.* (1996) noticed 100-1000 fold reduction in pest densities. De Bach and Rosen (1991) noted that of 4226 species of natural enemies introduced world-wide 1251 species have become established and the rates of success was 38%. As the targets for biological control is diverse and spread across taxonomically different groups, the opportunities are also wide-open.

Targets for biological control

There is a great deal of published information on the targets for biological control. The details of agents are as follows.

Invertebrates	-	300 species
Gastropoda	-	21 species
Arachnida	-	8 species
Insecta	-	290 species
Weeds	-	42 species

(*Source:* Ding Johnson, 2003)

Categorisation of results of biocontrol efforts

Results of biological control together with classical approaches (the mainform practiced earlier) are classified into three groups.

i. *Complete*: When the intervention with biocontrol methods is alone necessary to check the pest population, insecticide treatment becomes rare extent. Cottony cushion scale, *Icerya purchasi* controlled by vedalia beetle *Rodolia cardinalis*.

ii. *Substantial:* When the use of biocontrol method is primary and other forms of management are necessary to achieve economic returns. For example- *Menochilus sexmaculata* in groundnut ecosystem on *Aphis cracciyora*.

iii. *Partial:* When the use of biological control method does not give adequate economic control, it has to be integrated with other methods especially insecticides in areas where intensive cultivation is adopted. For example- Management of bollworm complex with *Trichogramma* sp. in Cotton.

Analysis of worldwide attempts at biological control of insect pests shows that more than 1300 cases of biocontrol have resulted in any one of the above categories of results. more than 135 species of insect pests and weeds have been controlled (substantially-completely) by introduced natural enemies. During the last one hundred years there has been 1.4 biological control successes per year (Johnson, 2000).

Perspectives of biological control

Biological control possesses unique attributes which other partners lack. The overall advantages of biological control is summarised below.

1. It can be directed against a single pest which may result in keeping its level below damaging proportion for many years.
2. It is the cheapest, safest and most efficient method of pest control.
3. The long-term use of biological control is ecologically sustainable and does not pollute the environment.

General limitations of biological control

Biological control though a distinct proposition also suffers from some setbacks. They are limiting the implementation of the programmes set forth in crops.

Non-availability of good quality biotic agents at the farm level on time is one of the major constraints faced in adopting the IPM practices.

1. Non-availability of good quality biotic agents at the farm level on time is one of the major constraints faced in adopting the biological control.
2. Residual population of pest is left behind. When this level is not acceptable economically, it has to be lowered by alternative means.
3. The action is slow and limits the use of chemical pesticides for the same or other pests in the same ecosystem.
4. It is intensive in terms of training and implementation. Hence, detailed agro-ecosystem analysis is needed to find the suitability. Taylor (1955) considered biological control suited only to tropics as the condition for life is more or less fairly uniform than temperate areas. However, Greathead (1986) was unable to find any significant differences in the performances. But the predictability of results is dependent on the local conditions which means analysis is required for individual locations for programme implementation.

5. As the natural enemies are specific the costs can be prohibitive when multiple insect species are to be tackled by inundative means.
6. The ultimate success of biological control programmes is dependent on the quality of laboratory bred natural enemies. The maintenance of quality control is essential in biological control programme.
7. Biological control agents can lead to extinction of species, for example Coconut moth, *Lerrvana iridiscens* in Fiji was controlled by the introduced tachnid, *Bessa remota* in 1925. In two years the pest became extinct (Howarth, 1991).

3.2 Benefits and Costs of Biological Control

Several case studies have shown that biological control can be competitive with chemical control or more profitable than it. In this context, it has been noted by several authors that biological control tactic with great capability to reduce the pest populations may not be economically feasible, by the same token, one that has modest impact on the target pest may be a cost effective control alternative. Successful biological control programmes report benefit: cost ratios from reduced pest impacts and decreased use of pesticides of 10:1 to 30:1 with some as high as 200:1 (Jetter *et al.,* 1994). However, the potential of biological control has largely remained untapped because it has bean underused, under exploited, under estimated and often untried and therefore scientifically unproven (Hokkanen, 1993).

3.2.1 Opportunities

The area under organic cultivation (crops) in India is estimated to be around 1,00,000 hectare. Besides, there are lakhs of hectare of forest area being certified as organic. Further, some states like Uttaranchal and Sikkim have declared their states as organic. Moreover, the area under organic crop cultivation may rise because of the growing demand of organic food, a result of increasing health consciousness among the people. This indicates that there is huge scope for growth of the biopesticide sector in India. At the same time increasing population can be fed by organic farming dependence is a big question and unless organic farming yield can be brought equal to that of conventional farming involving the use of agrochemicals etc, the organic farming may not be feasible at the moment. Analysts believe that there would be a greater development in the biopesticides sector (Desai, 1997). Due to its rich biodiversity India offers plenty of scope in terms of sources for natural biological control organisms as well as natural plant based pesticides. The rich traditional knowledge base available with the highly diverse indigenous communities in India may provide valuable clues for developing newer and effective biological control based methods.

The stress on organic farming and on residue free commodities would certainly warrant increased adoption of bioagents by the farmers. Increased adoption further depends on

1. Concrete evidences of efficacy of bioagents/ biopesticides in controlling crop damage and the resultant increase in crop yield,
2. Availability of high quality products at affordable prices,
3. Strengthning of supply chain management in order to increase the usage.

In this regard, an efficient delivery system from the place of production (factory) to place of utilization (farm) is quite essential.

The National Farmer Policy 2007 has strongly recommended the promotion of biopesticides for increasing agricultural production, sustaining the health of farmers and environment. It also includes the clause that biopesticides would be treated at par with chemical pesticides in terms of support and promotion. Further research and development of biological pest control methods must be given priority and people in general and agriculturists in particular must be educated about the handling and use of such control measures. All this will lead to a general understanding about the benefits of bioagents as green alternative. However, the need in the present day context is on IPM, INM, ICM and GAP and by practicing these quality of life and health will be assured.

3.3 Conclusions

Biological control is a natural process that plays an important role in the suppression of field crop pests in hill Eco-system. A lot of the best examples of biological control proceed completely unnoticed by the farmer, simply because both the pest and its natural enemies coexist at such low densities that there is no perceptible problem in the crop, even thought the pest is present. It is generally agreed that integrated pest management (IPM) is the preferred approach to sustainable pest control in agriculture. Whenever possible, IPM programs for field crops should be constructed on a foundation of biological control, with additional control measures applied only as needed and selected to conserve natural enemies and all non-target insects to the greatest extent possible. Crops genetically modified to express natural insect toxins have, to date, proved either neutral or favourable to biological control by virtue of reducing overall pesticide usage. Fallow areas around fields can serve as reservoirs of many natural enemies, especially if weed species are allowed to flower. If a pesticide treatment becomes necessary, leaving less-affected portions of a field untreated can provide a refuge for natural enemies and accelerate the field-wide restoration of biological control post-treatment. Some newer pesticide formulations are designed to be more selective for particular pests and will spare natural enemies. Whenever biological control has a role in pest population reduction, pest control decisions should be weighted by considerations of how natural enemies will be impacted and what tactics might be feasible to conserve them.

References

DeBach, P. (ed.). 1964. Biological Control of Insect Pests and Weeds. Reinhold Publishing Corporation, New York. 844 pp.

DeBach, P. and D. Rosen. 1991. *Biological control by natural enemies,* 2nd edition. Cambridge University Press, Cambridge, 440 pp.

Desai, S. T. 1997. Chemical industry in the post independence era: a finance analysis point of view. *Chemical Business,* 11(1): 25 - 28.

Dwyer, G., J. Dushoff and S.H. Yee. 2004. The combined effects of pathogens and predators on insect outbreaks. Nature 430: 341–345.

Franz, J.M. 1961. *Biological control of pest insects in Europe.* Ann. Rev. Ent. 6, 183-200.

Greathead, D J. 1986. Biological control in the tropics: Present opportunities and future. Pests in Tropical Africa Revue de Zoologie Africaine, v.100, n.1, 1986:85-96.

Howarth, F. G. (1991). Environmental impacts of classical biological control. Annual Review of Entomology 36: 485-509.

Howarth, F. G. 1983. Classical biological control: panacea or Pandora's box? Proceedings of the Hawaiian Entomological Society 24: 239-244.

Howarth, F. G. 1991. Environmental impacts of biological control. Annual Review of Entomology 36: 485-509.

Jetter, K., K. Klonksy, and C. H. Pickett. 1997. A cost/benefit analysis of the ash whitefly biological control program in California. Journal of Arborculture 23: 65-72.

Liebhold, A. and Kamata, N. (2000) Introduction -Are population cycles and spatial synchrony a universal characteristic of forest insect populations? Population Ecology, 42, 205–209.

Michaud, J.P. Sloderbeck, P.E. and Nechols, J.R. (2008). *Biological Control of Insect Pests on Field Crops in Kansas.* Kansas State University. Web at: *http://www.oznet.ksu.edu*

Myers, J. 1987. Population outbreaks of introduced insects: Lessons from the biological control of weeds. In: P. Barbosa and J. Schultz, [eds.], Insect Outbreaks. Academic Press, New York, 173-193 pp.

Paul DeBach (1964). Biological control of insect pests and weeds. Reinhold Publishing Corporation, 844 pages.

Samways, M. J. (1981). Biological Control of Pests and Weeds. Edward Arnold Publishers Ltd, London.

Smith, H. S. 1919. On some phases of insect control by the biological method. J. Econ. Ent. 12: 288-92.

Taylor, T. H. C. 1955. Biological control of insect pests. Ann. Appl. Biol. 42: 190-96. Van Driesche, R. G. and T. S. Bellows, Jr. 1996. Biological control. Chapman and Hall, New York. 539 pp.

Wiedenmann, R.N. & J.W. Smith, Jr. 1997. Attributes of natural enemies in ephemeral crop habitats. Biol. Control 10: 16-22.

Wissinger, S. A. 1997. Cyclic colonization in predictably ephemeral habitats: a template for biological control in an annual cropping system. Biological Control 10, 4-15.

Wood Mackenzie Consultants Ltd. 1995. (Personal Communication) 1996 Data on pesticides and animal health products. Edinburgh, United Kingdom.

Wood Mackenzie Consultants Ltd. 1996. (Personal Communication) 1934 Data on pesticides and animal health products. Edinburgh, United Kingdom.

INTEGRATED PEST MANAGEMENT UNDER PROTECTED CULTIVATION

4

Ajay K. Sood

4.0 INTRODUCTION

The warm, humid conditions and abundant food under protected conditions provide an excellent, stable environment for pest development. Often, the natural enemies that keep pests under control outside are not present under protected environment. For these reasons, pest situations often develop in the indoor environment more rapidly and with greater severity than outdoors (Fletcher, 1984; Enoch, 1996; Clarke *et al.* 1994; Bakker, 19995). The damage inflicted by arthropod pests on greenhouse crops varies with the pest and season. The level of damage that can be tolerated is greatly dependent on the type of crop. Producers of vegetable crops generally can accept a higher level of damage than those of ornamental crops that are produced for their aesthetic value.

Integrated pest management (IPM) is a systematic approach to manage pests that combines a variety of techniques and strategies to either reduce pest populations or lessen their economic impact. It is a site-specific strategy for managing pests that relies on correct pest identification and understanding the pest biology. With a long-term perspective it is easier to see that an investment in IPM can pay for itself in a higher-quality crop and a cleaner environment.

4.1 Insects and Mite Pests under Protected Environment

The growing conditions within the protected environment of greenhouse/ polyhouse are highly favourable to arthropod pests. In India, about twenty insect and mite species have been recorded to be associated with the crops under protected environment. Some of the important pest groups are aphids, caterpillars, leafminer, mites, thrips and whiteflies. The detailed account of the insect-pests associated with crops under protected environment is being given hereunder:

i. Aphids: Aphids or plant lice are small, soft-bodied, sedentary insects that form colonies on the leaves and stems of the host plants. They suck the cell sap from leaf or apical growing parts and prefer to feed on tender, young growth. Aphids multiply at rapid pace completing one generation in 7-10 days. They give birth to six to ten young ones per day with 50-100 in total life span. Thus, enormous populations can build up in a relatively short period.

   Feeding by aphids can cause leaves or stems to curl or pucker. This leaf distortion often protects the aphids from contact insecticides. They also secrete honeydew on which black sooty mold develop in the periods of high atmospheric relative humidity which interferes with photosynthetic activity. They also act as vector of many viral diseases.

ii. Caterpillars: Two species of caterpillars are associated with the crops under protected environment. These are tobacco caterpillar (*Spodoptera litura)* and tomato fruit-borer (*H. armigera*).

   Tobacco caterpillar: The fore wings of adult moths have beautiful grayish brown pattern with grey wavy lines. A single female lays about 300 eggs in clusters covered by brown hair. The full fed larva, upto 5 cm in length, is velvety black with yellowish green dorsal stripes and lateral white bands. They pupate in soil.

   They are mainly foliage feeders, but occasionally damage fruits. Eggs are deposited by female moths that have made their way into the greenhouse and lay eggs in large masses. Heavy infestations may occur in isolated areas within the greenhouse, and timely spot treatments can often provide effective control.

   Tomato fruit-borer: Tomato fruit-borer is one of the most significant insect-pests of field-grown crops. They are less common in greenhouses, but when infestations do occur, they can cause serious losses. This is because, as their name implies, this caterpillar feeds directly on the fruit. This insect attacks many other crops. These are stout bodied, green or brown striped caterpillars about 3-4 cm long when fully grown. The moths are light brown with a small brown spot on each forewing.

   Tomato fruit borer feed on blooms, as well as immature and fruits. The individually laid eggs are most often deposited on leaves near upper bloom clusters. If not controlled, heavy infestations of these caterpillars can destroy the crop. Because large larvae that have already bored into fruit are very difficult to control, it is important to detect infestations and begin treatment when caterpillars are small.

iii. Leafminers; Leafminers are larvae of small flies. They damage plants by feeding between the upper and lower surface of the leaf by forming mines. The mines increase in width as the larva grows. When fully grown, the larva emerges from the leaf and fall to the ground to pupate. Each female fly lays 50 to 100 eggs by making punctures in the leaf tissue.

Damage is caused by larvae mining the leaves. If the infestation is severe, photosynthetic activity is reduced affecting plant vigour. Because the damaging stages of the insects occur entirely inside the leaf, control with contact insecticides is ineffective once the damage appears. Infestations can be avoided through the use of good cultural practices, hand romoval and disposal of infested leaves and use of chemical controls when necessary.

iv. Mites: Mites are sap-sucking pests which attack a wide range of greenhouse plants. Two species, the two-spotted spider mite *(T. urticae)* and the yellow mite (*P. latus*), are known to be persistent problems. These mites feed by piercing tissue with their mouthparts and sucking out cell contents.

Two-spotted spider mites are light to dark green with two distinctive black spots on the abdomen. Eggs are spherical and clear when laid. After hatching, the larva has three pairs of legs, but later stages have four pairs and are known as proto- and deuto-nymphs. Heavy infestations of the two-spotted spider mite produce fine webbing which may cover the entire plant. Generally they feed on the undersides of leaves, giving the upper leaf surface a speckled or mottled appearance. Leaves of mite-infested plants may turn yellow and dry up, and plants may lose vigour and die when infestations are severe. Females can lay 100 to 150 eggs, and during hot, dry weather the life cycle may be completed in seven days. The pest can complete upto 30 generations in a year. In the extreme winters the females are reported to overwinter in the plant debris and cracks and crevices.

Yellow mites are minute, elliptical, semi-transparent, greenish mites. These mites thrive when the temperature is mild with humid environment and can complete their life cycle in about two weeks. Depending on the type of plant attacked, they may infest the entire plant or be concentrated around the buds. Infested leaves become distorted and often curl inward; foliage may become darker than that of healthy leaves. Because of their small size, infestations often go undetected until the damage is severe. Usually it is the nature of the injury, not the mites themselves, that alerts greenhouse managers to yellow mite infestations. Often, it is better to discard infested plants or parts than to attempt to control the problem with pesticides.

v. Thrips: Thrips are tiny, slender insects about 1-1.5 mm long. They range in color from light brown to black. They have four wings, each fringed with a row of long hairs that are held flat over their back. Female thrips insert eggs into slits in the tissue. Eggs hatch in two to seven days. Nymphs feed much like adults and molt four times during development. They pupate in debris or soil. Winged adults are carried into the greenhouse on contaminated plant material, or they fly in.

   Thrips infest the leaves, flowers, buds and young fruits of a crop. They feed by rasping the plant surface and sucking up the exuding sap. Heavily infested leaves have a mottled or silvery appearance.

vi. Whiteflies; Whiteflies are serious pests in the greenhouse. Two species of whiteflies are prevalent under protected environment namely, greenhouse whitefly (*T. vaporariorum)* and cotton whitefly (*B. tabaci*). The former is prevalent in temperate regions whereas the other is a serious pest in tropical and sub-tropical parts of the country.

   Whiteflies are small snowy white in colours, about 2 mm in length, flutter from the undersides of leaves when the plants are disturbed. The lower surface of the leaves is infested with all life stages of whiteflies. The female of these sap-sucking insects may lay upto 150 eggs at the rate of 25 per day. The entire life cycle takes 21-36 days, depending on the greenhouse environment. They can complete more than 12 generations in a year.

   Both the nymphs and adults of whitefly feed on phloem cell sap and causes chlorotic spots. The leaves dry-up prematurely and plant growth is affected. Nymphs also secrete a sticky substance known as honeydew, which covers leaf surfaces and flowers on which the sooty growth takes place under humid weather conditions (Zippori *et al.*, 1988).

Management strategies

Since greenhouse conditions allow rapid development of pest populations, successful control of insect pests on greenhouse vegetables and ornamentals depends on several factors. The IPM programme for protected cultivation can be described as a pyramid constructed having three key components namely, Avoidance of problem; Sampling and early detection; and Curative measures.

With "Avoidance" as the foundation, virtually all the non-curative preventive management components (e.g. physical, mechanical and cultural means) can be fitted. Confronted with a pest crisis, short-term function depends on the upper two levels of the pyramid. However, sustainable, long-term strategies must depend on the development of this solid foundation. At the same time, a pyramid-strategy developed for one pest must be compatible with like strategies in place for all pests of a system.

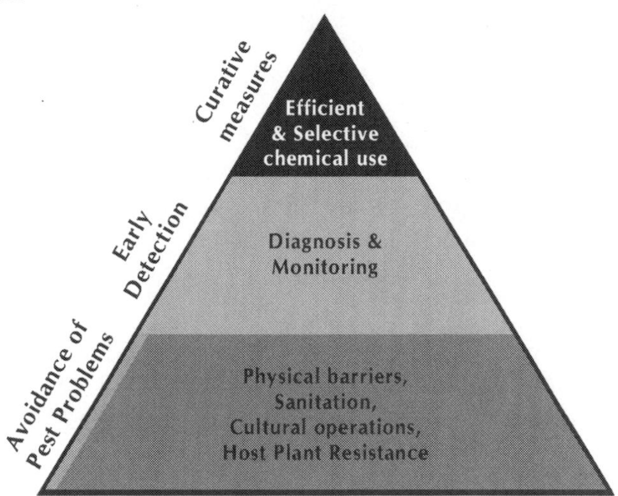

Although physical and cultural measures do not have a curative effect after heavy infestations of greenhouse pests, their preventive and delaying effect on the rapid increase of pests is of great importance in integrated pest management.

1. Avoidance

i. Use of physical barriers

Use of insect-proof screens: One strategy particularly suited to greenhouse IPM programmes is the use of physical barriers to exclude insect-pests. Screening removes an important variable from a grower's pest management program - the movement of pests into the greenhouse from outside. This includes common greenhouse pests such as thrips, aphids, leafminers and whiteflies, but also some less common pests such as fruit borers. Screening vents and doorways can greatly limit the movement of insect pests into the greenhouse. However, selection of proper screen size mesh and assuring adequate airflow are more important. Selection of screen with particular mesh size is site specific and depends on the prevalent pests of the crop in the locality. The smaller the holes in the mesh, the more reduction there will be in airflow. This reduction can be counteracted by increasing the surface area through which air flows. Screening can be stretched from gutter to gutter to increase surface area.

Provision of double door: Limited access to screened areas is beneficial since insects may come in the protected structure on clothing or be swept in with the wind. Building a screened foyer to create a double-door entry partially solve the problem of wind-carried insects. Special efforts must be put in for repairing holes or tears immediately, and cleaning the screens to maintain airflow.

ii. Sanitation

Many of the methods incorporated in an IPM strategy are logical greenhouse operating procedures and basic horticultural practices (Jarvis, 1992; Giocomelli and Roberts, 1993; Wittwer and Castilla, 1995). The following methods produce a healthier crop, prevent many pest problems, and isolate pests to smaller areas in the greenhouse. A basic component of cultural practices is sanitation. Infestations are easier to prevent than to cure. It has been observed that insects can enter the protected structure in following three ways:

- Introduction of infested seedlings/ planting material
- Infestation from other plants within the protected structure
- Infestation from host plants outside but near the protected structures

Before introducing a new crop into the greenhouse, it is extremely important to eliminate pests from the previous crop. Remove all plant debris and weeds from the greenhouse. Many pests also occur on other crops or broadleaf weeds. For this reason, it is important to avoid growing other crops next to the greenhouse and to prevent heavy growths of broadleaf weeds around the outside edges of the greenhouse. Under protected environment monoculture is suggested, however, if one has to go for polyculture then avoid staggered planting.

A fallow period of two to four weeks reduces the pest load considerably. To determine the presence of thrips, whiteflies, leafminers, or other insects, set up yellow sticky cards and indicator plants after watering. Observe for any insects that are trapped on the cards after two days and continue till the activity is ceased and only thereafter the decision regarding plantation of new crop be made.

iii. Balanced use of fertilizer

Fertilization schedules based on balanced use of nutrients should be followed. Nitrogen should be applied only as needed for optimal growth. Periodic heavy applications set up nitrogen surpluses that cause excessive growth, which favour the population growth of aphids, and other pests. Application of potassium at desired levels has been found to reduce the incidence of insect-pests.

iv. Pinching and Pruning

Pinching-off damaged plant parts, flowers, and spotted leaves (and those with insect larvae or egg deposits) can be a very effective way of reducing the spread of pests in the greenhouse. The plant debris should be placed immediately in a covered container before being disposed-off. This practice can be helpful in reducing the pest population of all the targeted pests.

Pruning lower leaves after harvesting lower fruit clusters is helpful measure in removing large numbers of developing leafminers and whiteflies.

2. Early Detection

Scouting and early detection are critical to manage the insect infestation successfully. Monitoring or scouting is the regular, systematic inspection of the plants and exteriors to identify and assess pest problems. It includes inspection of foliage and flowers; and the use of insect traps. Many insect infestations begin in isolated spots within the greenhouse. Timely crop monitoring identifies situations where pests are absent or are at levels well below those necessary to cause damage, thus preventing unnecessary control applications and expenditures.

i. Scouting

Scouting procedures for most greenhouse-grown crops are based on visual observations and are used to provide estimates of the pest population in protected environment. The common pests that attack greenhouse crops do not distribute themselves evenly throughout the crop. Therefore, it is imperative to scout the entire greenhouse in a consistent, uniform pattern. Inspect the entire plant, including the soil surface, for the presence of arthropod pests. Look at the plant systematically each time. Begin at the bottom and work up. Look at the older leaves, the young, tender leaves, and the flush growth. When the crop is young, it is important to check all the leaves on the plant. Because a majority of arthropod pests associated prefer the underside of a leaf, it is important to turn the leaves over to check for pests.

Scout the crop on a regular basis and at least 1-2% of the total plants should be inspected at weekly interval. A thorough greenhouse inspection reveals the location and severity of any current pest problems. One should use a field data sheet to record the identification, location, and severity of all pests present, and record the effectiveness of any treatments.

ii. Monitoring

It is a relative method of insect population estimation where no direct observations on the plants for the presence of insect-pests are needed.

However, the pest population is estimated with the help of attractant traps. For whiteflies, aphids, thrips and leafminer adults, yellow sticky cards (4"x12" or 6"x12") are an excellent supplement to pest observation in the protected environment. Additionally, for thrips blue coloured sticky traps can also be used. Though the sticky traps are available commercially, however, they can also be fabricated indigenously at growers level by using pieces of fluorescent yellow chart paper and pasting castor oil/ mustard oil or the commercial grade adhesive, the polybutene, on both the surfaces. The traps are placed in a grid pattern and 1-2 yellow sticky cards per 100 square meter of floor area are used. If the target is mass trapping, then number of traps can be increase to ten or more. Hang the yellow sticky cards/ traps in the crop with the help of strings about 4" to 6" above the plant canopy. As the crop grows, cards can be moved up. Designate the location of each sticky card on a map of the greenhouse. Check the sticky cards every scouting visit (twice a week if possible) and record the total number of whiteflies, thrips, winged aphids, and leafminer flies from each card on the field data sheet. Change the cards when more than 60-70% of the area is covered by trapped insect.

Apart from this, sex pheromone baited traps can also be used for detecting moths of tobacco caterpillar and tomato fruit borer. These traps allow visualizing population trends and can be used to time the application of pesticides or release of bioagents.

3. Curative Measures

i. Biological Control

Practically every crop pest has its natural enemies in the form of parasitoids or predators. The biological control involves a large scale multiplication and liberation of such agents, or creating conditions under which the naturally occurring agents can act effectively (van Lenteren and Woets, 1988; Shipp *et al.* 1991; Wearing, 1988). In field situations, some very outstanding successes have been achieved by using biocontrol agents (Gullino, 1992; Hanan *et al.*, 1978; Quinlan, 1988), but the method suffered a set-back owing to the large-scale and indiscriminate use of insecticides.

Use of specific predators and parasitoids in greenhouse ecosystem to manage pests forms the basics of biological control. But to be effective, biological control must be well planned and begun when the target populations are low. In Western Europe and North America, the bioagents are commercially available and are being used successfully for the management of pest problems under protected situations. However, under Indian conditions this technique though is having potential but have following limitations

- Temperature extremes and the effect of chemical pesticides affecting the capabilities of natural enemies.
- Low tolerance of pest damage amongst consumers especially in ornamentals
- Lack of a system for supplying natural enemies of good quality

ii. Chemical Control

The insecticides are curative in action and are one of the important tools to check the flaring pest populations. Insecticides belonging to different groups namely, botanicals pesticides, microbial pesticides, Insect Growth Regulators (IGR), synthetic chemical pesticides etc. are in use. Some of the serious limitations of the insecticides have been highlighted in recent years. Out of these, the problems of insecticides residues on crops have been agitating the minds of people in developed and developing countries of the world due to indiscriminate use of pesticides. However, the sensible approach suggested and generally accepted is for need based judicious and safe use of insecticides.

Under protected environment, in order to avoid the contamination of produce from pesticide residues, use of pesticides having less persistence and adopting suggested waiting period between last insecticidal application and harvesting can be followed.

Other challenge with pest managers is to minimize the development of resistance in pest species to the pesticides. This can be checked or delayed by avoiding repeated application of same insecticide and also from same group. Incorporating botanicals, microbials and others in management schedule proves better in Insecticide Resistance Management Programme (IRM).

Apart from this, the safety of the pesticide applicator should also be taken into consideration as the chemical pesticides become more volatile under high temperature regimes of the protected environment. One should apply the pesticides by taking all safety measures including the protective kit to avoid the direct effect of pesticides to the applicator. A re-entry period of atleast 12 hr should be observed after application of pesticides. Insecticides with fumigant action need to be avoided under protected structures.

References

Bakker, J.C. 1995. *Greenhouse Climate Control: An Integrated Approach,* Wageningen Press.

Clarke, N.D., Shipp, J.L., Jarvis, W.R., Papadopoulos, A.P. and Jewett, T.J. (1994) Integrated management of greenhouse crops – A conceptual and potentially practical model, *Horticultural Science* 29, 846–849.

Enoch, H.Z. 1986. Climate and protected cultivation, *Acta Horticulturae* **176**, 11–20. Fletcher, J.T. (1984) *Diseases of Greenhouse Plants,* Longman, London.

Giacomelli, G.A. and Roberts, W.J. 1993. Greenhouse covering systems, *Hort Technology* 3, 50–58.

Gullino, M.L. 1992. Integrated control of diseases in closed systems in the sub-tropics, *Pesticide Science* 36, 335–340.

Hanan, J.J., Holley, W.D. and Goldsberry, K.L. 1978. *Greenhouse Management,* Springer-Verlag, Berlin.

Jarvis, W.R. 1992. *Managing Diseases in Greenhouse Crops,* American Phytopathological Society Press, St Paul, Minn.

Quinlan, R.J. 1988. Use of fungi to control insects in glasshouses, in M.N. Burge (ed.), *Fungi in Biological Control Systems,* Manchester University Press, Manchester, pp. 19–36.

Shipp, J.L., Boland, G.J. and Shaw, L.A. 1991. Integrated pest management of disease and arthropod pests of greenhouse vegetable crops in Ontario: Current status and future possibilities, *Canadian J. of Plant Science* 71, 887–914.

van Lenteren. J.C. and Woets, J. 1988. Biological and integrated pest control in greenhouses, *Annual Review of Entomology* 33, 239–269.

Wearing, C.H. 1988. Evaluating the IPM implementation process, *Annual Review of Entomology* 33, 17–38.

Wittwer, S.H. and Castilla, N. 1995. Protected cultivation of horticultural crops worldwide, *Hort Technology* 5, 6–23.

Zipori, L, Berlinger, M.J., Dayan, E., Dahan, R., Shmuel, D., Mordechi, Sara and Aharon, Y. 1988. Integrated control of *Bemisia tabaci* in greenhouse tomatoes from early planting, *Hassadeh* 68, 1710–1713.

INTEGRATED DISEASE MANAGEMENT FOR SUSTAINABLE AGRICULTURE

5

Vishal Gupta and Vijay K. Razdan

5.0 INTRODUCTION

Integrated pest management (IPM) is an approach which maximizes natural control of pest populations, based upon knowledge of each pest, in its environment and its natural enemies, farming practices modified such as changes in the planting or sowing schedules, to affect the potential pests adversely and to aid natural enemies of the pests. The term Integrated Pest Management was first given by Smith and van den Bosch (1967), based on the integrated control concept given by Stern *et al.* (1959), entomologists from University of California, who defined it as applied pest control which combines and integrates biological and chemical control and supplemented with the plantation of the resistant varieties. There is need for regular scouting to determine the levels of pests and the environmental factors. When the disease is above the threshold level only then the suppressive measures should be taken with most suitable technique or combinations of techniques, such as biological control, genetically resistant hosts, environmental modifications and when necessary and appropriate, selective use of pesticides, while causing minimum disruption of natural enemies. This approach is markedly different from the traditional application of pesticides on a fixed schedule. IPM strategies generally relay first upon all other concepts of disease management before chemically altering the environment. Thus IPM is the optimization of pest control in an economically and ecologically sound manner. This is accompanied by the use of multiple tactics in a compatible manner to maintain pest damage below the economic injury level while providing protection against hazards to humans, animals, plants and the environment. A successful disease control programme depends on a crop production system which is closely aligned with the goals of pest management. One must start with the selection of appropriate varieties, an irrigation system that minimizes leaf wetness, a fertilizer programme that results in optimum plant growth, bed preparation, plant density, and canopy management that affords optimum air circulation and pesticide coverage when needed, a transplant programme which minimizes transplant shock, a clean seedling production programme, effective pest monitoring during the season, and finally, a harvest and transportation procedure which maximizes shelf life. The five components of an IPM programme are prevention, monitoring, correct disease and pest diagnosis, development and use of acceptable thresholds and optimum selection of management tools. The management strategies available include genetic control, cultural control, biological control and chemical control. In general, weeds are most effectively controlled with cultural and chemical practices. Diseases are controlled most by host-plant resistance, cultural practices, and chemicals. Insects and nematodes are controlled with all four strategies. For disease management, it is important to understand the potential of a pathogen to infest and spread in the crop. The three parameters of disease progress are 1) the initial amount of inoculum, 2) the rate of disease increase and 3) the time the crop is grown. These parameters may interact in different ways to produce a rapid pathogen population increase. The rate of disease increase over time is dependent upon the interactions of the pathogen, plant host and the environment. For disease management purposes, we are most concerned with the interactions of the pathogen and host at the population level of organization. However, environmental conditions play a critical role in determining the nature of plant disease epidemics. In plant pathology, this set of interactions is known as the disease triangle. Understanding the biology of the pathogen, host-pathogen interactions, and the effect of environmental factors on this dynamic process in time and space (disease epidemiology), is critical for planning and implementing effective and efficient management strategies. These strategies can affect particular aspects of the growth of the pathogen population. Host-plant resistance can affect all three parameters by 1) reducing the amount of inoculum via resistance to particular strains of the pest, 2) by slowing the rate of pathogen buildup by reducing its reproductive capacity and 3) by reducing the total period of exposure in short-season varieties. Cultural control is aimed at reducing the primary inoculum (sanitation) or reducing the rate of disease increase by modifying the crop environment. A good example of the latter is the use of drip irrigation rather than overhead to reduce free water on foliage. The total time that a crop is exposed to a pathogen may be minimized by optimizing plant growth, thus reducing the time to harvest. Biological control usually affects the rate of pathogen buildup. Finally, chemical control can affect the amount of inoculum available at the beginning of the season (i.e. soil fumigation) and/or reduce the rate of disease development by killing a portion of the pathogen involved in later stages of the epidemics.

IPM has established as the desirable pest control strategy from the perspective of both national policy and science based pest control. Broadly IPM can be defined as: the careful consideration of all available pest control techniques and subsequent integration of appropriate measures that discourage the development of pest populations and keep pesticides and other interventions to levels that are economically justified and reduce or minimize risks to human health and the environment. IPM emphasizes the growth of a healthy crop with the least possible disruption to agro-ecosystems and encourages natural pest control mechanisms (FAO, 2002). Inclusion of IPM in the plant pathology vernacular was only after the formal involvement of plant pathologists with entomologists, nematologists and weed scientists in IPM programmes under Huffaker Project, named after Carl Huffakar, Entomologist from University of California. Plant pathologists embraced integrated disease management by applying fundamental information on loss potential and pathogen biology, ecology and epidemiology, and applying the basic concepts of plant disease management. The principles of plant disease management should always be based on the integration of basic concepts such as avoidance, exclusion, eradication, protection, resistance and therapy.

5.1 Accurate Diagnosis of Diseases

Proper disease identification is critical for making correct disease management decisions. This will save time, money, and the environment. Effective use of fungicides and other pesticides depends on correct identification of the problem. The accuracy of any diagnosis depends upon the information supplied, the specimen material selected and the condition of the specimen when it arrives at a clinic. Digital images of the fresh specimen with symptoms and field-view images of the problem might be useful in some cases. In order to apply disease management practices, there should be knowledge of which pathogens are present or are likely to appear in a particular field or season. Descriptive and pictorial manuals are helpful for identification of diseases commonly found. It is important to know the common diseases of a given crop specific to your area.

5.1.1 Monitoring of Pathogens

As mentioned, monitoring is a critical component of an effective IPM programme. Monitoring can be direct (looking for the pathogen or disease) or indirect (recording environmental conditions which affect disease development). Direct monitoring of diseases can be based on symptoms or signs of the pathogen. Identification of pathogens is generally difficult, because pathogens usually are microscopic and can be detected typically after the disease process has begun. Most monitoring is actually for disease symptoms, with the control strategy aimed at reducing further spread. Even when visible symptoms are evident, levels of disease may be so low as to make detection very difficult. To optimize the chances of detection, one should concentrate on those areas where disease is most likely to occur, for example in low areas or areas of lush growth. If this is not possible, an array of sampling designs may be used, such as a diagonal across the field, a random walk, a stratified design where each subsection of the field is sampled, or a stratified random design where a random sample is taken in each subsection of the field. The appropriate sampling design will depend on the level of disease expected, the distribution of the disease and sampling schemes already in place for other pests. Disease distribution within a field is dependent, in large part, on the source of inoculum for the pathogen. If the disease is seed borne, in many cases the first diseased plants will be more uniformly distributed in the field. If the disease is soil borne, it may often be found in clusters in the field. If it is transmitted by insects, the distribution may be more random, or a field edge effect may be apparent. Thus, it is important to understand the biology of the disease when developing an appropriate sampling strategy for it. Indirect monitoring of disease most often involves data commonly gathered which includes temperature, relative humidity and leaf wetness. The models are often developed from controlled environmental chamber experiments where the minimum, maximum and optimum temperatures and relative humidity for fungal growth, germination and/or disease development are identified. Leaf wetness, either monitored directly or by prediction of dew point based on the relative humidity and temperature conditions, is used if the pathogen requires free water for germination. The prediction of disease events through environmental monitoring has been very successful in a few cases and is used widely for those crops and diseases where sufficient research exists.

5.1.2 Prevention and Management Practices

Site Selection and Preparation: Soil borne diseases remain a major limiting factor for the production of vegetables. It is important to start with clean soil and proper sites for crops. Plowing and disking will reduce pathogen carryover in old crop refuse. The longer the fallow period, the more pathogen populations are reduced. It is also essential to follow the latest recommendations for soil fumigation, cultural practices and biological control options to eliminate or reduce initial inoculum of soil borne pathogens. It is important that soil compaction is avoided, since this interferes with root growth, encourages soil moisture retention and promotes root diseases. Preparation of raised beds generally allows for better drainage. Prior to planting, soil should be tested for nutrient levels and nematode populations (and other pathogens if tests are available). Planting times can be altered to avoid or reduce development of certain diseases.

5.2 Host Resistance

Development of resistant varieties through hybridization is a cheap method of disease management for the farmers. With the development in biotechnology it is possible to introduce distant resistance genes in the plant by manipulation, genetic modification and multiplication of plants through techniques such as tissue culture and genetic engineering. Creation of transgenic plants in which resistance genes from sources other than that particular plant species or incorporation of avirulence genes of the pathogen to impart resistance is now possible. Induction of Acquired resistance, in which the plant acquires localized or systemic resistance during through the effect of chemicals or microorganisms, using phosphates and carbonates as foliar therapy has been found to turn on the resistance genes. Rhizobacteria are also known to induce systemic acquired resistance in the foliar parts against many diseases simultaneously. Even biochemical resistance of non-genetic nature can be developed in plants by chemotherapy or host nutrition, but this type of resistance is temporary. Temporary physiological resistance in plants can be developed through chemotherapy, when systemic fungicides and antibiotics applied to the foliage or through the roots persist in the plants and while their toxic level is maintained the pathogen cannot invade the tissue. Nutrition cannot change a susceptible variety to a resistant variety, but making major-and micro-nutrients available there is strengthening of the tissue that can ward off invasion of the pathogen. It is very important to choose cultivars with multiple pathogen and nematode resistance whenever possible. Control of many diseases of vegetables (Fusarium wilt, Verticillium wilt, and gray leaf spot for tomato) is achieved primarily by this method. Recently, varieties resistant to *Tomato Yellow Leaf Curl Virus* (TYLCV) and *Tomato Spotted Wilt Virus* (TSWV) have been identified. Firm-fruited, crack-resistant tomato fruit can escape infection of some pathogens.

5.3 Irrigation Management

High soil moisture enhances the development of soil borne pathogens including *Phytophthora*, *Pythium* and the bacterial wilt pathogen. Excess water damages roots by depriving them of oxygen and creates conditions that favor infection by certain soil borne pathogens. Irrigation management, based on plant needs, will help to create an environment unfavorable for pathogen survival and disease development. Use of tensiometers or other devices for irrigation scheduling and avoidance of low areas can help in disease management.

5.4 Soil and Fertilizer Management

Plant nutrition and soil pH can also impact some diseases. Fertilizers with a higher proportion of nitrate nitrogen (NO_3) than ammoniacal nitrogen (NH_4) will help to reduce the incidence of Fusarium wilt on tomato. Increasing soil pH by liming is a good management strategy to reduce Fusarium wilt incidence as well as Botryis gray mold severity. Optimum calcium nutrition and higher soil pH may reduce the incidence of bacterial wilt in the field. Adequate calcium is necessary to minimize blossom end rot and to provide for overall healthy growth. Avoiding excessive nitrogen leads to less dense canopies, thus improving air movement in the canopy.

5.5 Cultural Practices

Cultural practices serve an important role in prevention and management of plant diseases. The benefits of cultural control begin with the establishment of a growing environment that favors the crop over the pathogen. Reducing plant stress through environmental modification promotes good plant health and aids in reducing damage from some plant diseases. Sanitation practices aimed at excluding, reducing or eliminating pathogen populations are critical for management of infectious plant diseases. It is important to use only pathogen-free transplants, especially for late blight, bacterial spot, viral diseases and early blight. In order to reduce dispersal of soil borne pathogens between fields, stakes and farm equipment should be decontaminated before moving from one field to the next. Reduction of pathogen survival from one season to another may be achieved by destruction of volunteer plants and crop rotation. Removal of cull piles and prompt destruction of crops should be applied as a general practice. Avoid movement of soil from one site to another to reduce the risk of moving pathogens. For example, sclerotia of *Sclerotinia sclerotiorum* and *Sclerotium rolfsii*, are transported primarily in contaminated soil. Minimizing wounds during harvest and packing reduces post-harvest disease problems. Depending on crops and other factors, sanitation of soil can be achieved to some degree by solarization.

5.6 Crop Rotation

Crop rotation is a very important practice, especially for soil borne disease control. For many soil borne diseases, at least a 3-year-rotation using a non-host crop reduces pathogen populations. This practice is beneficial for Phytophthora blight of pepper and Fusarium wilt of watermelon. Land previously cropped to alternate and reservoir hosts should be avoided whenever possible. Vegetable fields should be located as far as away as possible from inoculum and insect vector sources.

Weed control is important for the management of viral diseases. Weeds may be alternate hosts for several important vegetable viruses and their vectors. Elimination of weeds might reduce primary inoculum. Cover crops help to reduce weed populations that may harbor pathogens between seasons. For this purpose use cover crops that grow fast and provide maximum biomass. Non-host cover crops help to reduce weed populations and primary inoculum for soil borne pathogens.

Excessive handling of plants such as in thinning, pruning and tying may be involved in spread of pathogens, particularly bacteria. Because some pathogens can only enter the host through wounds, situations which promote plant injury should be avoided. During pruning process and harvest, workers should periodically clean their hands and tools with a disinfectant, such as isopropyl alcohol. If applicable, plants can be staked and tied for improved air movement in the foliar canopy. A more open canopy results in less wetness, discouraging growth of most pathogens.

Soil aeration and drying can be enhanced through incorporation of composted organic amendments in the soil. Build up of inoculum can be reduced by removing all plant materials (infected and apparently healthy) after harvest. Between-row cover crops reduce plant injury from blowing sand. Polyethylene mulch can be used as a physical barrier between soil and above-ground parts of plants. This is an important practice for fruit rot control in the field. Highly UV-reflective (metalized) mulches repel some insects. It is beneficial to use metalized mulch during certain times of the year when insect vectors of some viral diseases are prevalent. *Tomato spotted wilt virus* (TSWV) incidence and associated vector thrips populations have been demonstrated to be effectively reduced by using metalized mulches on tomatoes.

5.7 Biological Control

Biocontrol agents for use in vegetable disease management are increasing in use especially among organic growers. These products are considered safer for the environment and the applicator, than conventional chemicals and are mainly used against soil borne diseases. Examples of commercially available biocontrol agents include the fungi *Trichoderma harzianum* and *T. virens*, an actinomycete *Streptomyces griseoviridis* and a bacterium *Bacillus subtilis.*

5.8 Chemical Control

Fungicides and bactericides are an important component of many disease management programmes. It is important to remember that chemical use should be integrated with all other appropriate tactics mentioned above. Information regarding physical mode of action of a fungicide is helpful in timing of fungicide applications. Physical modes of action of fungicides can be classified into four categories: protective, after infection, pre-symptom, and anti-sporulant (post-symptom). Protectant fungicides include the bulk of the foliar spray materials available. In order to be effective, protectant fungicides, such as copper compounds, mancozeb, etc., need to be on the leaf (or plant) surface prior to arrival of the pathogen. Systemic (therapeutic) fungicides, based on their level of systemically are active inside of the leaf (can penetrate at different rates through the cuticle). Systemic fungicides may stop an infection after it starts and prevent further disease development. If necessary, fungicides must be used based on recommended fungicide resistance management strategies. A new strategy to chemically manage plant diseases without direct interference with the pathogen is the triggering of plant defense reaction. Chemicals must be used at recommended rates and application frequencies. Besides selection of the most efficacious material, equipment must be properly calibrated and attention paid to the appropriate application technique. As always, the key to effective disease management is correct diagnosis of the problem. Always read the pesticide labels and follow the instructions carefully. Effective management of whiteflies, thrips, and aphids should be practiced to reduce the incidence and secondary infections of viral diseases vectored by these insects.

References

Abawi, G. S., and Barker, K. R. 1984. Effects of cultivar, soil temperature, and populations of *Meloidogyne incognita* on root necrosis and Fusarium wilt of tomatoes. Phytopathology 74:433-438.

Abdul-Baki, A. A., and Teasdale, J. R. 1993. *A no-tillage tomato production system using hairy vetch and subterranean clover mulches.* Hort Science 28:106-108.

Abdul-Baki, A. A., and Teasdale, J. R. 1997. *Snap bean production in conventional tillage and in no-till hairy vetch mulch.* Hort Science 32:1191-1193.

Adeniji, M. O., and Chheda, H. R. 1971. Influence of six varieties of *Cynodon* on four *Meloidogyne* spp. J. Nematol. 3:251-254.

Beaty, E. R., and Tan, K. H. 1972. *Organic matter, N, and base accumulation under Pensacola bahiagrass.* J. Range Manag. 25:38-40.

Brenneman, T. B., Sumner, D. R., Baird, R. E., Burton, G. W., and Minton, N. A. 1995. *Suppression of foliar and soilborne peanut diseases in bahiagrass rotations.* Phytopathology 85:948-952.

Burton, G. W. 1954. R*oot penetration, distribution and activity in southern grasses measured by yields, drought symptoms and P32 uptake.* Agron. J. 46:229 233. Cantliffe, D. J., Hochmuth, G. J., Locascio, S. J., Stansly, P. A., Vavrina, C. S., Polston, J. E., Schuster, D. J., Seal, D. R., Chellemi, D. O., and Olson, S. M. 1995. *Production of solanaceae for fresh market under field conditions: Current problems and potential solutions.* Acta Hortic. 414:229-244.

Chase,C. A., Sinclair, T. R., Chellemi, D. O., Olson, S. M., Gilreath, J. P., and Locascio, S. J. 1999. *Heat-retentive films for increasing soil temperatures during solarization in a humid, cloudy, environment.* HortScience 34:10851089.

De Ceuster, T. J. J., and Hoitink, A. J. 1999. Prospects for composts and biocontrol agents as substitutes for methyl bromide in biological control of plant diseases. Compost Sci. Util. 7:6-15.

De Ceuster, T. J. J., and Pauwels, F. 1995. Soil disinfestation in the Belgian horticulture–A practical view. Acta Hortic. 382:37-50.

Dickson, D. W., and Hewlett, T. E. 1989. Effects of bahiagrass and nematicides on *Meloidogyne arenaria* on peanut. J. Nematol. 21(suppl.):671-676.

FAO, 2002. Analysis of an FAO survey of postharvest crop losses in developing countries. *AGPP Misc.* 27. FAO Rome.

Vanachter, A. 1975. *Fumigation against fungi.* Pages 163-184 in: Soil Disinfestation. D. Mulder, ed. Elsevier Science Publishing Co., New York.

Yamazaki, H., and Hoshina, T. 1995. Calcium nutrition affects resistance of tomato seedlings to bacterial wilt. Hort. Science 20:91-93.

INTEGRATED DISEASE MANAGEMENT IN PROTECTED ENVIRONMENT

6

Deepika Sud

6.0 INTRODUCTION

The crop pests and diseases pose a major constraint to increase food production. Insect pests, diseases and weeds together cause a total loss of approximately 30% of total crop produce. The rapid change in the agro-ecosystem leaves a snag in the establishment of harmony and disturb ecosystem due to wide usage of chemical pesticides, fertilizers, and synthetic plant growth regulators. Diseases are a fact of life in most greenhouse vegetable operations. Implementing an effective disease management program may well determine the success or failure of a greenhouse business. Disease management strategies aim to prevent the establishment of diseases in greenhouse plantings, as well as to minimize the development and spread of any diseases that may become established in a crop.

Integrated Disease Management (IDM) is a concept derived from the successful Integrated Pest Management (IPM) systems developed by entomologists for insect and mite control. In most cases IDM consists of scouting with timely application of a combination of strategies and tactics. These may include site selection and preparation, utilizing resistant cultivars, altering planting practices, modifying the environment by drainage, irrigation, pruning, thinning, shading, etc., and applying pesticides, if necessary. But in addition to these traditional measures, monitoring environmental factors (temperature, moisture, soil pH, nutrients, etc.), disease forecasting, and establishing economic thresholds are important to the management scheme. These measures should be applied in a coordinated integrated and harmonized manner to maximize the benefits of each component. For example, balancing fertilizer applications with irrigation practices helps promote healthy vigorous plants. However, this is not always easy to accomplish, and "disease management" may be reduced to single measures exactly the same as the ones previously called "disease control." Whatever the measures used, they must be compatible with the cultural practices essential for the crop being managed.

6.1 Component of Integrated disease management

i. *Cultural practices*
   - Sanitation.
   - Tillage and cultivation.
   - Crop rotation.
   - Management of temperature and moisture.
   - Management of fertility and pH.
   - Organic soil amendments and mulches.
   - Modification of the biological environment.

ii. *Preventive measures*
   - Regulatory mechanisms to prevent introduction and spread of plant pathogens.
   - Use of disease-free seed and planting material.
   - Seed treatment.

iii. *Biological control*

iv. *Plant resistance*

v. *Physical control techniques* (Steam, soil solarization, fire and flooding)

vi. *Disease control chemicals* (Fungicides, Nematicides, Bactericides, Viricides etc.)

A. Important Diseases of Vegetable Crops and their Integrated Management

I. Tomato *(Lycopersicon esculentum Mill.)*

Tomato is prone to a number of diseases. The causes may be fungus, bacteria, virus, nematodes or abnormal environmental conditions. Some of the important diseases and their Management are described below:

1. Damping-off *(Phytophthora, Pythium, Rhizoctonia, Fusarium, Sclerotium* sp.*)*

Symptoms: This disease occurs in the nursery bed. In pre-emergence damping off, the young seedlings are killed before they reach the surface of the soil. Since this happens under the soil surface, the disease is often not detected except for the resulting poor stand. The post-emergence damping off is very conspicuous. This phase of the disease is characterized by the toppling over of infected seedlings, any time after they emerge from the soil until the stem has hardened sufficiently to resist invasion. As the disease advance, the stem become constricted at the base and plant collapse.

Management

i. *Cultural Practices*

- Use disease free seed from disease free areas
- Raised nursery beds to drain off excess water
- Thin sowing to avoid overcrowding.
- Light sandy soils for nurseries or use pure fine sand sawdust mixture for raising seedlings.
- Use well-decomposed manure.
- Light but frequent irrigations
- Sterilization of soil by burning a 30cm thick stacks of farm trash on the nursery bed.
- Sow the seeds in lines 5 cm apart and cover the nursery beds with grass or polythene sheet and remove it with the emergence of seedlings
- Soil solarization for 6-8 weeks during summers

ii. *Biological Control:* Seed coating with spores of *Trichoderma harzianum* and *Penicillium oxalicum* and cells of *Pseudomonas cepacia* and *Pseudomonas fluorescens* have been found very effective bio-control methods of damping off.

iii. *Chemical Control*

- Treat the nursery beds with formalin (1:7) water at least a week before sowing.
- Treat the seeds with fungicides like Thiram /Captan/ Bavistin (carbendazim) / vitavex power (carboxin+thiram) @ 2 g/ kg seed or Raxil (Tebuconazole) @ 1g/ kg seed.
- Drench the nursery beds with Indofil M-45 (0.25%) + Bavistin (0.1%) solution or Companion /Saaf (mancozeb) + carbendazim) fungicide @ 0.2%.
- Select healthy plants for transplanting and always transplant seedlings in the afternoon.
- Before transplanting, the uprooted seedlings should be dipped in the solution of 0.25% Dithane M–45 (mancozeb) and 0.1% Bavistin (carbendazim). Fix the plants and water with mixture of Indofil-M45 (mancozeb) (0.25%) and Bavistin (carbendazim) (0.1%).

- Spray the nursery with a mixture of Ridomil (metalaxyl + mancozeb) @0.2% or Indofil M-45 @ 0.25% / Bavistin (carbendazim) @0.1%.
- For good healthy seedlings, urea spray of 3g/l is given when the plants are 8-10 cm tall.

2. Early blight *(Alternaria solani)*

Symptoms: Initial disease appears as brownish-black spots on older foliage. Spots expand beyond 1/4 in becoming concentric with a yellow halo. Stem lesions appear small, dark and sunken, expanding to elongate, zoned lesions, girdling may occur at the soil line. Fruit spots occur at the calyx end of ripe or green fruit appear leathery, concentric and velvety black.

Management

- Use disease free seed.
- Diseased plants should be uprooted and burnt just after detection in the field.
- Remove lower leaves after first fruit sets; remove affected leaves
- Use *Bacillus subtilis* and *Trichoderma harzianum, T. viride* and *Pseudomonas fluorescens* as soil application.
- Seed treatment with Thiram (3 g/kg).
- Spraying the crop with Bordeaux mixture (1%) at weekly interval before onset of disease.
- Spray Indofil Z–78 (zineb) (0.25%) / Score (difenoconazole) (0.5%) /Indofil M-45 (mancozeb) (0.25%)/ Blitox (copper oxychloride) (0.3%).

3. Late Blight *(Phytophthora infestans)*

Symptoms: Late blight affects both the leaves and fruits of tomato. Late Blight is the disease responsible for the Irish Potato Famine. Late Blight spreads rapidly. Greasy looking, irregularly shaped gray spots appear on leaves. A ring of white mould can develop around the spots, especially in wet weather. The spots eventually turn dry and papery. Blackened areas may appear on the stems. The fruit also develop large, irregularly shaped, greasy gray spots.

4. Buckeye rot *(Phytophthora parasitica)*

Symptoms: Brown spots appear on green and ripe fruit, often at the blossom end. The spots have bands of dark and light brown rings. These spots remain firm and smooth, although internally the rotted tissue turns mushy and can cover half the fruit. Young green fruit, when infected, usually become mummified. Fruit touching or near the soil are most likely to become infected. Severely affected roots become necrotic and decayed. The leaves become bronze and later dieback from the tip.

Management

- Stacking of plants and prune tomatoes to promote air movement and reduce leaf wetness. Do not stake and prune when wet, as diseases can be spread by people and tools.
- Destroy the diseased plant and rotten fruits. Maintain proper drainage.
- Removal of leaves upto 15 cm from soil surface.
- Avoid over head watering.
- Some of the bioagents i.e. *Penicillium aurantiogriseum, P. veridicatum, Aspergillus terreus, T. viride* etc. applied in the soil have an antagonistic effect on Phytophthora infestans.
- Spray Ridomil MZ / Moximate (cymoxanil + mancozeb) @0.25% and Indofil M-45 (mancozeb) Antracol (propineb)/ Kavach (chlorothalonil) @0.25%.

5. Septoria leaf spot *(Septoria lycopersicae)*

Symptoms: Septoria can infect all above-ground parts of the plant other than the fruit but infections are most obvious and extensive on the foliage. The infections are characterized by small (1/8"), circular lesions with dark borders and greyish centres. Close inspection reveals tiny black or brownish dots within the lesions (pycnidia of the fungus).The leaves eventually wither and die. The disease progresses up the stem and total defoliation of the plant may finally occur.

Management

- Remove all plant debris.
- Staking increases air circulation and helps to dry the leaves-this reduces favourable conditions for infection.
- Mulching acts as a barrier between the soil (containing infected fragments of plant debris) and the tomato and prevents splashing of spores onto the lower leaves.
- Spray Dithane M-45 (mancozeb) @ 0.25% or Bavistin (carbendazim) @0.1%.

6. Fusarium wilt *(Fusarium oxysporum f. sp. lycopersicae)*

Symptoms: Infestation often occurs on mature plants after flowering and at the beginning of fruit set. The initial symptom is the yellowing on the lower leaves and a slight wilting on the infected shoots. Yellowing begins to appear on one side of a leaf and then all leaflets become yellow on the other half of the leaf. As the disease progresses, one side of the plant wilts and this process spread to the other side as the infection worsens. An infected plant often dies before maturing

Management

- Apply farmyard manure where soil is acidic

- Use certified, disease-free seeds
- Use *Bacillus subtilis* and *Trichoderma harzianum, T. viridae* and *Pseudomonas fluorescens* as soil application.
- Seed treatment with Bavistin (carbendazim) @ 2.5g/kg is effective to some extent.
- Seedling dip and drenching of seedlings or plants with Bavistin (carbendazim) @ 0.1%

7. **Powdery mildew** *(Oidiopsis taurica/ Oidium neolycopersici)*

Symptoms: Symptoms of the disease occur only on the leaves. Symptoms initially appear as light green to yellow blotches or spots that range from 1/8 - ½ inches in diameter on the upper surface of the leaf. The spots eventually turn brown as the leaf tissue dies. The entire leaf eventually turns brown and shrivels, but remains attached to the stem. A white, powdery growth of the fungal mycelium is found on the top of leaves.

Management

- Removal of diseased leaves.
- Restrict visitor access to the greenhouse.
- Follow strict greenhouse hygiene throughout the growing season.
- Control outdoor weeds surrounding the greenhouse.
- Conduct a through year-end clean up and dispose of all crop debris off-site or by burning or burying in a landfill.
- Keep ornamentals and imported tropical plants out of the greenhouse and immediate area.
- Improve greenhouse climate to reduce relative humidity and increase air circulation.
- Spray Contaf/Sitara (hexaconazole) @0.05% or Karathane (dinocap) @ 0.1% or Tilt (propiconazole) @ 0.05% / Score (difenoconazole) @ 0.05% or Bayleton @0.05%.

  Other fungicides like Benomyl, Carbendazim @ 0.1% are also effective.

8. **Bacterial wilt** *(Ralstonia solanacearum)*

Symptoms: Wilting first appears on the youngest leaves of plants during hot daytime temperatures. The infected plants may recover, temporarily, in the evening, when temperatures are cooler. A few days later, a sudden and permanent wilt occurs. The roots and lower portion of the stem have a browning of their vascular system. The invaded roots may rot due to infection from secondary bacteria. Diseased stems that are cut and placed in a small container of water will show yellowish or grayish bacterial ooze coming from the cut end. When conditions are less favorable for disease development (for example, cool and dry), the infected plants may only show signs of stunting, and adventitious roots may develop on the main stems. The lower leaves will turn yellow before wilting symptoms occur.

How to identify the bacterail wilt: Plants wilt, first on the lower leaves during a hot day and then recovering at night. A few days later, a sudden and permanent wilt occurs. Look for brown discoloration of the vascular system. Ooze will flow from a cut, infected stem.

Management

Avoid contaminated land. Suitable rotations can only be determined through local experience because of the diversity of *R. solanacearum strains* and *races,* and the many agro-climatic zones where reports occur. Eradicate weed hosts. Remove wilted plants, root debris and volunteer hosts, and burn them to reduce spread of the disease from plant to plant. Disinfest tools when used in an infested field. Wash with water or bleach or sterilize by flame. Wash the soles of shoes after working in an infested area. Work in the infested portion of a field after working in the noninfested areas. There are tomato varieties with some tolerance to bacterial wilt. Variations in race and strain of *R. solanacearum* make it difficult to utilize these varieties in some regions.

9. Bacterial spot *(Xanthomonas campestris pv. vesicatoria)*

Symptoms: Xanthomonas campestris is the widespread disease on the leaves, symptoms appear as small (about 3 millimeters), irregular, black, greasy lesions. Leaves with many lesions may turn yellow, may appear ragged or may fall off. Infection of flower parts usually results in serious blossom drop.

Management

Control of the disease depends on the use of bacteria free seed and seedlings, resistant varieties, crop rotations.

II. Capsicum *(Capsicum annum)*

Capsicum is prone to a number of diseases. The causes may be fungus, bacteria, virus, or abnormal environmental conditions. Some of the important diseases and their Management are described below:

1. Powdery Mildew (*Leveillula taurica*)

Symptoms: In general, pepper crops become more susceptible to this powdery mildew as they mature. Older plants and lower leaves are the first to show evidence of powdery mildew infection. Pepper powdery mildew needs living host plant tissue to grow and survive. The fungus only infects the leaves not the fruit or stems of pepper plants. Check for pepper powdery mildew by closely inspecting the underside of older leaves for the first signs of the disease. Look for fluffy, white patches of powdery mildew on the underside of leaves. These patches may turn brown rather than remaining white. The top surface of the leaf may appear normal or have diffuse, yellow patches which correspond to the mildew colonies on the lower surface. Early powdery mildew infections can be seen more easily by holding the leaf up to the light and looking for developing mildew colonies. Severely infected leaves wither and drop off causing plants to die.

Management

- Removal of diseased leaves
- Restrict visitor access to the greenhouse
- Follow strict greenhouse hygiene throughout the growing season.
- Control outdoor weeds surrounding the greenhouse
- Improve greenhouse climate to reduce relative humidity and increase air circulation
- Spray Contaf/Sitara (hexaconazole) @0.05% or Karathane (dinocap) @ 0.1% or Tilt (propiconazole) @ 0.05% / Score (difenoconazole) @ 0.05% or Bayleton @0.05%. Other fungicides like Benomyl, Carbendazim @ 0.1% are also effective.

2. **Phytophthora blight and fruit rot** (*Phytophthora nicotianae* var. *nicotianae)*

Symptoms: Phytophthora infects roots, crowns, stems, leaves, and fruit, causing seedling damping-off, stem lesion, stem blight, leaf spot, and fruit rot. The first symptom on pepper in the field is commonly crown rot. A lesion girdling the base of the stem causes rapid collapse and death of the plant. This phase of the disease usually occurs in low-lying areas of the field. Following rainstorms, black, girdling lesions form on the stem and in the exiles of pepper branches, resulting in wilting of leaves and branches. The affected plants gradually die. It is not uncommon for an entire field to exhibit plant blight. Leaf symptoms are less common. When leaves are infected, half-moon-shaped, tan lesions form at the margin. Infected fruit develop dark, water-soaked lesions, which are commonly covered with white mold. The water-soaked lesion may not be immediately obvious; however, the soft, infected tissues may slip off, or can be easily punched when touched during picking. Affected fruit wither, but remain attached to the plant.

Management

- Use disease free seed
- Grow crop on raised and well drained beds
- Avoid sprinkler irrigation
- Destroy rotten fruits and proper drainage

- Use of biocontrol agents such as *Bacillus subtilis, B. vallismortis, Paenibacillus* and mycorrhizal inoculants.
- Pepper seed treatment with biomass of *Gliocladium virens* and *Burkholderia cepacia.*
- Seed treatment with Ridomil MZ (2g/kg seed)
- Spray Ridomil MZ / Moximate (cymoxanil + mancozeb) @0.25% and Indofil M-45 (mancozeb) Antracol (propineb)/ Kavach (cholrothalonil) @0.25%.
- Silica amendment of nutrient solutions.

3. Anthracnose/Fruit rot (*Colletotrichum capsici*)

Symptoms: Colletotrichum is capable of causing disease on virtually all parts of the pepper plant during any stage of plant growth. Fruit symptoms initially begin as water-soaked lesions that become soft, slightly sunken, and become tan. The lesions can cover most of the fruit surface and multiple lesions occur. The surface of the lesion becomes covered with the wet, gelatinous spores from salmon-coloured fungal fruiting bodies (acervuli) with numerous black spines (setae). Concentric rings of the acervuli are common within the fruit spots. In some cases, the lesions are brown, not orange, and then black from the formation of setae and sclerotia (a dark, fungal survival structure).

Management

- Use disease free seed
- Hot water seed treatment at 52o C for 30 min.
- Destruction of crop debris
- Seed treatment with Thiram/ Vitavax (carboxin) (2.5g/kg seed)
- Spray Dithane M-45 (mancozeb) (0.25%)/ Bavistin (carbendazim) (0.1%)/ Companion (mancozeb) + carbendazim) @ 0.25%.

4. Stem Rot *(Sclerotium rolfsii)*

Symptoms: The fungus attacks the stem of the plants near or at the soil line and forms a white mold on the stem base. Later in the season, small, round brown bodies (sclerotia) appear in the mold. Infected plants wilt and slowly die. Vascular discoloration can be observed in stem tissues above the lesion

Management

- Avoid water logging.
- Use *Pseudomonas* fluorescens, *Bacillus* subtilis and *Trichoderma harzianum* or *T. viridae.*
- Drench the plants with Captan @ 0.25% or Tilt (propiconazole) @ 0.1% or Blitox (copper oxychloride) @ 0.3%.

5. Bacterial wilt

Symptoms: Wilting of leaves, sometimes only few branches of the plant, occurs during hot days followed by recovery throughout the evening and early hours of the morning. The wilted leaves maintain their green color and do not fall as disease progresses. Under favorable conditions complete wilt will occur. The vascular tissues in the lower stem of wilted plants show a dark brown discoloration. These symptoms are very similar to those of Phytophthora blight, induced by *Phytophthora capsici*. Furthermore, a cross section of the stem of a plant with bacterial wilt produces a white, milky strand of bacterial cells in clear water.

Management

Because it is caused by a genetically diverse soilborne pathogen with a wide host range, bacterial wilt is very difficult to control after it is established in the poly house. No single measure totally prevents losses caused by the disease. Cultural practices, if judiciously used, can reduce disease incidence and severity, allowing the disease to be manageable. Seedlings must be free from infection by *R. solanacearum*. It is mandatory that commercial seedling producers use irrigation water not contaminated with the pathogen. Fields should not be over-irrigated, because excess soil moisture favors disease build-up. Crop rotation with non-susceptible crops reduces soilborne populations of the bacterium.

III. Cucumber *(Cucumis sativus)*

Cucumber is prone to a number of diseases. The causes may be fungus, bacteria, virus, or abnormal environmental conditions. Some of the important diseases and their Management are described below:

1. Downy mildew *(Pseudoperonospora cubensis)*

Symptoms: Downy mildew symptoms first appear as small yellow spots on the topside of older leave. The centre of the lesion will eventually turn tan or brown and die. The yellow spots sometimes take on a "greasy" appearance. They usually lack a distinct border. During prolonged wet periods the disease may move onto the new growth. As the disease progresses, the lesions expand and multiply, causing the field to take on a brown and "crispy" appearance. On most cucurbits the lesions appear irregular but on cucumber, they often are angular and confined by the leaf veins. Under humid conditions, the lesion often develops a downy growth immediately on the underside of the light yellow lesions observed on the top of the leaf. This downy growth is particularly noticeable in the mornings after a period of wet weather or when conditions favour dew formation. The downy growth on the underside of the lesions is frequently speckled with dark purple sporangia (spore sacks) which can be observed with a hand lens.

Management

- Avoid over watering and remove infected plants
- Wider spacing and application of potassic fertilizers helps in reducing the disease levels.
- Spray of neem oil is also effective
- Use of biofungicides such as *Bacillus subtilis*
- Spray Ridomil MZ/ Moximate (cymoxanil + mancozeb) @0.25% and Indofil M-45 (mancozeb) Antracol (propineb)/ Kavach (chlorothalonil) @0.25%.

2. Powdery mildew *(Sphaerotheca fuliginea/ Erysiphe cichoracearum)*

Symptoms: White, powdery fungal growth develops on both leaf surfaces, petioles, and stems. The growth is primarily asexual spores called conidia. It usually develops first on crown leaves, on shaded lower leaves, and on leaf undersurfaces. Yellow spots may form on upper leaf surfaces opposite powdery mildew colonies. Older plants are affected first. Infected leaves usually wither and die. Plants may senesce prematurely. Fruit infection is rare on watermelon and cucumber. cleistothecia are dark brown, minute structures (about 0.003 inches in diameter), barely discernible without a hand lens, that develop late in the growing season and protect the sexual spores within from adverse conditions.

Management

- Removal of diseased leaves.
- Sufficiently space the plants.
- Restrict visitor access to the greenhouse.
- Follow strict greenhouse hygiene throughout the growing season.
- Control outdoor weeds surrounding the greenhouse.
- Conduct a through year-end clean up and dispose of all crop debris off-site or by burning or burying in a landfill.
- Keep ornamentals and imported tropical plants out of the greenhouse and immediate area.
- Improve greenhouse climate to reduce relative humidity and increase air circulation.
- Application of conidial suspensions of the hyperparasite *Ampelomyces quisqualis, Sporothrix rugulosa, Tilletiopsis minor* and *Verticillium lecanii* to greenhouse cucumbers inoculated with *Sphaerotheca fuliginea* increased cucumber yield compared with an untreated control.
- Spray Contaf/Sitara (hexaconazole) @0.05% or Karathane (dinocap) @ 0.1% or Tilt (propiconazole) @ 0.05% / Score (difenoconazole) @ 0.05% or Bayleton @0.05%.

- Other fungicides like benomyl, carbendazim @ 0.1% are also effective.

3. Angular leaf spot *(Pseudomonas syringae pv. lachrymans)*

Hosts: All members of the Cucurbitaceae family; susceptibility can vary between genera and species

Symptoms: Initial symptoms appear as small, water-soaked lesions on the leaves and are often visible shortly after crop emergence. These spots expand until they are confined by the veins. This gives them an angular appearance for which the disease is named. Under conditions of high humidity, the spots may ooze droplets of liquid containing bacteria.

Management

- Practice good sanitation (e.g., use clean equipment and tools, avoid work in poly house when wet, locate cull piles away from production fields). Use pathogen-free seed or treat seed to eliminate the pathogen. Use certified disease-free transplants. Practice crop rotation (at least 2 years without cucurbits). Avoid overhead irrigation. Use resistant varieties, when available. Pesticide sprays (e.g., copper compounds)

4. Bacterial leaf spot *(Xanthomonas campestris pv. cucurbitae)*

Symptoms: Spots first appear on squash and pumpkin leaves as small water- soaked or greasy areas on the underside of leaves, and as indefinite yellow areas on the upper side of leaves. In about five days the spots become round to angular with thin, brown, translucent centres and a wide, yellow halo. The spots enlarge up to about 7 mm diameter. Occasionally young stems and petioles are attacked. Young fruit may also be affected. Fruit may produce a light brown ooze from small, water-soaked areas, which can extend into the seed cavity, causing seed infection.

Management

- Use disease-free seed. Do not work in the crop when it is wet. Practice rotation and destroy plant debris by burning, if possible, or ploughing in. Use recommended chemical treatments, but often it is too late after fruit symptoms are noticed.

5. Bacterial wilt *(Erwinia tracheiphila)*

Cucumber, cantaloupe, squash, and pumpkin are susceptible, whereas water- inside xylem vessels of leaf vein of black rot-infected cabbage. *Symptoms:* Affected plants develop sudden wilting of foliage and vines and finally die. Affected squash fruit develop a slime rot in storage. The severity of the disease varies from an occasional wilted plant to destruction of 75 to 95% of the crop.

Management

- The bacterial wilt of cucurbits can be controlled best by controlling the cucumber beetles, especially the early ones, with insecticides. To avoid squash rot in storage, only fruit from healthy plants should be picked and stored in a clean, fumigated warehouse. Resistant cucurbit varieties should be preferred to more susceptible ones.

Viruses associated with Vegetables

(Tomato, Capsicum, Cucumber)

- Mosaic : *Tobacco mosaic virus* (TMV), *Cucumber mosaic virus* (CMV), Potato virus Y (PVY), Tobacco virus etch (TEV), *Pepper mottle virus* (PeMV), Pepper mild mottle virus (PMMV), *Pepper severe mosaic virus* (PSBMV)
- Tospo viruses: *Tomato spotted wilt virus* (TSWV), Capsicum cholorotic virus (CaCV), *Groundnut bud necrosis virus* (GBNV)
- Leaf curls: *Tomato leaf curl virus* (TLCV), *Tomato yellow leaf curl virus* (TYLCV), *Chili leaf curl virus* (CLCV)

1. Capsicum mosaic complex: The mosaic complex is caused by many viruses like Tobamoviruses: TMV, *Pepper green mottle virus; Cucumber mosaic virus, Potato virus Y, Tobacco etch virus, Pepper mottle virus, Pepper severe mosaic virus, Pepper mild mottle virus.* These viruses spread among the plant through aphids and contact. Viruses like *Tobacco mosaic* and *Pepper mild mottle virus* spread through contact and as seed contaminant but not by insects. Whereas *Potato virus Y, Tobacco etches, Pepper vein mottle virus* spread both by aphid vectors as well as contact. In contact spread, the cutting and pruning knives used for trailing the plants under polyhouse plays an important role. This type of symptoms is caused by many viruses and is generally called as mosaic complex.
2. Capsicum mosaic: *Cucumber mosaic virus*: The symptoms appear as light and dark green areas on leaves (mosaic) of infected plants. There is reduction in leaf lamina, plant remains stunted and fruits of infected plants are deformed. The virus is transmitted by aphids in non- persistent manner may also be through seed. Diseased plants remain stunted and do not bear normal fruits after one or two pickings.
3. Tobamoviruses (*Tobacco mosaic and Pepper mild mottle*)*:* Leaves develop a mild systemic mosaic and plant also show crinkling. Severe symptoms such as distortion, rings, lines and necrotic spots develop on fruits of infected palnts. The virus is transmitted through contact and seed contaminant but not by insects. Main source of primary

inoculum includes infected seed, crop debris, work benches and humans working in the polyhouse.

4. *Potato virus Y, Tobacco etch, Pepper veinal mottle:* These viruses cause mosaic, severe mosaic, shoe stringing, stunting of plants and fruit deformation. Transmitted by aphids in non-persistent manner. The virus survives on weeds, volunteer plants.
5. *Tomato spotted wilt virus:* In general, tospovirus symptoms causes consist of chlorotic or necrotic rings, lines pattern, or spots on leaves, fruits and stems, along with necrotic streaks on stems. There is bronzing, curling, and wilting of leaves. The infected plants remain stunted and necrosis of parts or whole plants. The symptoms vary with the type of host, organ affected and age of plant or organ at the time of infection. Tospoviruses are transmitted by many species of thrips *(Frankliniella occidentalis), tobacco thrips (F. fusca), common blossom thrips (F. schultzei), onion thrips (Thrips tabaci),* and *melon thrips (T. palmi).* TOSPO viruses have wide host range. Tomato plants show poor growth, leaf rolling, browning of leaves and wilting. Spots ringspots on fruits are very common.
6. Geminivirus

i. *Tomato leaf curl virus*: Tomato leaf curl virus belongs to a group of viruses which cause a range of destructive plant diseases worldwide. Tomato leaf curl virus (TLCV) and tomato yellow leaf curl virus (TYLCV) are the most damaging. Plants are stunted or dwarfed. Leaflets are rolled upwards and inwards. Leaves are often bent downwards and are stiff, thicker than normal, have a leathery texture and often have a purple tinge to the veins on the underside. Young leaves are slightly chlorotic (yellowish). Flowers appear normal. Fruit, if produced at all, are small, dry and unsaleable. Affected plants tend to be distributed randomly or in patches. TLCV can be confused with several other tomato disorders such as tomato big bud, tomato yellow top, physiological leaf roll and phosphate and magnesium deficiency.

ii. *Tomato and pepper leaf curl:* There is leaf curling, reduction in leaf size, and thickening of leaves of the diseased plants, followed by bunchy growth of the branches or severe stunting of plants. Early infections of the plants may result in complete yield loss. The virus is transmitted by whitefly (*Bemisia tabaci*) in semi- persistent manner.

Virus Management Strategy

1. Plant, Polyhouse and surrounding hygiene (removal of infection foci.)
2. Raise healthy nursery or procure seedlings from reliable resources (reduction in initial inoculum).

3. Establish a minimum 2 month crop free period during the summer, preferably from mid-June to mid-August.
4. Avoidance of virus & vector infestation in the polyhouse.
   a. Use of sterilized knife or scissors during pruning and training of plants to avoid virus spread (contact transmitted).
   b. Remove and destroy all infected plants.
   c. Monitor vector population with yellow/ blue sticky traps. d. Use of insecticides at regular interval to avoid vector
   e. Use of resistant varieties if available.

References

Arneson, PA 2001. *Plant Disease Epidemiology*. http//www.apsnet.org/edcentre/ advanced/ topics/epidemiology

Charles W A, Jean B. R. and Jonathan G. S. 2000. Disease Management for Vegetables and Herbs in Greenhouses Using Low Input Sustainable Methods collage of Agriculture and life sciences Vegetable Disease Information Note 2 (VDIN-002)

Hull R. 2005. Methods for Assay, Detection and Diagnosis. *In:* Matthew's Plant Virology (eds 4th Academic Press) pp. 983.

Hye Yeon HY, Park M, Lee HB and Kim K. 2008. Outbreak of *Cucumber mosaic virus* and *Tomato spotted wilt virus* on Bell Pepper Grown in Jeonnam Province in Korea Plant Pathol. J. 24: 93-96.

Jacobsen B 2001. Disease Management. *In:* Encyclopedia of Plant Pathology, O.C. Maloy and T.D. Murray, eds. Wiley, New York. p351-356

Jarvis, W.R. 1989. *Managing Disease in greenhouse crops.* Plant Disease: 73: 190-194

TECHNOLOGICAL INTERVENTIONS IN CULTIVATION OF SAFFRON

7

C.S. Kalha, Vishal Gupta and V. K. Razdan

7.0 INTRODUCTION

Jammu and Kashmir region of India enjoys the monopoly of growing saffron crop. Cultivation of saffron is mostly concentrated around Pampore and adjoining Karewa belts of Pulwama district. Cultivation on small scale is done in Kishtwar District where the importance of saffron besides its export value is related to rural employment, high net profit compared to other crops and religious importance. The word saffron originated from the 12th-century old French term *safran*, which derives from the Latin word *safranum*. *Safranum* is also related to the Italian *zafferano* and Spanish *azafran* (Harper, 2001). *Safranum* comes from the Arabic word *asfar* which means "yellow" (Katzer, 2001; Kumar, 2006). In Sanskrit it bears the name of Kum-Kum and Kesar. It is introduced in China by Mongel invaders, as depicted in China Materia Media "Pun-Tsaou" compiled in 1552-78. As Kashmir is concerned, the saffron cultivation started in the reign of King Laltaditya during 550 A.D. In "Tozaki-Jahangiri" some reference has been made regarding saffron cultivation in Kishtwar. Famous Kashmiri Vaids namely Vegbhatta and Sushtra used saffron as an ingredient in Ayurvedic medicines. Historically, it has been used as an ingredient of fragrance, as a dye and in herbal medicine but the high cost of the spice currently stops its widespread use in these areas. Saffron contains more than 150 volatile and aroma- yielding compounds. The stigma of the saffron flower contains number of chemical substances. Carbohydrates, minerals, mucilage, vitamins (especially riboflavin and thiamine) and pigments including crocin, anthocianin, carotene, lycopene, zigzantin, flavonoids, aminoacids, proteins, starch, gums and other chemical compounds have also been described in saffron (Abdullah, 1993; Abdullah, 2002 Rios *et al.*, 1996; Winterhalter and Straubinger, 2000). Crocin is responsible for the colour of saffron, picrocrocin for its bitter taste and safranal for aroma. The shelf life of saffron is greatly related to the fate of its pigments (Raina *et al.*, 1996). Safranal is the main aroma factor which comprises of about 60% of the volatile components of saffron. It has insecticidal and pesticidal properties, and may comprise up to 4% of dry saffron. (Abdullah, 1993; Abdullah, 2002; Rios *et al.*, 1996; Winterhalter and Straubinger, 2000; Tarantilis and Polissiou, 1997).

7.1 Diseases

The crop suffers heavy losses because a number of diseases affect the yield and most prominent among them is corm rot. Corm rot of saffron is presently found to be the major constraint in the production aspects, which affect the saffron production and simultaneously stops the production of daughter corms. Management of corm rot can successfully be done by applying systemic chemicals as corm dipping treatments. Corm treatment with Bavistin 50 WP @ 0.2% for 15-20 min. before sowing is widely adopted for management of corm soil diseases.

7.1.1 Symptoms

Corm rot disease is the major one, which increases drastically year after year in fields. Diseased corms show brown to darker brown sunken and irregular patches below corm scales. The rot lesions are usually 1mm deep having raised margins. Symptoms are mostly located in root and bud regions. In some cases, light green, black or white fungal mass is observed. Severely infected corms have dry foliage. Drying starts from tip to downwards. No spots are observed on leaves. Yellowish foliage could easily be pulled out from the infected field. Patches of diseased foliage are scattered throughout field. In severe cases the entire corm turns into black or powdery mass. Incidence of rot is less in newly planted field and maximum in old planted fields. The rot causing pathogens get multiplied year after year in the fields because of longer planting cycle and poor agronomic practices. Injury to the corm enhances disease incidence. Corm rot of saffron, drought, poor cultivation practices along with the nutrient deficient soil played major hurdles in the successful cultivation of saffron and responsible for reduction in production and productivity of this important commercial crop.

The major precautions adopted for the raising of saffron crop are as:-

Uniform size of corms: It is mandatory to grade the saffron on the basis of their size before sowing. Optimum weight of corms (8-10 g) should be preferred for raising of saffron crops whereas, the undersized, small ones should be sorted out and opted for raising the nursery of saffron corms which ensures the enhancement of corm size and better production of daughter corm for future crops. Planting of corms without sorting is the main reason for spread of corm rot inoculums. Sorting of corms ensures flowering in first year with check on corm rot disease.

Transportation: During transportation, storage and handling of corms utmost care should be taken to avoid any injury, damage or wound to the planting material which predisposes the saffron crop to complex diseases. Proper aeration should be adopted during packing and transportation of corms. During storage, the heaps or stacking of corms should be avoided and the corms must be placed in layers so as to reduce the chances of injury to corms and proper circulation of air. Any type of injury/wound/damage not only reduces the vigor of the corms but also aggregates the chances of infection in the seed lot. Care should be taken during lifting of corms from one place to another. Corms should be distributed immediately after transportation for sowing purposes, as poor storage conditions results in rotting of corms.

Land preparation: Well coordinated work plan is required to produce a good saffron crop. Continuous cultivation of saffron on same field should be avoided. Crop rotation helps in reducing the diseases/ pest problem and also maintain the physical, chemical and biochemical properties of soil. Field preparation starts in the month of May and is continued till the time of planting i.e., last week of July. During this period land should be ploughed with mould board plough/ tractor for 3-4 times and clods be broken by spade or manually so that field become loose and free from any stubble or any kind of weeds. It is important to prepare raised beds, with a slope with all round drains of convenient depth and width giving it the shape of vegetable nursery with spacing: R × R = 20 cm, C × C = 15 cm along with depth of 10-15 cm. Planting depth affects corm production, more buds sprout from shallow planted corms than from deep planted corms, resulting in more daughter corms. Corms must be planted by scraping the scales in such a manner that the apical portion is located upward and corm base are in contact with the bottom of the furrow. The diseased, injured and wounded corms should be discarded. Injured corms (including scales) should be collected and buried. Spreading or placing the damaged or diseased corms on the cultivated field (a common practices) result in spreading the propagules from one field to another field and chances of infection. Drainage is an essential factor for the better growth and development of corms. About 50 q corms are required for one hectare. Sowing is done with the mould board plough by putting the corms in the furrow and then manually presses the corms for optimum depth and finally covered by soil with a hoe. Under Kishtwar conditions, the sowing of corms is recommended in I_{st} fortnight of August which helps in promotion of good crop; early and late sowing of corms affected by various pest and diseases along with flower abortion and germination failure, which ultimately result in crop losses. Corm treatment with carbendazim 50 w.p. @ 4g/kg of corms for 15 minutes and drying of corms before sowing reduced or suppress the chances of infection.

Soil health: Soil health is the basic and fundamental aspect in increasing the production of any crop. Soil analysis brings forth the condition of the soil. On the basis of soil testing the proper doses of inorganic fertilizers should be applied as the concept of nutrient management was lacking in Kishtwar and the saffron soils were deficient in nutrients. About 10-15 tons/ha of decomposed FYM should be applied. Small heaps of farm yard manure are scattered to the whole saffron field before plouging and it can mix in the soil by mould board plough for proper distribution. It improved the soil quality and also provides nutrients. However, use of integrated plant nutrient system (FYM @ 175 kg/ha + inorganic fertilizer @ 30:20:15 kg/ N, P and K, respectively) resulted enhancement in the saffron yield.

Intercultural operations: The first hoeing is done during June-July. The operation is considered vital to ensure that dormant corms receive proper aeration in the soil else would result in corm rot disease during the hot summer season. Subsequently second hoeing is done in early September by hand harrow to avoid injury to the buds. This improves moisture retention capacity and enhance growth and ultimately development of corms takes place. During the first year, saffron field require hoeing and weeding, just before the buds appear on the corms. After the flowering light hoeing is recommended to promote daughter corm production. Collaborative and improved cultural practices along with soil application of fungal bio-agents, *Trichoderma viride* and *T. harzianum* @ 2.5 kg/ha mixed with well rotten farmyard manure (100 kg) recommended to reduce the soli borne inoculums. Corm dip treatment with carbendazim 50 % W.P. @ 0.2% for 25-30 min is highly effective in managing corm rot disease. Use of healthy corms followed by the application of carbendazim @ 0.2% in the subsequent years appears as the best management strategy for combating the disease.

Irrigation: Spring rain is considered favorable for corm production. Light precipitations in September are beneficial to initiate plant growth activity and promotes flowering yield (stigma). Rain or cold during flowering spoils the saffron yield and high temperature encourages diseases. In absence of rains, two light irrigation should be applied at the intervals of ten days at the critical stage of the crop (35 days after sowing). Irrigation has significant impact in increasing the yield of saffron.

Flowering: Sprouting starts in the first week of October like pin-pointed shoots. The corms sprouts grow fast and within a fortnight attain a proper size with three fleshy leaves resembling pine-needles. Afterwards flowers emerge during the morning hours from the middle of these leaves (15 Oct. to 15 Nov). Flowers are picked up early in the morning, after the dew disappears as the frosted flowers decay easily. The crop is very meager during the first year but increases gradually and again declines after the fourth year. Slow drying (under natural conditions) gives a poor quality product. High temperature (Solar/Hot air) drying adds the brightness of color, and preserves the quality of product. Drying at 40±5 °C for 4-6 hours improves the quality of the product.

Corm lifting: The corms are extracted from the 4- year old saffron field with kundali and gathered in wooden basket in the last week of July. The rotten and disease corms should be separated from healthy ones during digging. The extracted corms are placed over the cemented floor and stored in well ventilated room until the sowing season starts. Storing corms for a longer period may reduce flowering potential.

References

Abdullah, F.I. 1993. *Biological effects of saffron.* Biofactors 4: 83-86.

Abdullah, F.I. 2002. Cancer chemopreventive and tumoricidal properties of saffron (*Crocus sativus* L.). Exp. Biol. Med. Maywood 227: 20-25.

Abdullah, F.I. and Frenkel, G.D. 1992. The effect of saffron on intracellular DNA, RNA and protein synthesis in malignant and non-malignant human cells. Biofactors 4: 43-45.

Abdullah, F.I., Caballero, O.H., Riveron, N.L., Pereda, M.R., Hernandez, J.M. and Perez, L.J. 2003. Use of *in vitro* assays to assess the potential genotoxic and cytotoxic effects of saffron (*Crocus sativus* L.). Tox. *In vitro* 17: 731-736.

Cappelli, C., Donato, G., and Di Donato, G. 1994. Attacks of *Fusarium oxysporum* Schlecht in crops of Saffron in Abruzzo. Informatore Agrio 50 (25): 75-76.

Carta, C., Fiori-M. and Franceschini, A. 1982. Charcoal rot of saffron (*Crocus sativus* L.) bulbs. Studi-Sassaresi 29: 193-197.

Deng, Y., Guo, Z.G., Zeng, Z.L. and Wang, Z. 2002. Studies on the pharmacological effects of saffron (*Crocus sativus* L.): a review. Zhongguo Zhong Yao Za Zhi 27:565-568.

Grigg, D.B. 1974. *The Agricultural Systems of the World*, Cambridge University Press, ISBN 0-521-09843-2.

Gulam, H. and Devi, L.S. 2003. Corm rot diseases of saffron in Kashmir valley. Indian Phytopathology 56 (1): 122.

PEST MANAGEMENT PROBLEMS AND STRATEGIES IN HILL FARMING

8

Uma Shankar and D. P. Abrol

8.1 INTRODUCTION

The increasingly precarious situation facing the world's hill regions has been the focus of a growing level of awareness over the past two decades. The high altitude regions offer a challenge to agricultural development agencies which is shared by many hill areas of the world characterized by marginality, extreme diversity, environmental fragility and complex smallholder farming systems. There is justifiable concern relating to the conservation of hill and mountain ecosystems, to the extent that the concept of sustainability has been adopted as a broad slogan by the environmental movement (Lele, 1991)

Conditions specific to mountain regions, such as inaccessibility and diversity have placed a high premium on large families, encouraging population growth. Children are essential to family subsistence strategies offering free labour, and they are consequently considered to be a source of wealth. Population control measures in these regions have ignored this scenario (Sharma and Banskota 1990, cited in Jodha, 1991).

Increased population generates an increased demand for cropland, and for firewood, both resulting in deforestation. As the distance required to walk to collect firewood increases, so the critical threshold is reached when households resort to the use of animal manure as fuel, so depriving the cropping systems of their major source of fertilizer, weakening soil structure and decreasing yields (Ives and Messerli, 1984). This situation, allied to the reduction in ground cover and the stabilizing influence of tree roots leads to surface erosion, landslides, mass wasting, flooding and sedimentation. Subsequent decline in the overall systems productivity is inevitable.

This oversimplification ignores the complexity of the reasons that cause deforestation, and avoids the more recent proposals that the effects of deforestation are probably less severe than originally thought (Metz, 1991). However, there is little doubt that the mountain systems of subsistence agriculture are undergoing a process of rapid change that relate to persistent negative trends in the variables relating to the health of the natural resource base, the productivity of land based activities and management practices of the region (Jodha, 1991). The migration of crop production into more marginal and forest cleared areas, as well as the uncontrolled grazing of ruminant animals beyond the carrying capacity of the land (Subedi, 1990) are merely examples of these negative trends, resulting to a large degree from the inherent characteristics of mountain environments.

Conventional answers to this problem would lie in the use of inorganic fertilizer and high yielding varieties (Balogun *et al.*, 1988). Whilst this has been the professed Government approach, the emphasis of the strategy is skewed towards road accessible areas which have access to irrigation, and where there is scope for large increases in production. In the hills, the farmers will continue to rely upon traditional sources of nutrients for the maintenance of soil fertility.

The economy is so heavily dependent upon natural resources, that serious environmental threats that tend to degrade this base, notably deforestation and soil erosion are now being afforded the attention they deserve. Given this situation of rapid environmental degradation and consequent declining production levels, the unsustainability of the mountain environment (Jodha, 1991) and the development of sustainable agricultural systems are a priority issue.

Because land available for new cultivation is now very limited, then most of this increase will have to derive from improved yields. The basic needs for agricultural development, programme are as follows:

1. Emphasis should be upon production of food grain crops in irrigated areas, with vegetables, fruits, eggs, meat and milk production concentrated along the highways and near urban areas. Programmes in the hills should be undertaken on the basis of micro-climatic areas and local requirements.
2. Strengthening on farm and farmer participatory research.
3. Improving productivity per unit area of crops, rather than increasing the area cropped.
4. Developing new technologies for rainfed agriculture while not neglecting technology development for irrigated land.

5. Considering research needs of the whole farming system rather than focusing on individual crops or other farm enterprises.
6. More closely integrating the activities of agricultural research, extension and training services.
7. Strengthening the agricultural technical support services so that they more effectively reinforce the production programmes.
8. Generate appropriate technologies for crop, livestock and horticultural production under Hill Farming conditions that maintain a sustainable resource base. This will emphasise agroforestry and soil conservation.

The limitations to development in the mountains are the harsh agro-ecological conditions of the fragile environment, the widely scattered population settlements, poor transportation and communication facilities, and the almost total lack of markets for produce. In the hills, the complex farming systems with the individual components being integrated and interdependent, inadequate plant nutrients for the achievement of high yields, and lack of irrigation facilities to exploit the land to best advantage are identified restrictions. Added to these, weather conditions are often adverse and can result in total crop failure from hail, flood or landslide. Under both these environments, lack of farmer knowledge about improved technology, and scarcity of labour and animal draught power at peak field operation periods, further erodes the potential to optimize production and yields.

Concepts of agricultural development in the hills

During the past decade, the concept of "Farming Systems Research" and adoption of a "Multidisciplinary Approach" towards solving smallholder farming problems have become the prevailing philosophy. These procedures rely upon people of different specialist subject matter disciplines working together as a team with the active participation of the farmers concerned to identify the technical, economic, social, political and other constraints which they are faced with.

8.1 Problem identification and research organization

To identify the problems, firstly visit as many farmers as possible in the time available and question them about their constraints to production, resource availability and use. Secondly, the overall group is subdivided into smaller units consisting of mixed subject disciplines, both for ease of interviewing farmers without intimidation, and obtaining wide coverage. The composition of these groups is changed at regular intervals, and at minimum on a daily basis. Thirdly, as much in-field discussion as possible is promoted, combined with extensive note-taking, and a draft report is written in the field, the salient points of which are discussed with the farmers again, so that the priorities can be provisionally assigned before returning to the Centre. Based upon this report, priorities for research are assigned, and programmes for their investigation are developed.

8.2 Sustainability

Achieving sustainable development in the hills and mountains is a daunting prospect. The negative trends as indicators of unsustainability are presented in Table 8.1 and the indicators of sustainability could be considered as the inverse of those changes presented above. Ecological sustainability indicators should relate progress with regard to the conservation of ecological life support systems and biodiversity should ensure that renewable resources are sustained and the depletion of non-renewable resources minimized, and should ensure that the carrying capacity of supported ecosystems is not exceeded. Logical indicators to the sustainability of animal production systems include yield and production trends, levels of genetic diversity within domestic species, increased or stable levels of diversity of enterprises, limited extraction from locality, stabilised inputs with increased/stable output, increased areas of regenerated growth and cultivated land, decreases in eroded, overgrazed and degenerated land, and increased availability of agricultural products.

Table 8.1. Negative Changes as Indicators of Unsustainability of Mountain Agriculture

	Changes Related to:		
Visibility of Change	Natural Resource Base	Production Flows	Resource Use/ Management Practice
Directly visible changes	Increased landslides and other forms of land degradation; abandoned terraces; fragmentation of land; changed botanic composition of forest/pasture; reduced water-flow for irrigation and domestic use.	Prolonged negative trend in crop, livestock yields; increased input per unit output; increased time and distance to gather food, fuel and fodder; lower per capita availability of agricultural products	Reduced fallowing, crop rotation, intercropping, diversified enterprises; cropping of sub-marginal land; unbalanced and high intensity of input use.
Changes concealed by responses to changes	Substitution of: cattle by sheep/goat, deep-rooted by shallow rooted crops; shift to non-local inputs; substitution of water flow by fossil fuel for grinding mills; and substitution of manure by chemical fertilizers.	ncreased seasonal migration; introduction of externally supported public distribution system; intensive cash cropping on limited areas.	Shifts in cropping pattern and composition of livestock; reduced diversity; increased specialization in monocropping.

Source: Jodha (1991).

8.3 Technical strategies for sustainable animal production

Increasing production from hill environments necessitates the intensification of resource use without resource degradation (Schroeder, 1985; Jodha, 1991). On the demand side, consideration must be given to balancing the pressure on mountain resources with their carrying capacity. Since agriculture is the major activity at the human-environment interface, attempts at operationalising ecological sustainability have focused on agricultural development (Lele, 1991). However, to date most development strategies have paid more attention to the supply side, with little consideration for the demands upon the system.

For watershed stability, the misuse of livestock is particularly damaging (Pereira, 1989). The importance of animals as nutrient recyclers has led to an increased population of ruminant stock on grazing and forest land that has long exceeded its carrying capacity. Overgrazing rapidly eliminates palatable, nutritious grass and shrub species. Forested areas, vital for the stability of steep mountain slopes, have been depleted as they have been harvested beyond their regenerative capabilities. Several options are available to counter the damage caused by overgrazing, including a reduction in livestock numbers through increased marketed offtake, sterilization and culling, or controlling livestock movements by switching to stall-fed systems for milk and meat production.

Resource use intensification for higher productivity without resource degradation will involve the introduction of improved breeds which will allow farmers to maintain output levels with fewer animals. Similarly, improved animal health reduces mortality rates and allows fewer animals to be maintained. Although this may imply a reduction in manure output, the gradual transition of large ruminant husbandry to stall feeding systems will not only improve production but will also enable larger quantities of manure to be collected and composted. This will relieve pressure on fodder resources. Small ruminants, especially sheep which are more suited to extensive grazing and migratory systems will then be able to take more benefit from the relatively fragile marginal areas.

Livestock management adapting traditional methods and working in concert with the natural ecosystem will not only be more likely to be successful, but will also enhance the natural resource base upon which it depends and in so doing contribute to a sustainable agricultural system.

8.4 Influence of human population on soil fertility

Increases in human population in the past twenty years have converted large parts of the forest areas into agricultural land for food production.

More intensive cultivation, while maximising the exploitation of soil nutrients in the form of food grains, has meant that less has been put back in return to realise sustainable production. Soil fertility under the traditional farming system has been maintained by repeated addition of various amounts of organic compost/manure, ranging from three to 21 mt/ha/annum. Large amounts of compost are produced from a mixture of livestock manure, forest leaf-litter and farm waste. However, over-exploitation of forests is limiting the supply of fodder, litter and humus, and unless this trend of diminishing natural resources is checked in time, and the traditional practices of maintaining soil fertility are modified with available or improved technologies, the hill farming systems will no longer remain sustainable. Use of chemical fertilizer to cope with these losses is costly, and is generally impractical for hill farmers in any case, both because of a lack of service infrastructure to provide it, and its cost.

8.5 Soil fertility status in the hills

Hill farming systems have always been sustained by using traditional practices for maintaining soil fertility. However, overall yields of almost all crops in the hills have decreased over the last twenty years due to a decline in soil fertility. Soil is being lost from the whole land use system through erosion on the one hand, and on the other, fertility is also lost because of intensified cropping patterns, and through other factors such as faulty agricultural practices.

The estimated soil loss per unit area for different land use systems for a typical watershed in the middle mountains (Table 8.2) reveals that maximum soil loss occurs from degraded range land, and poorly managed sloping terraces.

Table 8.2. Estimated soil loss from different land use in the mid-hills of Nepal.

Type of land	Estimated soil loss (mt/ha/annum)
Well managed forest land	0 – 10
Well managed rice terrace	0 – 10
Well manged maize terrace	5 – 15
Poorly managed sloping terrace	20 – 100
Degraded range land	40 – 200

Source: Laban (1978, cited by LRMP, 1986)

Severe soil loss due to erosion, and nutrient extraction from the soil by cereal cropping, indicate the need for soil conservation as a first step in conjunction with maintaining its fertility status, before planning any programmes to increase productivity. Special attention is needed to check soil loss by improving agricultural practices, and care should be taken to utilize manure efficiently by proper composting and mixing with the soil.

8.6 Nutrient recycling in the hill agricultural system

Soil provides the medium for plants to grow, a surface for animals to live on, and food to enable human beings to survive. Forage plants that are grown in the soil are the "inputs" consumed by animals which act as the "factory" to convert plant material into a usable "output", manure. This can then be put back onto the soil in the form of compost, to replenish fertility, and constitutes the traditional nutrient cycling system of hill agriculture. In addition, the use of forest litter. farm waste, and the growing of grain legumes, green manures and bio-fertilizers such as Azolla spp. with minimal use of chemical fertilizers, has also helped to maintain the soil fertility on farm land. Plants (including herbs, shrubs, and trees) protect the soil from the direct effect of rain and wind, and minimise surface run-off by checking top soil erosion, and causing water to slowly percolate into the earth. Their root systems and decomposing plant parts improve the moisture-holding capacity of the soil, and further enhance plant growth. Forage grasses and shrubs that are growing on or around farm land, also help to bind soil particles through their thick root systems, again preventing loss. Natural decay of plant parts including agricultural residues, are mixed with the soil by decomposition processes.

Though animals have only a limited direct effect on restoring soil fertility, they are the major means by which plant produce is converted to manure through digestion of crop residues and fodder/forage. Addition of their dung helps to improve soil texture, and to decompose litter more easily. With a traditional feeding system, and farm yard manure (FYM) preparation method, an adult large ruminant animal provides about four tonnes of FYM (Table 2) or about 10–13kg of nitrogen per annum. However, six large ruminants are required to provide sufficient FYM for one hectare of farm land under a rice - maize - wheat system for one year (Subedi and Gurung, 1991).

8.7 Methods of maintaining soil fertility

Construction, maintenance and modification of terraces, drains, irrigation waterways and tunnels, and the use of mulch and cover crops lends support to the concept that farmers are skilled in soil and water conservation. The poor and declining fertility status of hill slopes is usually corrected by application of compost and manure. Amounts added very greatly, from 10 to 60 mt/ha, with 15–20 mt/ha the average for the mid-hills. Farmers and researchers have identified the maintenance of soil fertility as an important factor that requires more attention in order that present hill farming systems can be sustained.

8.8 Soil conservation

No improvement in soil fertility can be contemplated until soil conservation methods are practiced. Soils of hills are lost through detrimental agronomic practices such as slicing terrace risers every year, excessive tillage and hoeing in the rainy season and severe grazing pressure on pasture and forest lands. In order to first check mass soil erosion, improvements to the management of grazing land and degraded forest land are essential. Use of minimum tillage methods, and preventing the practice of slicing tall bariland risers should be adopted to reduce further soil losses.

8.9 Increased Productivity of Forest and Grazing Land

The major reason for declining soil fertility is the need to use the land more intensively because of increasing human population, coupled with a reduction in manure production, so that nutrients extracted by food crops are not adequately replaced. This is the result of a reduction in animal populations in some areas, but is mostly the result of depletion of the animal feed resources from the forest and grass lands, which means that livestock are not realising their full potential the year round.

8.10 Improved Animal Management Systems

Animal populations already overburden the hill farmer, and it is essential to consider complete stall-feeding in order to use the available feed effectively and maximize manure production. The wastage of valuable urine can be prevented and utilized by improving drainage and constructing a store pit at the animal shed. Losses of manure due to rain and sun could be minimised by providing some kind of simple shelter over the compost heap/pit. Similarly, animal production could be improved by the timely supply of feed and water, without wastage. Straw as a livestock feed can be improved in quality by treatment with urea, and by the practice of ensiling or otherwise preserving the summer surplus of grass and agricultural crop by-products. These could then be consumed during the food scarcity period of winter.

8.11 Improvement in Cropping Practices

In order to supply food grain for a steadily increasing human population from a fixed or limited land resource, improvements to existing farming practices are inevitable. From the soil conservation and fertility standpoint, intercropping of grain legumes within the major cropping systems should be encouraged whenever possible. Similarly, planting grasses and legumes on terrace risers, on farm boundaries and on irrigation bunds should be practised more widely. Legume crops such as cowpea and crops such as oats and berseem can be grown after the rice is harvested using zero tillage, with broadcasted seed while the ground is still moist. Such practices would provide substantial amounts of forage with a minimum of labour, and render the soil more fertile. Improved crop varieties will give more return over local varieties, particularly where intensive cultivation, and irrigation facilities, or other input supplies are available. However, to achieve this in the hills, government subsidies in addition to technical information may be necessary.

8.12 Insect pest problems

Hill agriculture is comparatively more vulnerable to insect pest infestation due to the occurrence of varied climatic conditions. Locational specificity in hills is determined by factors like altitudinal differences, steepness of slopes, directions of mountain and valley etc. Microclimatic specificity of some hilly pockets results in development of hot spot for certain pests. Because of rich organic matter, light textured soil and major area (about 90%) under rainfed condition, white grubs have become key pest of the region. In addition, the intensive agricultural technologies in recent past have added to pest related problems. Intensive vegetable farming especially in valley areas of hills has witnessed increased pest build up during recent past and the management of pests has been mostly relied on the chemical means, which in turn poses several ill effects such as killing of natural enemies, environmental pollution, health hazard etc. Moreover, some of the states like Uttaranchal is pressing hard for organic farming in the region. The major challenge in the event of this farming would be the management of insect pests and diseases. Hence, for higher productivity and production of the crops, it is vital to adopt the Integrated Pest Management (IPM) strategy with a minimum use of pesticides, which are considered as one of the major factors for the burgeoning pest problem.

8.13 Pest Management Strategies

Integrated pest management was practiced knowingly or unknowingly by human being since time immemorial. However, IPM as an organized concept was evolved since its introduction in the late 1950s, when the focus was on combining suitable methods to limit pests in a crop, within the framework of sustainable agricultural development. The concept of IPM has been defined in different ways by different workers. Stern *et al.* (1959) defined applied pest control, which combines and integrates biological and chemical control. Chemical control is used as necessary and in a manner, which is least disruptive to biological control. Integrated control may make use of naturally occurring biological control as well as biological control affected by manipulated or induced biotic agents. Smith and Reynolds (1966) defined integrated pest control as a pest population management system that utilizes all suitable techniques in a compatible manner to reduce pest populations and maintain them at levels below those causing economic injury. FAO,

1967 defines integrated control as a pest management system that in the context of the associated environment and the population dynamics of the pest species, utilizes all suitable techniques and methods in as compatible manner as possible and maintains the pest populations at levels below those causing economic injury. While FAO, 1980 defines it as an interdisciplinary approach incorporating the judicious application of the most efficient methods of maintaining pest populations at tolerable levels.

Ideally, an integrated pest management program considers all available pest control actions, including no action, and evaluates the potential interaction among various control tactics, cultural practices, weather, other pests, and the crop to be protected. Pedigo, 1989 defined integrated control as control of pests that emphasizes selective use of insecticides so as to conserve natural enemies in the agro ecosystem. Allen and Rajotte, 1990 defined IPM as a systematic approach to crop protection that uses increased information and improved decision-making paradigms to reduce purchased inputs and improve economic, social, and environmental conditions on the farm and in society. Moreover, the concept emphasizes the integration of pest suppression technologies that include biological, chemical, legal, and cultural controls.

Different pest management strategies or practices are being used by the farmers to reduce the pest infestation. The most important tool for evaluating an insect problem is its economic threshold. The economic threshold is the pest density at which management action should be taken to prevent an increasing pest population from doing economic damage to the crop. The economic threshold takes into consideration the market value of the crop, injury and damage, insect density, lag time before implementation occurs, and cost of management. This threshold information is important in distinguishing the difference between minor plant injury and true economic damage. Economic threshold information is available for some, but not all insect pests. When economic threshold information is not available, other expertise must be used to determine damage potential of a given pest to a crop. Although the concept of the economic threshold can be applied to the use of many tactics, it is primarily used in association with insecticidal control. The most common strategies used for insect control are chemical and cultural practices. There are different insecticides available in the market with different modes of action. Insecticides have the ability to inflict high mortality quickly, and are the most common and effective form of rescue treatment. These chemicals vary in toxicity to humans or other animals, and also in their potential impact on the environment.

Cultural practices are the manipulation of the cropping environment to reduce rates of pest increase and damage. There are many different cultural practices that can help to reduce pest impact. Some of the more common techniques include improving plant vigor, tillage practices that disrupt the insect's life cycle and destroy crop residue, changing planting dates to minimize insect impact, and crop rotations that include non-susceptible crops. Some cultural controls may also increase or decrease pests' natural enemy populations. Sound cultural control methods often become stable farming practices that serve multiple purposes. Use of resistant crop varieties is another useful practice in managing insect pests. By using resistant varieties, other insect control measures may not be needed because the crop is less likely to be economically injured by the insect pests.

Biological control is the use of any biological agent to reduce insect pest populations. Predators, parasitoids, and pathogens are common and effective biocontrol agents. Biological control agents may occur naturally in the cropping environment or may be introduced as needed. This practice usually has low environmental impact but often requires a long time before natural enemy populations are established and control is achieved. The IPM principle does not preclude the chemical pesticide use, instead it uses it as one of the tools in the management package, to be used prudently integrated with other tools. Thus, the concept of integrated pest management of insects contains three basic elements. These include: maintaining insect populations below levels that cause economic damage, the use of multiple tactics to manage insect populations, and the conservation of environmental quality.

8.14 Sustainable Agriculture

Sustainable agriculture is a system of agriculture that is ecologically sound, economically viable and socially acceptable in the short as well as long term. It avoids specific set of farming practices, rather represents the goal of developing a food production system that: i) yields plentiful, affordable, high-quality food and other agricultural products, ii) does not deplete or damage natural resources (such as germplasm base, soil, water, wildlife etc.), iii) promotes the health of the environment, iv) supports a broad base and diversity of farms and the health of rural communities.

IPM and sustainable agriculture share the goal of developing agricultural systems that are ecologically and economically sound. IPM may be considered as a key component of a sustainable agriculture system. The hill ecosystem is considered as very fragile and the conventional agriculture is followed in most part of the hilly region. However, there is a bright scope for organic farming in the hills due to less usage of inorganic chemicals. Primary strategy in organic farming for the management of insect pests and diseases is the prevention through proactive strategies which include healthy, biologically active soils (increasing belowground diversity), habitat for beneficial organisms (increasing aboveground diversity) and appropriate plant cultivars and maintenance of crop diversity. Hence, ensures minimal losses from the pests.

8.15 Factors affecting pest population

The insect pest population is drastically influenced by any change in the environmental factors. The environmental factors are broadly grouped into abiotic and biotic factors.

Abiotic factors: The population density of pest fluctuates with the prevailing weather conditions such as temperature, moisture, light, wind, edaphic factors etc. Since, insects are poikilothermic animals i.e. they do not posses temperature regulation system of their body, their body temperature, more or less, follows the surrounding temperature. Exposure to temperature beyond the favourable range, whether low or high, may retard growth and development of .the insects or may even cause their death. The species adapted to live in temperate climate, hibernate in a particular developmental stage. A constant supply of moisture is essential for metabolic reactions as well as for the dissolution and transport of salts. The water content in insects varies from less than 50% to more than 90% of the total body weight. Excessive moisture can prove harmful to the insect population by affecting normal development and by encouraging disease-causing microorganisms. Edaphic factors such as structure, texture and composition of soil, drainage, topography etc. also affect the distribution and abundance of the insects. Example, the cut worm, *Agrotis ipsilon* larva live in light texture soil in which they move around freely in response to daily or seasonal temperature and moisture changes. Major topographic factors that affect biography of insects include height of mountain chains, steepness of slope, directions of mountains and valleys and exposure of the slope. These factors also affect the climate of an area, thus influences the distribution of certain insects. However, their impact on the status of the major insect pests has yet to be quantified. Mountain ranges also act as physical barriers to the spread of insects (Yazdani and Agarwal, 1997).

Biotic factors: The growth and development of insects is largely influenced by biotic factors such as natural enemies and food. The natural enemies of insect pest population include predators, parasitoids and disease causing microorganisms such as fungi, bacteria, viruses etc. The natural enemies are also influenced by various environmental factors such as weather and hyper parasitism. Weather is known to determine the effectiveness of the natural enemies. Example, in the control of apple wooly aphid with *Aphelinus mali*, the prevalent temperature in spring would determine the effectiveness of the parasitoid. Similarly, quality and quantity of food play important role in insect's survival, longevity, distribution, reproduction, speed of development etc. (Dhaliwal and Arora, 2001, Atwal and Singh, 1990).

Hill agriculture and insect pest problem

With the increase in the productivity and production of the agricultural crops, by intervention of high yielding varieties, agro-chemical, irrigation, monoculture etc., there have been tremendous changes in the pest scenario of the crops (Dhaliwal and Arora, 2001, Atwal and Singh, 1990). Agronomic recommendations for exploitation of full potential of HYVs has certainly helped in enhancing crop production but at times it has provided congenial condition for many insect pests. Close spacing of paddy has resulted in the increased population of BPH, gall midge and leaf folder. In North West Himalayan Region, cultivation of spring rice is the most common practice which provides availability of the rice crop for a quite longer duration in the field, hence promotes pest build up in the region. Routine crop rotation in some cases may provide congenial environment for insect pest multiplication. Cultivation of off-season vegetables in hills provides food for many insect pests round the year. The tobacco caterpillar, *Spodoptera litura* and tomato fruit borer, *Helicoverpa armigera* were found to damage cabbage, cauliflower, broccoli, tomato, chilli, capsicum etc. round the year. Destruction of forests for cultivation of agricultural crops led to the migration of insect pests of forest, mainly white grubs to agricultural fields for egg laying and subsequent damage to the crops. So far, authors have recorded about 40 species of white grubs from the region. Among which, many of them are considered as major pests of agricultural crops.

Human preference for certain agricultural crops like rice, wheat, maize prompted farmers to grow more of these crops. Despite the rich biodiversity in agriculturally important crops, wheat, maize and rice accounts for 60% of plant derived calories in human diet (Hawtin, 1998). In NWHR, some of the less preferred crops like finger millet and barnyard millet are now gaining momentum because of their nutritive value. Monoculture of certain crops is considered as one of the major factor for increased insect pest problem. Mono-cropping of vegetable crops especially in the valley areas prompted the chances of outbreaks of many insect pests.

Nutritional requirement of the crops are mostly met with the application of FYM in traditional farming of the hills. However, application of compost, which is not well decomposed, in the rainfed conditions also helps in increasing the white grub problem in the region. Modern agriculture has undoubtedly contributed a lot in increasing the crop productivity as well as crop production all over the country, but at the same time it has provided congenial conditions for multiplication of many of the insect pests. Therefore, before advocating any practice for crop production it has to be thoroughly tested from insect pest point of view.

8.16 Major insect pests in hilly areas

Insect pests have been the major constraint in agricultural crop production in the hilly tracts of the country. Due to varied climatic conditions of the hilly areas, many pockets have become the hot spot for certain insect pests. Crop wise important insect pests of NWHR are discussed hereunder along with available management strategies. (Garg, 1992, Bhatt and Sushil, 2004)

8.16.1 Rice

Rice is grown both under rainfed and irrigated conditions. Under rainfed condition, the crop is generally sown in March-April and harvested in September-October. Under irrigated conditions the crop is transplanted in June to July and harvested by October. Therefore, rice crop is available in the fields from March to October. The insects are reported to attack at all the stages of crop growth. Although major borer in hilly area is pink stem borer, *Sesamia inferens*, but during recent surveys, attack of yellow stem borer, *Scirpophaga incertulus* was also recorded in some parts of Kumaon hills. Leaf folder, *Cnaphalocrosis medinalis* is also an important pest of rice in hills especially during August- September. Other important pests recorded on rice in hills are *Leptocorisa acuta*, *Dolycoris indicus, Nezara virudula*, *Sogatella furcifera*, *Cletus* sp., different species of white grubs like *Anomala dimidiata*, *Holotrichia seticollis*, grasshopper, crickets, earwigs, *Elaunon bipartitus*, *Nephotettix nigropictus*, *Agrotis* sp. *Chaetocnema basalis*, *Atherigona oryzae* etc. At some locations under favourable conditions, these pests may become important. In H.P., rice hispa was also recorded as important pest of rice.

8.16.2 Maize and small millets

Maize and small millets are generally less attacked by insect pests but some times sporadic pest attack was observed. On maize, finger millet and barnyard millet major attack of *Sesamia inferens* was recorded. On maize, however, occurrence of *Chilo partellus*, *Atherigona soccata*, blister beetle, some species of white grubs and cutworms were also recorded. In moisture stress area, occurrence of cutworm (*Agrotis flammatra* and *Agrotis segetum*) was found very serious. *Atherigona falcata* was found serious in barnyard millet at early stage of the crop.

8.16.3 Pulses

Major pulse crops of the hill region are black gram (urd), green gram (moong), horse gram (gahat), pigeon pea (arhar), french bean (rajmash) and soybean in *kharif* and pea, lentil and gram in *rabi*. Yield of most of these crops are drastically influenced by the attack of a number of insect pests. Due to heavy attack of *Helicoverpa armigera* and blister beetle, farmers show reluctance in growing pigeon pea.

8.16.4 Vegetables

Vegetables have enough potential under hill agro-ecosystem especially in valley areas wherever irrigation facility is available. Major vegetables grown in hills are brinjal, cabbage, cauliflower, onion, pea, capsicum, french bean, chillies, squash etc. besides several off season vegetables such as tomato, capsicum, cauliflower etc. Heavy infestations by the insect pests have been observed on vegetable crops due to prevailing congenial conditions for the growth of the insects in the hilly areas.

8.17 Conclusion and future strategies

In conclusion, insect pests have been the major constraint in agricultural crop production in the hilly tracts of NW Himalayan region of the country. To combat the burgeoning problem, relying on the pesticides, is not desirable in the present day of global consciousness towards health hazards and environmental safety. One of the strongest alternatives for the management of the insect pests would be integrated pest management with emphasis on biological control using potential insect parasitoids and microbial pathogens.

Rich biodiversity and fast changing cropping pattern toward commercial agriculture invite vibrant monitoring and surveillance system. Due to emphasis on organic farming in the region, there is a need to stress upon exploitation of potent local bioagents and botanicals against major insect pests, besides conservation and augmentation of exotic as well as native natural enemies. In addition, already existing proven IPM technologies available in different parts of the country need to be validated under hill condition. As the hills posses varied climatic conditions due to several factors like topography, steepness of slope, variations in soil structures, face of mountain etc., niche based IPM module has to be worked out to ensure better effectivity of the various management tools. In the event of shift in cropping system, particularly perennial crops, a critical vigilance would be of paramount importance in order to prevent emerging pest problems. Overall general awareness about IPM among the inhabitants coupled with government and institutional support would be necessary in the geographically challenged region of the country.

References

Allen, W. A. and Rajotte, E.G. 1990. *Ann. Rev. Entomology*. 35: 379-97.

Atwal, A.S. and Singh, B. 1990. Pest population and assessment of crop losses, Indian Council of Agricultural Research, New Delhi. 131 p.

Bhatt, J.C. and Sushil, S.N. (Eds) 2004. A compendium on diseases, insects and weeds of rice, maize and wheat crops in Indian Himalayas. 118 p.

Dhaliwal, G.S. and Arora, R. 2001. Integrated pest management, Ludhiana, Kalyani Publishers. 427p.

FAO. 1980. Research Summary. Integrated pest management. EPA-600/8-80-044. 28 pp.

Garg, D.K. 1992. Insect pests of agricultural crops in UP hills and their management. Almora VPKAS, 25 p.

Hawtin, G. 1998. Conservation of agro-biodiversity for tropical agriculture. In: Crop Productivity and Sustainability: Shaping the Future, edited by V.L.Chopra, R.B. Singh and A. Varma. Oxford & IBH, New Delhi. pp. 917-925.

Ives, J.D. and Messerli, B. 1984. Stability and instability of mountain ecosystems: lessons learned and recommendations for the future. Mountain Research and Development 4, 63–71.

Jodha, N.S. 1991. Mountain Farming Systems: Coping with Unsustainability Prospects. Paper presented at the 11th FAO Session of FAO Farm Management Commission for Asia and Pacific, Kathmandu, December 4–6, 1991.

Laban, P. 1978. Field Measurements on Erosion and Sedimentation in Nepal. Department of Soil Conservation and Watershed Management. FAO/UNDP. IWM/WP/05 (cited by LRMP, 1986).

Lele, S.M. 1991. Sustainable development: a critical review. World Development 19, 607–622.

LRMP. 1986. Land Resource Mapping Project. Soil Report. His Majesty's Government of Nepal/Government of Canada.

Metz, J.J. 1991. A reassessment of the causes and severity of Nepal's environmental crisis. World Development 19, 805–820.

Pedigo, L.P. 1989. Entomology & Pest Management. Macmillan Pub. Co., Inc. N.Y. 646 pp.

Pereira, H.C. 1989. Policy and Practice in the Management of Tropical Watersheds. Westview Press, Boulder, Colorado.

Schroeder, R.F. 1985. Himalayan subsistence systems, indigenous agriculture in rural Nepal. Mountain Research and Development 5, 31–44.

Sharma, P. and Banskota, M. 1990. Population dynamics and sustainable agricultural development in mountain areas. In Sustainable Development of Mountain Agriculture. N.S. Jodha, M. Banskota, T. Partap (eds). IBH Publ. Co. Pvt. Ltd., New Delhi, India, (cited in Jodha, 1991).

Smith, R. F. and Reynolds, H.T. 1966. Principles, definitions and scope of integrated pest control. Proc. FAO Symposium on Integrated Pest Control 1: 11-17.

Stern, V. M., Smith, R.F., van den Bosch, R and Hagen, K.S. 1959. The integrated control concept. *Hilgardia*, 29: 81-101.

Subedi, A. 1991. Agriculture and environment: Some issues in the Nepalese context. LAC Discussion Paper 5/91, Lumle Agricultural Centre, Kaski, Nepal.

Subedi, K.D. and Gurung, G. 1991. Soil fertility thrust towards sustainable agriculture: experiences of Lumle Regional Agricultural Research Centre. In: Soil Fertility and Erosion Issues in the Middle Mountains of Nepal pp 61–82. IDRC, Ottawa, Canada.

Yazdani, S.S. and Agarwal, M.L. 1997. Elements of insect ecology. Narosa Publishing House, New Delhi. 210 p.

BIORATIONALS IN INTEGRATED PEST MANAGEMENT

9

D. P. Abrol and Uma Shankar

9.0 INTRODUCTION

Disproportionately growing human population has resulted in increased demand for production of food supplies. In ancient times, when the population was low, the traditional agriculture could meet the demand of food supplies. With the increase in human population demand for food supply increased which was met by increasing the area under cultivation. As the land resources became limited, efforts were made to increase productivity by fighting the losses inflicted by insects through the discovery of pesticides. At the same time, they proved to be more dangerous due to their indiscriminate and excessive use, contaminating the food (milk, honey, cereals, vegetables and fruits), environment (ground water, soil etc.), and resulting in pest resistance, pest resurgence and pest outbreaks.

Around late mid of the last century, Stern *et al.* (1959) gave the concept of "Integrated Control" when control of insects depended largely on broad spectrum insecticides such as organochlorines, organophosphates (OPs), and carbamates, all neurotoxic. Their concept closely resembled the present day Integrated Pest Management (IPM) where chemical and biological control could work within an ecological frame work. Furthermore, the broad spectrum insecticides resulted in severe adverse effects in the form of environmental degradion, resistance to insecticides, pest resurgence and public interests have driven the demand for alternative pest control strategies. Development of biorational pesticides have now made it really possible to realize the concepts of Stern and colleagues (1959) which advocated the vision of more holistic biological and chemical controls. The advances in tools of physiology, toxicology, and biotechnology has made it easier to realize the goal in most effective way.

The US Federal definition states that there is paradigm shift towards "risk reduction" as a way of placing focus on the consequences and impacts of pests and pest management tactics rather than just pesticides. It defines "IPM as a long-standing, science-based, decision-making process that identifies and reduces risks from pests and pest management related strategies. It coordinates the use of pest biology, environmental information, and available technology to prevent unacceptable levels of pest damage by the most economical means, while posing the least possible risk to people, property, resources, and the environment. IPM provides an effective strategy for managing pests in all arenas from developed agricultural, residential, and public areas to wild lands. IPM serves as an umbrella to provide an effective, all encompassing, low-risk approach to protect resources and people from pests" Anonymous (2004). Reducing the risks associated with pest management tactics like pesticides in plant protection is the challenge of pest control for the twenty-first century.

Biorational agents and approaches shall be useful to meet the societal challenges before us (Ishaaya 2003; Horowitz and Ishaaya 2004; Ishaaya *et al.* 2005) to provide safer and environmentally friendly approaches to manage the pests and reducing pesticide resistance problems. Disrupting the activity of specific biochemical sites such as neuropeptides, ecdysone and juvenile hormones, tyramine, octopamine and GABA receptors and use of natural plant products, semiochemicals and exploit genetically modified-plants, insects, and symbionts in the combat against insect pests and disease-borne vectors are some of the approaches which hold promise for future pest control strategies.

Biorational integrated pest management help to produce safe food and addresses the ecological concerns from the negative impact of pesticide residues and their pressure on an ecosystem (Abrol and Shankar, 2014). Biologically rational control is a major component of Integrated Pest Management (IPM), which plays a crucial role in management of most of the crop ecosystem. The tactics utilized in IPM programs include chemical, cultural, physical, and biological control. Most management programs rely heavily upon the use of insecticides applied when periodic scouting indicates that pests have exceeded a pre-determined threshold.

Thus, the concept of IPM of insects contains three basic elements: (i) maintaining insect populations below levels that cause economic damage; (ii) the use of multiple tactics to manage insect populations; and (iii) the conservation of environmental quality. Excessive and indiscriminate use of pesticides endangers the health of farm workers and consumers of agricultural products worldwide (Goodell, 1984). People do not want have reliance on chemicals and look for alternate strategies for pest control such as cultural, biological and biorational methods. According to Rola and Pingali (1993), this has been necessitated as a result of the negative impact of pesticides on biodiversity and food and water quality, and the high cost of pesticides and the development of resistance in pests. The Government of India adopted IPM as a cardinal principle of plant protection in 1985.

Insecticides provide quick control of pests but often require repeated applications to provide long term management. It has long been recognized by researchers and, more recently, by IPM practitioners that the integration of biological control (mortality induced by natural enemies including parasites, predators, and pathogens) into IPM programs is essential for long term, stable management of pests.

Although this information has been collected for at least the last 50 years for some insecticides, the term "biorational" has only recently been proposed to describe those insecticides that are efficacious against the target pest but are less detrimental to natural enemies. The term at times has been used to describe only those products derived from natural sources, i.e. plant extracts, insect pathogens, etc. However, we choose to define a biorational pesticide as "any type of insecticide active against pest populations, but relatively less harmful to non-target organisms, and, therefore, non-disruptive to biological control" (Stansly *et al.* 1996).

Biorational pesticides are those synthetic, organic, or inorganic pesticides that are both low toxicity and exhibit a very low impact on the environment. They also have minimal impact on species for which they are not intended (called *non-target species*). Biorational pesticides include oils, insecticidal soaps, microbials (such as *Bacillus thurengienesis* and entomopathogenic nematodes), botanical (plant-based) and insect growth regulators. Biorational pesticides should therefore be first choice whenever a pesticide is needed. Biorational pesticides are a distinct group, inherently different from conventional pesticides. They are comprised of two major categories: the biochemical pest control agents (e.g. pheromones, hormones, natural plant growth regulators and enzymes) and the microbial pest control agents (e.g. microorganisms). Pesticides to be included in these categories must be naturally occurring, or if man synthesizes the chemical, then it must be structurally identical to a naturally occurring chemical. Minor differences between the stereochemical isomer ratios (found in the naturally occurring compound compared to the synthetic compound) will normally not rule out a chemical being classified as a biorational unless an isomer is found to have significantly different toxicological properties from those of another isomer.

During the past three decades efforts have been made to reduce the risk of human exposure to pesticides, especially insecticides. There has been a great demand for safer and more selective insecticides for natural enemies and non-target organisms. In view of the above, many conventional pesticides have been replaced by newer biorational or low risk insecticides. Hara (2000) defined, biorational insecticides as the synthetic or natural compounds that effectively control insect pests, but at the same time have low toxicity to non-target organisms and the environment. Consequently, new insecticides have been developed which are more selective than conventional insecticides, safer, and fit well into integrated pest management (IPM) programs (Casida and Quistad 1998).

Most of the old conventional insecticides were nerve poisons that focused mainly on four target sites in the insect's nervous system viz γ-aminobutyric acid (GABA) gated chloride channel (endosulfan), octopamine receptor (amitraz), acetylcholine enzymes (organophosphates [OPs] and carbamates) and the voltage gated sodium channel (DDT and pyrethroids) (Holloway *et al.* 1999; Ishaaya 2001).

The limited number of target sites exploited by conventional insecticides has created problems with resistance to these insecticides (Georghiou 1983). Efforts have been made to find biorational insecticides with novel modes of action, which have no cross-resistance with the old insecticides. New classes of insecticides include: neonicotinoids, avermectins, phenlpyrazoles, spinosyns, pyrroles, oxadiazines and various groups of insect growth regulators (IGRs). Most are still nerve poisons acting at specific target sites in the insect's nervous system. Although some new insecticides act similarly to the old nerve poisons, they have higher affinity to insect receptors than to mammalian. The other insecticides affect specific systems, such as the molting processes, metamorphosis, and the insect endocrinology system.

Most of the newer insecticides are preferable to the conventional insecticides because of their specificity to target pests, their effectiveness at low rates and their non-persistent characteristics in the environment (Table 9.1). Their selectivity renders many of these insecticides suitable for IPM programs (Casida and Quistad, 1998).

Table 9.1. List of biorational insecticides and miticides

(*Source:* https://nevegetable.org/about-guide)

Insecticides	Trade Name	Target Pests	Comments
Azadiractin	Amazin Plus 1.2%ME^{OG}; Aza-Direct^{OG}; AzaGuard^{OG}, Azatin O^{OG}; XL; Azatrol EC^{OG}; Azera^{OG}(M); Ecozin Plus 1.2%ME^{OG}; Molt-X^{OG}; Neemix 4.5^{OG}, Ornazin 3%EC	Aphids, caterpillars, leafhoppers, leafminers, thrips, whiteflies, beetles, and other insects	An insect growth regulator extracted from the seeds of the neem tree, works by contact or ingestion against immature stages, and has antifeedant properties.
Bacillus thuringiensis subsp. aizawai	Xentari^{OG}	Caterpillars, as listed for *Bt kurstaki*, as well as cross-striped cabbageworm, armyworms, cutworm	Use in rotation with Bt kurstaki products to prevent resistance. May be used in greenhouse or field.
Bacillus thuringiensis subsp. Israelensis	Gnatrol WDG^{OG}	Fungus gnats	For use in greenhouse.
Bacillus thuringiensis subsp. Kurstaki	CoStar^{OG}; Deliver^{OG}; Dipel ES, Dipel DF^{OG}; Javelin WG^{OG}	Many caterpillars, including cabbage looper, diamondback moth, imported cabbageworm, European corn borer, hornworm	Must be ingested. Apply when caterpillars are actively feeding. Safe on beneficial insects. For some species, more effective on smaller larvae.
Buprofezin	Talus 70DF	Leafhoppers, mealybugs, planthoppers, whiteflies	Insect growth regulator.
Burkholderia spp. (heat killed) Strain A396 and spent fermentation media	Venerate XC^{OG}	Caterpillars, cutworms, aphids, mites, thrips, whiteflies	Works by contact and ingestion to disrupt insect exoskeletons and interfere with molting.

Insecticides	Trade Name	Target Pests	Comments
Capsicum oleoresin extract, garlic oil, soybean oil	Captiva	Caterpillars, mites, thrips, leafhoppers, whiteflies	Has anti-feedant, anti-egg laying, and irritant activity, and weakens cuticles of immature stages of insect and mite pests.
Chenopodium ambrosioides extract	Requiem	Aphids, whiteflies, thrips, leafminer, mites	Contact insecticide/ acaricide. Can be applied for ovicidal control. Start when populations are at threshold. Compatible with beneficial insects.
Chlorantraniliprole	Coragen	Caterpillars, Colorado potato beetle, leafminers	May be used as soil or transplant water treatment at planting, in drip or as foliar. Avoid run-off in surface waters. Non-toxic to bees.
Chromobacterium subtsugae strain PRAA4-1	Grandevo^{OG}	Caterpillars, aphids, whiteflies, mites, leafhoppers,thrips	A selective biological insecticide/miticide containing fermentation solids. Works as a stomach poison upon ingestion. No systemic activity. Toxic to bees exposed to direct treatment or residues on blooming crops or weeds.
Clarified hydrophobic extract of neem oil	Triact 70^{OG}, Trilogy^{OG}	Primarily labeled for (but not limited to) mite control	Can be used to control mites. For best results use when population levels are low to prevent build-up. Repeat applications are needed.
Clove oil, thyme oil and *cinnamon oil*	Ecotrol G^{OG}	Root maggots, wireworm, cutworm, Symphylans	Soil-incorporated volatile plant oils on a corn cob carrier, applied before, during or after planting. Exempt from registration and reporting under Federal EPA standards.

Insecticides	Trade Name	Target Pests	Comments
Cottonseed oil	Sea-cide[OG]	aphids, mites, thrips,	A broad-spectrum, fish oil based, horticultural oil (91% fish oil, 3% cottonseed oil) that acts as an insecticide and fungicide. Works on contact.
Cryolite	Prokil Cryolite 96	Colorado potato beetle (larvae only), cucumber beetle, flea beetle, cutworms, some caterpillars	Stomach poison, must be ingested. Sodium aluminofluoride mineral, applied as spray or dust. Reapply to cover new growth. May cause extensive wear of nozzles.
Cyromazine	Trigard	Leafminer, Colorado potato beetle	An insect growth regulator.
Diflubenzuron	Dimilin	Armyworms, pepper weevils	A selective insect growth regulator that disrupts the molting process of insect larvae.
Etoxazole	Zeal	Mites	A mite growth regulator that works as an ovicide and larvicide.
Flonicamid	Beleaf	Aphids, plant bugs, whiteflies	A pyridinecarboxamide that works by contact and ingestion. Feeding stops rapidly and mortality will follow.
Flubendiamide	Belt SC; Synapse WG	Caterpillars including armyworms, loopers, hornworm, corn earworm, European corn borer	A diamide, active by insect larval ingestion, leading to a rapid cessation of feeding followed by death. Labeled for use as a foliar spray on Brassicas and turnip greens, leafy vegetables, sweet corn, cucurbits, fruiting vegetables, and legumes to control caterpillars.

Insecticides	Trade Name	Target Pests	Comments
Flupyradifurone	Sivanto	Aphids, leafhoppers, whiteflies, Colorado potato beetles, squash bugs	A broad-spectrum insecticide in a new class of chemistries, the Butenolides. Acropetally systemic, and translaminar. No toxicity to bees in lab tests.
insecticidal soap (Potassium salts of fatty acids)	Des-X[OG], M-Pede[OG]	Aphids, leafminers, mites, thrips, whiteflies	Works on contact. Can be phytotoxic to some crops, test on small plot. Avoid treatment when plants are stressed. May also harm some beneficials. Also active against powdery mildews.
Iron phosphate	Sluggo: Snail and Slug Bait[OG]	Snails, slugs	Bait which causes feeding to cease. Death occurs over 3-5 days. Exempt from tolerance and has a zero hour reentry interval due to low toxicity to people and wildlife.
Kaolin	Surround WP[OG]	Flea beetle, striped cucumber beetle, leafhopper, thrips	Interferes with insects' ability to recognize their host; particles cling to cuticle. Plant leaves turn white but growth is not inhibited. Reapply after heavy rain. To avoid lumps in mixing, form a slurry, then dilute. Maintain agitation.
Metaldehyde	Deadline Bullets	Slugs, snails	A toxic bait. Not for direct application to or contamination of edible portions of the plant; to be applied as a band treatment between rows after formation of edible parts. Long lasting; resistant to rain and watering.

Insecticides	Trade Name	Target Pests	Comments
Methoxyfenozide	Intrepid	Many species of caterpillars	Mimics molting hormone; causes premature molt and death.
Novaluron	Rimon	Caterpillars, beetles, leafminers, squash bugs, thrips, whiteflies	An insect growth regulator for use on immature stages.
Petroleum oil (mineral, parrafinic oils)	JMS Stylet OIl, Organic JMS Stylet Oil[OG], Suffoil X[OG]	aphids, leafminers, beetle larvae, mites, thrips, leafhoppers, whiteflies	A horticultural oil (80% petroleum oil) insecticide, miticide and fungicide. Kills eggs, larvae and nymphs of insects and mites and adults of soft-bodied insects through suffocation. Registered for use as a foliar spray on a variety of crops. Also labeled for certain diseases.
Pymetrozine	Fulfill	Aphids	Works by contact and ingestion, translaminar with long residual.
Pyrethrin	PyGanic EC 5.0[OG], PyGanic ED 1.4[OG], Pyrenone (includes piperonyl butoxide)	Asparagus beetle, blister beetle, cucumber beetle, flea beetles, Mexican bean beetle, potato leafhopper, many caterpillars, aphids, stink bugs	Botanical insecticide with broad-spectrum activity. Contact toxin with rapid knockdown but short period of activity (non-persistent). Highly toxic to fish. Derived from a chrysanthemum species cultivated in Africa. Some formulations are approved for organic crops.
Pyriproxyfen	Esteem, Distance IGR, Knack	Thrips, whiteflies shoreflies, fungus gnats	Label may include greenhouse or field use. An insect growth regulator that suppresses development of embryo within the egg, immature,and pupal stages of the insect. No activity against adult insects.

Insecticides	Trade Name	Target Pests	Comments
Rosemary Oil, Peppermint Oil, other essential plant oils	Ecotrol EC^{OG}, Ecotec^{OG}	Aphids, beetles, bugs, early stages of caterpillars, leafminers, mites, thrips	A broad-spectrum, works on contact as an insecticide and miticide. Greenhouse and field use. Thorough coverage is needed. Exempt from registration under Federal EPA standards.
Soybean Oil	Golden Pest Spray Oil^{OG}, Captiva	Primarily soft-bodied insects	A soybean-derived horticultural oil that works as a contact insecticide, as well as a feeding and oviposition deterrent.
Spinetoram	Delegate, Radiant	See spinosad, below.	A second-generation spinosyn, similar to spinosad, below.
Spinosad	Blackhawk, Entrust^{OG}, Seduce Insect Bait^{OG}, GF-120 Naturalyte^{OG}	Caterpillars, Colorado potato beetle, asparagus beetle, fleabeetle, leafminers, thrips. Baits labeled for cutworms, earwigs (Seduce) or fruit flies (GF-120)	Acts both as a contact and stomach toxin. Derived from soil bacterium *Saccharopolyspora spinosa*. Non-disruptive to most predator insect species and some parasites. Rotate with other selective biorationals to prevent resistance.
Spiromesifen	Oberon	Whiteflies, nymphs & pupae, mites	Contact insecticide and miticide.
Sulphur	Microthiol Disperss^{OG}	Mites	Micronized wettable sulfur, also labeled as a fungicide.
Tebufenozide	Confirm	Caterpillars	Selective insect growth regulator.

Most of the biorational insecticides have low toxicity to bees and beneficial insects. Most of them have dermal and oral LD50 values over 2,000 mg/kg. The symbol OG indicates a pesticide that has been listed by the Organic Materials Review Institute (OMRI) as compliant with the National Organic Standards and therefore approved for use in organic production.

9.1 Neonicotinoids

The neonicotinoids are among the most effective group of insecticides, used especially for controlling sucking insects such as whiteflies, aphids and leafhoppers, and various coleopteran pests such as Colorado potato beetle. The neonicotinoids represent a class of chemicals, nicotine mimics, in which the biochemical target is the nicotinic acetylcholine receptor (nAChR) of both the central and peripheral nervous systems (Bai *et al.* 1991). Similar to the original nicotine, the neonicotinoids bind to these receptors, resulting in excitation and paralysis followed by insect death. The structure-activity relationships based on binding affinity to the receptor and insecticidal activity has been summarized by several investigators (Tomizawa and Yamamoto 1992, Liu and Casida 1993, Yamamoto *et al.* 1995, Nauen *et al.* 2001, Tomizawa and Casida 2003). The neonicotinoids have a broad insecticidal spectrum, exhibiting systemic and translaminar properties, and high residual activity (Elbert *et al.* 1990, 1998; Takahashi *et al.* 1992). Imidacloprid was the first commercial neonicotinoid to successfully control agricultural pests. It is widely used as a seed dressing and in soil applications, as it is considered a relatively polar material with good xylem mobility.

Because of their systemic and translaminar properties the neonicotinoid insecticides are particularly effective in controlling sucking pests such as aphids, leafhoppers and whiteflies.

Horowitz *et al.* (1998) made comparative bioassays of two neonicotinoids, acetamiprid and imidacloprid, against the whitefly *Bemisia tabaci* (Gennadius), using foliar and systemic applications on cotton seedlings and found that both under laboratory and field conditions, the ovicidal and nymphicidal activities of foliar applications of acetamiprid on cotton seedlings were much higher than those of imidacloprid. According to LC50 and LC90 values, acetamiprid was 10- and 18-fold more potent than imidacloprid. Both compounds were effective against whitefly adults when applied to soil; however, the potency of imidacloprid was somewhat higher than that of acetamiprid. In experimental cotton plots, when 60 g a.i. /ha acetamiprid and 210 g a.i. /ha imidacloprid were applied foliarly, the residual activity of acetamiprid on whitefly adults lasted approximately ten days, as compared with three days for imidacloprid.

However, in other assays,they found activity different host plants. The translaminar residual activity of imidacloprid on cabbage leaves against *Myzus persicae* (Sulzer) was superior to that of acetamiprid; against *Aphis gossypii* (Glover) on cotton it was inferior to that of acetamiprid. Feeding bioassays revealed no significant difference in activity between the two neonicotinoids (Buchholz and Nauen 2002).

Neonicotinoids are flexible insecticides that can be used as foliar, drench or seed dressing treatments, and they act as contact and stomach poisons. As a foliar treatment these insecticides would have adverse impact on beneficial insects, especially predatory hemipterans. The drench and seed treatments are safer to most natural enemies, for instance.

Ellsworth and Martinez-Carrillo (2001) found that Imidacloprid – used in vegetables and melons against the whitefly, *B. tabaci*, was a key chemical in the southwestern United States and was fully integrated with sampling plans and action thresholds, as part of resistance management guidelines. The effect of imidacloprid as a foliar agent (Provado) on beneficial insects was assessed. This insecticide exhibited a relatively mild detrimental effect on common predatory bugs (McNally and Mullins 1996).

Similarly, Marquini *et al.* (2002) found that the use of imidacloprid in bean plots in Brazil showed no apparent impact on beneficials including spiders; however, Tanaka *et al.* (2000) found that imidacloprid was toxic to some species of predators and parasitoids. In field trails for controlling the Italian locusts, *Calliptamus italicus* (L.) in Kazakhstan, imidacloprid had no adverse impact on non-target organisms (Wilps *et al.* 2002). In comparative experiments of various conventional insecticides vs. imidacloprid, Figuls *et al.* (1999) found that imidacloprid is moderately persistent for the two mirid predators, *Dicyphus tamaninii* Wagner and *Macrolophus caliginosus* Wagner. In other tests, imidacloprid was harmful in a semi-field test to the anthocorid predator, *Orius laevigatus* (Fieber) (van de Veire *et al.* 2002). The following example demonstrates the compatibility of neonicotinoids with various biocontrol agents for controlling agricultural pests: In the Colorado potato beetle (CPB), *Leptinotarsa decemlineata* (Say) pest management control was combined effectively by pheromone-mediated augmentation of the predatory bug, *Podisus maculiventris* (Say) along with a treatment of imidacloprid (Aldrich and Cantelo 1999), although this predator is susceptible to direct contact with imidacloprid (De Cock *et al.* 1996; Boyd and Boethel 1998). Neves *et al.* (2001) have examined the compatibility of various entomopathogenic fungi (*Beauveria bassiana*, *Metarhizium anisopliae* and *Paecilomyces* sp.) under laboratory conditions with different neonicotinoids (acetamiprid, imidacloprid and thiamethoxam). Their conclusion was that in most cases, the neonicotinoids at recommended formulations and concentrations had no adverse effect, and can therefore be used in IPM programs.

In greenhouse and field studies, imidacloprid and the entomopathogenic nematodes *Heterorhabditis bacteriophora* and *Steinernema glaseri* and other nematode species interacted synergistically against third-instars of the Japanese beetle, *Popillia japonica* (Newman) (Koppenhofer *et al.* 2002), thiamethoxam interacted synergistically but to a lesser extent. Synergistic combinations of nematodes and neonicotinoid insecticides could be used for curative treatments of white grub infestations, especially against *Exomala orientalis* (Waterhouse) and *P. japonica.* This combination may allow spot-treatment of turf areas that exceed damage thresholds, thereby limiting the environmental impact of the insecticide applications. It can be concluded that although the neonicotinoids have mild effect on beneficials, especially as foliar agents, their efficacy for controlling insect pests and their versatile use render them important components in IPM and IRM programs.

9.3 Insect Growth Regulators (IGRs)

Several investigators have summarised the information on selectivity and importance of growth regulators in IPM programs (Retnakaran *et al.* 1985; Ishaaya 1990, Ishaaya and Horowitz 1998; Smagghe and Degheele 1998; Oberlander and Smagghe 2001; Ishaaya *et al.* 2002a).

9.3.1 Chitin synthesis inhibitors

Among the inhibitors of chitin synthesis are the benzoylphenyl ureas (BPUs), such as diflubenzuron, chlorfluazuron, lufenuron and novaluron. During the past three decades, the BPUs have been developed as commercial compounds for controlling agricultural and forest pests, especially lepidopterans. BPUs act on insects by inhibiting chitin formation (Ishaaya and Casida 1974) thereby causing abnormal endocuticular deposition and abortive molting (Mulder and Gijswijt 1973). BPUs are compounds with selective properties, affecting the larval stage. They act mainly by ingestion, but in some species they suppress fecundity and exhibit ovicidal and contact toxicity (Ishaaya and Horowitz 1998). The first commercial compound in this group was diflubenzuron, but subsequently more potent BPUs have been developed (Ishaaya 1990). They are very effective in controlling insect pests of cotton, corn and vegetable crops such as *Spodoptera* sp., and heliothine sp. A new BPU insecticide, novaluron, has somewhat more contact and translaminar activity as compared with other BPUs, thereby affecting in addition to lepidopteran and coleopteran larvae, whiteflies and leafminers (Ishaaya *et al.* 1996, 2002a).

BPUs generally affect the larval stages of insects, which synthesize chitin during their molting processes. Hence, the adults of beneficial species, predators and parasitoids, are seldom affected. For this reason, BPUs are considered important components in IPM programs. Buprofezin is a thiadizine-like compound with long residual activity that also acts as a chitin synthesis inhibitor. It has both contact and vapor activity, and acts on nymphal stages of sucking insects such as leafhoppers, plant hoppers, scales and whiteflies (Ishaaya *et al.* 1988; De Cock and Degheele 1998). Its mode of action resembles that of BPUs, although its structure is not analogous. The compound inhibits incorporation of 3H-glucose and N-acetyl-D-3H-glucosamone into chitin (Izawa *et al.* 1985; Uchida *et al.* 1985). As a result of chitin deficiency, the procuticle of the whitefly nymph loses its elasticity and the insect is unable to molt (De Cock and Degheele 1998). Similar to the BPUs, buprofezin is effective on immature stages and not on adults. The compound has a mild effect on natural enemies (De Cock and Degheele 1998) and is considered an important component in IPM programs for controlling whiteflies in cotton, vegetables and ornamentals.

9.3.2 Ecdysone agonists and juvenile hormones

Since various physiological and biochemical processes in insects are regulated by molting and juvenile hormones, novel insecticides which target ecdysteroid and juvenile hormone (JH) receptor sites have been developed during the past two decades (reviewed by Dhadialla *et al.* 1998). The Rohm and Hass Company (Pennsylvania, USA) has developed several substitutes for dibenzoyl hydrazines (diacylhydrazines), which act as ecdysone agonists. Two of these compounds, tebufenozide and methoxyfenozide, have been commercialized for controlling lepidopteran pests (Dhadialla *et al.* 1998; Smagghe and Degheele 1998). These compounds disrupt insect hormonal systems, and bind to the ecdysteroid receptors, thereby accelerating the molting process (Wing 1988; Palli *et al.* 1996). Tebufenozide is very effective against lepidopteran pests such as *Spodoptera* sp., *Helicoverpa zea* (Boddie) *and Manduca sexta* (L.) in cotton, cereal, maize, rice and vegetables (Wing *et al.* 1988; Chandler *et al.* 1992; Smagghe and Degheele 1992; Ishaaya *et al.* 1995). Methoxyfenozide is five- to tenfold more potent than tebufenozide (Ishaaya *et al.* 1995), and it also controls lepidopteran larvae, but with more activity against budworm/bollworm and diamondback moth (Le *et al.* 1996; Dhadialla *et al.* 1998; Smagghe and Degheele 1998). Both compounds have no appreciable cross-resistance with conventional insecticides (Ishaaya *et al.* 1995), but the risk for development of resistance to both compounds is apparent (Moulton *et al.* 2002). These diacylhydrazines are considered highly selective with no harm to natural enemies (Dhadialla *et al.* 1998; Smagghe and Degheele 1998) and as such fit well in IPM and IRM programs.

Among the JH mimics, fenoxycarb and pyriproxyfen effectively control agricultural, veterinary and human health pests. Fenoxycarb was the first commercial JH compound for controlling agricultural pests (Dorn *et al.* 1981; Masner *et al.* 1987), and pyriproxyfen is a fenoxycarb derivative in which a part of the aliphatic chain was replaced by pyridyl oxyethylene. Pyriproxyfen is a potent JH mimic affecting the hormonal balance in insects, thereby resulting in strong suppression of embryogenesis, metamorphosis and adult formation (Itaya 1987; Langley 1990; Koehler and Patterson 1991). It is considered a leading insecticide for controlling whiteflies (Ishaaya and Horowitz 1992, 1995; Ishaaya *et al.* 1994) and scale insects (Peleg 1988). Although high resistance to pyriproxyfen in *B. tabaci* has evolved in some areas (Horowitz and Ishaaya 1994; Horowitz *et al.* 1999, 2002), it remains an important component for controlling whiteflies in cotton fields (Dennehy and Williams 1997; Denholm *et al.* 1998ab).

In general, compounds that act on insect hormonal receptors, or on particular insect processes not present in mammals, are considered important insecticides for controlling insect pests with minimal harm to humans, beneficial arthropods, and the environment; however, some natural enemies (especially predators) can be affected by IGRs. In addition, there is some concern regarding BPUs with respect to their potential effect in the environment on crustacean species (Ishaaya 1990; Ishaaya and Horowitz 1998).

Among other mechanisms, selectivity of IGRs can result from penetration through the insect cuticle (Medina *et al.* 2002). This can explain the inhibition of egg hatch in the predator *Chrysoperla carnea* (Stephens), after exposure to diflubenzuron, and the nontoxic effect observed in the ecdysone-agonist, tebufenozide. In comparative assays, tebufenozide was harmless to various predators of cotton pests (Elzen 2001), but it was found moderately toxic to predators of the diamondback moth, *Plutella xylostella* (L.) (Liu and Sengonca 2002). In laboratory and field trials, tebufenozide was not toxic to the coccinellid predator, *Stethorus punctum* (LeConte); but other IGRs, including fenoxycarb and teflubenzuron, were harmful to some stages of this predator (Biddinger and Hull 1995). Fenoxycarb was also toxic (as an ovicide) to the predator *C. carnea* (Bortolotti *et al.* 2000).

The introduction of IGRs to whitefly management resulted in optimization of control of these severe pests. Two IGRs, pyriproxyfen and buprofezin, which have been introduced as essential components in whitefly management for cotton in Israel (Horowitz *et al.* 1994) and Arizona (Dennehy and Williams 1997; Denholm *et al.* 1998ab; Ellsworth 1998), have replaced the broad spectrum insecticides, pyrethroids and organophosphates. These changes have had an impact on populations of natural enemies and allowed cotton growers to manage whiteflies with fewer disruptions to the environment and non-target organisms (Ellsworth 1998; Naranjo *et al.* 2002).

In greenhouse poinsettias, the use of IGRs (especially buprofezin) reduced parasitoid release rates for controlling whiteflies (Driesche *et al.* 1999). In processing tomatoes, reduced risk of the newer pesticides, including IGRs, has provided alternatives to conventional pesticides for controlling whiteflies and other tomato pests (Zalom *et al.* 2001).

9.4 Insecticides from soil microorganisms

Over the past two decades, there have been attempts to produce insecticides derived from natural or biological sources. Various important classes of insecticides were formed mostly from soil microorganisms by fermentation procedures. Among these classes are avermectin, spinosyn and pyrrole insecticides. Another important group is the *Bacillus thuringiensis* (*Bt*) insecticides (and *Bt* crops) which are also covered in other chapters of this book.

9.4.1 Avermectins

The avermectins are a group of macrocyclic lactones isolated from fermentation of the soil actinomycete microorganism, a Japanese strain of *Streptomyces avermitilis* Burg. These compounds act as agonists for GABA-gate chloride channels (Mellin *et al.* 1983; Albrecht and Sherman 1987; Bloomquist 2001). They bind with high affinity to sites in the neuronal membranes of various insect species. It was suggested that avermectins may bind to a common site in all gated chloride channels (Arena *et al.* 1995).

Abamectin developed for agricultural use is a mixture of not less than 80% avermectin B1a and not more than 20% avermectin B1b (Fisher and Mrozik 1989). It is a broad spectrum pesticide and is highly toxic to many arthropods, including spider mites, agromizid leafminers, ants, cockroaches, and selected lepidopteran pest species (Dybas 1989, Lasota and Dybas 1991). Abamectin is considered a selective pesticide with relatively low toxicity to many non-target arthropods (Dybas 1989).

It is used at very low rates and degrades rapidly when exposed to light, especially when applied as a thin film on either inert surfaces or leaves (Wislocki *et al.* 1989). The half-life of a thin film of abamectin under laboratory light or sunlight is approximately 4 to 6 h regardless of surface or foliage type. Despite its rapid photodecomposition following application, abamectin provides residual activity in the field due to its translaminar action; rapid penetration of abamectin into cotton leaves within a few hours after application has been observed (Dybas 1989). In addition, various authors have reported that the use of mineral oil or surfactants in combination with abamectin extends its foliar residual toxicity, especially to phytophagous mites, under greenhouse and field conditions (Wright *et al.* 1985, Mizell *et al.* 1986, Dybas 1989).

The effects of abamectin applied alone or together with the mineral (paraffinic) oil ULTRA FINE® on sweetpotato whitefly, *B. tabaci*, were determined on cotton plants under laboratory (leaf dip bioassays) and field conditions (Horowitz *et al.* 1997). This study showed that a mixture of abamectin with mineral oil exhibits higher toxicity than abamectin applied singly. Abamectin alone was effective against the whitefly under laboratory conditions only if it had not been exposed to sunlight; the addition of oil is apparently essential to increase penetration into the leaves and reduce degradation by sunlight in the field. Although, several cases of resistance to abamectin in spider mites have been monitored (Eric Palevsky personal com.), it is still considered a leading compound for controlling mites in various agricultural systems.

9.4.2 Emamectin benzoate

The avermectins exhibit essentially the same mode of action but differ in their biological activity (Shoop *et al.* 1995). Among the new avermectins, the emamectin benzoate; exhibits excellent activity against lepidopteran pests and a wide range of insect species (Ishaaya and Horowitz 1998). Emamectin benzoate (Proclaim) is a novel macrocyclic lactone insecticide derived from the avermectins family. Emamectin, obtained from abamectin via a five step synthesis, was discovered after screening several hundred avermectin derivatives in an *in vivo* screen using tobacco budworm, *H. virescens,* and southern armyworm, *Spodoptera eridania* (Cramer). This compound was subsequently selected for further development in crop protection (Jansson and Dybas 1998). Later studies found that benzoate salt of emamectin had improved thermal stability and greater water solubility when compared with the original hydrochloride salt. As such, emamectin has broader spectrum insecticidal activity than abamectin.

Similarly to abamectin, emamectin affects the nervous system of arthropods by increasing chloride ion flux at the neuromuscular junction, resulting in cessation of feeding and irreversible paralysis. The compound acts on various lepidopteran pests such as *P. xylostella, Trichoplusia ni* (Hübner), *Heliothis virescens* (F.) and *Spodoptera exigua* (Hübner). It has translaminar activity, providing a relatively prolonged residual activity. While abamectin acts specifically on mites, with low efficacy on lepidopterans, emamectin showed considerable potency on diverse lepidopteran pests.

Wide-ranging bioassays were conducted to test the toxicity of emamectin against various important insect pests such as the lepidopteran pests *Spodoptera littoralis* (Boisduval) and *Helicoverpa armigera* (Hübner), the western flower thrips (WFT) *Frankliniella occidentalis* (Pergande), the whitefly *B. tabaci* and the leafminer *Liriomyza huidobrensis* (Blanchard) (Ishaaya *et al.* 2002b). Emamectin was found to be highly effective for controlling the western flower thrips *F. occidentalis* and the cotton bollworm *H. armigera* and to a lesser extent the Egyptian cotton leafworm *S. littoralis.* Although emamectin had some effect on the whitefly *B. tabaci* and the leafminer *L. huidobrensis*, further assays are required to evaluate its potential use for controlling these pests.

Although the direct activity of abamectin is toxic to non-target organisms (*e.g.* van de Veire *et al.* 2002; Childers *et al.* 2001), the speed with which it degrades on the surface of the leaves under field conditions renders this compound an important agent in IPM programs. Experiments for testing the side effects of abamectin on various species of predators and parasitoids (Shipp *et al.* 2000) showed high mortality of the majority of beneficial arthropods up to 6 days after application; thereafter, the residual toxicity of abamectin declined rapidly. It was concluded that this agent is compatible with predacious mites. As with abamectin, surface residues of emamectin decompose rapidly in sunlight, resulting in relatively low toxicity to beneficial insects (MacConnell *et al.* 1989; Feely *et al.* 1992; Jansson and Dybas 1998). Hence, these compounds are considered important components in IPM programs for controlling mites and other agricultural pests.

9.4.3 Spinosyns

The spinosyns, a new group of insecticidal macrocyclic lactones derived by fermentation of the soil actinomycete, *Saccharopolyspora spinosa*, are active on various insect pests, especially lepidopteran, thysanopterans and dipterans (Thompson and Hutchins 1999; Sparks *et al.* 2001). In addition, the spinosyns exhibit a favourable environmental and toxicological profile. Spinosyns initially cause spontaneous muscle contractions and tremors by exciting neurons in the central nervous system (Salgado 1998), and they excite this system directly when applied to insect ganglia. These effects are consistent with the activation of both nicotinic acetylcholine receptors and also GABA functions, which may increase their activity (Thompson and Hutchins 1999). However, the site of interaction on the acetylcholine receptors differs from that of neonicotinoids (Nauen *et al.* 2001).

The most effective insecticide of this group and the first entry into the market by DowElanco was Spinosad (Tracer®) (Sparks *et al.* 1996). Spinosad is a mixture of the two most abundant spinosyns, A and D (Kirst *et al.* 1992; Sparks *et al.* 2001). It acts primarily via ingestion, but also has some contact activity. This compound has translaminar activity, affecting hidden insect pests not targeted by the spray solution.

It is very active against a range of agricultural insect pests, especially lepidopterans such as the heliothine complex including *H. virescens*, and *H. armigera* and *H. Zea* (e.g. Sparks *et al.* 1996; Brickle *et al.* 2001). A recent report has shown that no reduction in susceptibility to spinosad was observed in Mississippi, USA, after three years of use against the tobacco budworm (*H. virescens*) and bollworm (*H. zea*) (Hardee *et al.* 2001). Spinosad became available for use in the Australian IPM-IRM strategy against *H. armigera* in the late 1990s (Holloway *et al.* 1999), but after 5 years of use, resistance levels to this insecticide increased 30-fold (Gunning 2002). Studies with the beet armyworm (*S. exigua*) populations have confirmed the vulnerability of spinosad to resistance (Moulton *et al.* 2000). Spinosad is among the most powerful insecticides for controlling the western flower thrips *F. occidentalis* and leafminers (Ishaaya *et al.* 2001) and was considered an effective and selective insect control agent for potential use in vegetables cultivated in greenhouses in Europe (Schoonejans and van der Staaij 2001). This compound is very effective in controlling cabbage lepidopteran pest complex such as *Pieris rapae* (L.), *P. Xylostella*, and *T. ni* (Hines and Hutchison 2001; Liu and Sengonca 2002). On the other hand, spinosad was not effective for controlling thrips in tomatoes or for suppression of the tomato spotted wilt virus (McPherson *et al.* 2002).

Spinosad is considered a biological control agent with relatively low mammalian toxicity and a favorable environmental profile (Sparks *et al.* 1996), occasionally this compound was found toxic to natural enemies. Spinosad was very effective against populations of onion thrips, but no natural enemies were found either in standard or in spinosad insecticide treated plots (Workman *et al.* 2002). In laboratory assays, spinosad was found toxic to two predators of spider mites, the thrips *Scolothrips takahashii* Priesner and the lady beetle *Stethorus japonicus* H. Kamiya (Mori and Gotoh 2001). Funderburk *et al.* (2000) demonstrated that *Orius* was compatible with spinosad treatments as compared with the broad spectrum pyrethroid, fenpropathrin, in pepper field experiments. In other experiments, the compatibility of spinosad and the predatory bug, *Orius insidiosus* (Say) was tested against *F. occidentalis* in greenhouse chrysanthemums (Ludwig and Oetting 2001). Their results indicated that in a restricted environment, spinosad may have an effect on *Orius* populations depending on activity of both the thrips and *Orius* in the plant canopy, and also the timing of insecticide applications and predator releases. In experiments conducted in Mexico (Cisneros *et al.* 2002), spinosad in a granular maize-flour formulation was toxic to the predaceous beetle, *Aleochara bilineata* Gyllenhal but larvae of *C. carnea* were not harmed (also Medina *et al.* 2001). They concluded that spinosad does not have an environmental safety profile similar to other biological insecticides. In cotton field trials, spinosad exhibited marginal to good selectivity to various natural enemies but was quite toxic to some parasitoids such as *Trichogramma* sp. and *Bracon* sp. (Ruberson and Tillman 1999; Tillman and Mulrooney 2000). On the other hand, in laboratory assays (Elzen 2001), spinosad did not harm two important predators of cotton pests, *O. Insidiosus* and *Geocoris punctipes* (Say), and similar results were reported in cotton fields (Duffie *et al.* 1998). It can be concluded that spinosad may harm several natural enemies in direct applications but with some precautions it can be used in IPM programs.

9.4.5 Pyrrole insecticides (chlorfenapyr)

Pyrrole insecticides are derived from a natural product, *dioxapyrrolomycin*, isolated from a strain of *Streptomyces*. Using *dioxapyrrolomycin* as a template, several pyrroles have been synthesized in Cyanamid's laboratory; chlorfenapyr has been commercially developed because of its broad spectrum activity against many species of Coleoptera, Lepidoptera, Acarina and Thysanoptera (Hunt and Treacy 1998). Chlorfenapyr is mainly a stomach toxicant, but has some contact activity (French *et al.* 1996). Field trials have demonstrated that foliar applications of chlorfenapyr was effective in controlling more than 70 insect pests and mites in cotton, vegetables, orchard trees, cereals and ornamentals (e.g. Miller *et al.* 1990). Chlorfenapyr acts at the mitochondrial level by uncoupling oxidative phosphorylation. The activated compound disrupts the proton gradient across mitochondrial membranes and impairs the ability of the mitochondria to produce ATP, which leads to cell destruction and death of the affected pest arthropod (Black *et al.* 1994; Hunt and Treacy 1998).

Chlorfenapyr was found moderately toxic to predatory arthropods in cotton (Duffie *et al.* 1998). In several bioassays, testing the residual effect of different insecticides including chlorfenapyr on the parasitoid *Diadegma semiclausum* (Hellén), fresh residue of this agent caused very high mortality; however, this toxicity did not persist long (Haseeb *et al.* 2000). The toxicity of selected insecticides (including chlorfenapyr) was evaluated on various hemipteran predators (Boyd and Boethel 1998) by contact using foliar residues and indirectly through the consumption of prey previously exposed to insecticides. The results demonstrated that although most of the newer compounds were more selective than older insecticides chlorfenapyr was as toxic as the older insecticides. In other laboratory bioassays, chlorfenapyr was found to have mild toxicity on some important beneficials of cotton pests (Pietrantonio and Benedict 1999). Under laboratory conditions, chlorfenapyr was harmful to several beneficials, but the level of selectivity increased under semi-field conditions (Leonard 2000). Under field conditions this insecticide has been reported to be safe to most beneficials, such as bumble bees and various predatory mites.

9.5 Other novel groups of insecticides

Among various new classes of pesticides the phenylpyrazole, pyridine and oxadiazine insecticides are demonstrated in this chapter.

9.5.1 Phenylpyrazoles (Fipronil)

The phenylpyrazoles comprise a new class of pesticides which exhibit herbicidal and insecticidal activities (Yanase and Andoh 1989; Klis *et al.* 1991). Fipronil, the first highly successful member of this class is active at the neuro-inhibitory GABA-gated chloride channels (Scharf and Siegfried 1999; Bloomquist 2001). Competitive bindings have demonstrated that this class of insecticides has greater affinity to the target site of insects than mammals, resulting in a high selectivity towards insects (Cole *et al.* 1993).

Fipronil exhibits broad activity against various insect pests such as soil insects (moderately systemic agent), foliar feeding pests such as the diamondback moth, (*P. xylostella*) *Spodoptera* sp., the Colorado potato beetle, the heliothine group, sucking pests such as thrips and *Lygus* sp. (but not aphids or whiteflies), and on various noncrop and household pests (Hamon *et al.* 1996). Fipronil, used in bait, was very effective in controlling the German yellowjacket wasp, *Vespula germanica* (F.) (Sackmann *et al.* 2001).

Early bioassays with fipronil in a cyclodiene-resistant strain of the housefly confirmed cross-resistance of about 90-fold to this compound (Bloomquist 1993). However, other recent studies have shown no resistance in major insect pests or lower resistance levels to this insecticide (Bloomquist 2001). It was concluded that cross resistance to fipronil in cyclodiene-resistant insects is less than expected; hence, the resistance situation does not preclude development of new compounds conferring the GABA receptor (Bloomquist 2001).

Fipronil is non-toxic to earthworms and various microorganisms (Hamon *et al.* 1996, Chaton *et al.* 2002) but is very toxic to bees. Laboratory residual insecticide bioassays (including fipronil) were conducted with the species *O. insidiosus* and *G. punctipes*, which are considered important predators of several economic pests of cotton (Elzen 2001), hence it was not considered compatible with IPM of cotton pests. In another study, fipronil was applied topically or as a residue to determine susceptibility of four natural enemies: *G. punctipes, Cotesia marginiventris* (Cresson), *Bracon mellitor* Say and *Cardiochiles nigriceps* Viereck (Tillman and Mulrooney 2001). When applied topically, all the insecticides were highly toxic to all beneficials, but when applied as residue, fipronil showed selectivity towards *C. nigriceps.* In other assays, in which fipronil was used as seed treatment in maize, mild to no effect was observed on the predator *O. insidiosus* (Al-Deeb *et al.* 2001). Hence, although fipronil was more toxic to various natural enemies than other new insecticides, it has little or no risk to beneficials when applied as a granule or seed treatment.

9.5.2 Pyridine insecticides (pymetrozine)

Pymetrozine, a novel azomethine pyridine insecticide, is highly specific against sucking insect pests (Fluckiger *et al.* 1992ab; Fuog *et al.* 1998). Its unique mode of action is not yet entirely understood, but it affects the nerves controlling the salivary pump and causes immediate and irreversible cessation of feeding due to an obstruction of stylet penetration, followed by starvation and insect death (Kayser *et al.* 1994). The compound is a powerful toxicant against aphids (such as *A. gossypii* and *M. persicae*), whiteflies (*B. tabaci* and *Trialeurodes vaporariorum* [Westwood]) and planthoppers, (*Nilaparvata lugens* Stål). Pymetrozine has systemic and translaminar activities and can be used as drench or foliar application (Fluckiger *et al.* 1992ab). It is also effective in lessening aphid-transmitted diseases caused by persistent viruses (Fuog *et al.* 1998); this action may be strengthened with an addition of mineral oil (Asjes and Blom-Barnhoorn 2002). As pymetrozine has shown effectiveness against different strains of resistant aphids with no evidence of cross-resistance, it can play an important role in IRM strategies for aphids (Ferguson *et al.* 1999; Foster *et al.* 2002).

Pymetrozine has very low mammalian toxicity and no appreciable effect on nontarget and beneficials. In laboratory experiments, various predators were exposed to this compound and it was found to be nontoxic (Jamses 2002). In other assays, no detrimental effect could be detected against predaceous coleopteran, heteropteran and neuropteran species (Sechser *et al.* 2002). In cotton fields, applications of pymetrozine did not affect predator and parasitoid populations (Ferguson *et al.* 1999; Sechser *et al.* 2002). It can be concluded that pymetrozine is a potential IPM component in controlling sucking insect pests.

9.5.3 Oxadiazines (indoxacarb)

It is a pyrazoline-type sodium-channel blocker. Indoxacarb was the first insecticide (DuPont) of the oxadiazine group to be commercialized. It is effective against lepidopteran pests as well as certain homopteran and coleopteran pests in vegetables, cotton, and other field and orchard crops (Harder *et al.* 1996; Wing *et al.* 2000). This product has been demonstrated against important insect pests such as: *Heliothis* sp., *Helicoverpa* sp., *Spodoptera* sp., *Plutella* sp., *Trichoplusia* sp., (lepidopterans); *Lygus* sp., *Empoasca* sp. (hemipterans) and also the Colorado potato beetle, *L. decemlineata* (Harder *et al.* 1997; Hammes *et al.* 1998; Michaelides and Irving 1998; Pluschkell *et al.* 1998; Sullivan *et al.* 1999; Wing *et al.* 2000).

Indoxacarb acts by inhibiting sodium ion entry into nerve cells, resulting in paralysis and death of target pest species. It is also a pro-insecticide that is bioactivated by esterases inside the insects following ingestion (Wing *et al.* 1998, 2000). The conversion of the pro-insecticide to the active compounds (S-enantiomers of the N-decarbomethoxyllated metabolites), which are powerful sodium channel blockers is correlated with the appearance of neurotoxic symptoms; *e.g.* several lepidopteran larvae can rapidly convert the pro-insecticide after ingestion. Various sucking insects can bioactivate indoxacarb but more slowly than lepidopterans (Wing *et al.* 2000). Thus, the rate of bioactivation is a critical factor in determining the toxicity of indoxacarb in different insect species. Since the sodium channel sites have already been exploited by DDT and pyrethroids, there is a fear of cross-resistance between the latter compounds and indoxacarb. However, its mode of action seems to be distinct from that of the pyrethroids and DDT (Wing *et al.* 2000). Consequently, no cross-resistance has yet been reported between indoxacarb and other insecticide groups (Holloway *et al.* 1999). Furthermore, negative cross-resistance has been observed between indoxacarb and pyrethroids (Gunning and Devonshire 2002) in *H. armigera.*

Indoxacarb was considered safe to beneficials. Several assays (Tillman *et al.* 1998; Ruberson and Tillman 1999; Ruberson *et al.* 2002) have been conducted to evaluate the toxicity of indoxacarb and other new insecticides to key natural enemies in cotton, such as *Chrysoperla rufilabris* (Burmeister), *G. punctipes, O. insidiosus*, *Trichogramma pretiosum* Riley and *C. marginiventris.* In these experiments indoxacarb exhibited good selectivity to all of the natural enemies tested. In other assays, the toxicity of indoxacarb was examined on *Anaphes iole* Girault (an egg parasitoid of *Lygus lineolaris* [Palisot de Beauvois] in cotton) (Williams *et al.* 2003). Although acute exposure of indoxacarb resulted in about 30% survival of the parasitoids, parasitoid wasps exposed subsequently increased their survival considerably.

The differential selectivity of indoxacarb against herbivorous versus insectivorous may be based on its modes of entry as direct spray or dried residues (Andaloro *et al.* 2000). Although the mechanisms of selectivity/safety of indoxacarb for a predator such as *G. punctipes* and its prey (*L. lineolaris*) is still under investigation, there may be a combination of differential feeding behavior of the predator and reduction in feeding by females exposed to indoxacarb-treated eggs of the prey (Tillman *et al.* 2002). The selectivity of indoxacarb to beneficials rendered this agent compatible with IPM programs in cotton fields.

9.6 Biopesticides

The latest developments of the biorational approaches in pest management has been the use of biopesticides for their specificity, safety to non-target organisms, particularly mammals, and utilization in low, sometimes in minute amounts have led to an intensive research program to discover and develop newer and safer pesticides. The three main groups including microbial insecticides in pest control, utilization of semiochemicals, and botanical insecticides hold promise for safety to the environment.

Until few years back, the biopesticides comprised little more than 1% of the total world pesticide market, but in recent years over 800 pesticides formulated in 21,000 different products and registered at the US Environmental Protection Agency (EPA). Of the more than 94% of the total market for biological control products, more than 90% of the total are based on the bacterium *Bacillus thuringiensis* Berliner (*Bt*) (Warrior, 2000). The chemical pesticides account for approximately 99% of the market, which are a cause of serious concern because of their adverse effects. The ideal pesticide should be toxic only to the target organisms, biodegradable, and should not drift into the groundwater. But this rarely happens and their widespread use in modern agriculture is cause of serious concern. On the other hand, the low level of acceptance of biopesticides is the result of several factors: inconsistent practical results, uncompetitive price in comparison to classical insecticides, inappropriate formulations and application to a limited range of pests. However, the expected advantages of biopesticides, that is their specificity, safety to non-target organisms, use in very limited amounts, which often decompose quickly, thereby resulting in lower exposures and largely avoiding the pollution problems caused by conventional pesticides, have led to numerous scientific works on new and safer pesticides, particularly in the last three decades (Koul and Dhaliwal 2002)

In addition to be inherently less harmful than conventional pesticides, biopesticides have been of immense value in specific integrated pest management (IPM) strategies and are very effective when they are produced and delivered correctly. In this case, the use of biopesticides can greatly decrease the use of conventional pesticides, while crop yields remain high. Another important attribute is the specificity of action on target pests and closely related organisms in contrast to broad spectrum conventional pesticides that may affect organisms as different as birds, insects, and mammals. In addition and due to the fact that, in many instances, for example in *Bt*, multiple toxins are involved in killing the insect, the possibility of creating resistance is greatly diminished, in spite of a large scale utilization over time.

9.6.1 Microbial pesticides

Microbial pesticides are those pesticides in which the active ingredient is a microorganism, bacterium, fungus, virus, or protozoan. These types of pesticides can control many different kinds of pests, although each separate active ingredient is relatively specific for its target pest(s); for example, there are fungi that control certain weeds and other fungi that kill specific insects. The most widely used microbial pesticides are subspecies or strains of *Bt*. Each strain of this bacterium produces a different mix of proteins, and specifically kills one or a few related species of insect larvae. While some *Bt* strains control moth larvae found on plants, other strains are specific for larvae of flies and mosquitoes. The target insect species are determined by whether the particular *Bt* produces a protein that can bind to a larval gut receptor, thereby causing the insect larvae to starve.

9.6.2 Bacteria

The most widely used insecticidal bacterium is *Bt*, a Grampositive soil bacterium that forms crystalline protein inclusion during sporulation (Hofte and Whitely 1989) During this process, this bacterium produces d -endotoxin, an insecticidal crystal protein that is encoded on bacterial plasmids. The formation of crystal proteins (usually referred to as Cry proteins) is an essential feature that distinguishes *Bt* from a very close related species *B. cereus*, which can produce emetic and diarrheal toxins that cause food poisoning (Sudakin 2003) Cry proteins are produced as protoxins that are proteolytically converted into a combination of up to four smaller toxins upon ingestion. These proteins bind to specific receptors in the larval midgut epithelium causing formation of large cation-selective pores that increase the water permeability of the cell membrane. A large uptake of water then causes cell swelling and eventual rupture of the midgut. The detailed mode of action of these toxins, however, is still unknown although a number of recent studies have added new data to disclose their mechanism of action (Charles *et al.*

1996, Manceva *et al.* 2004, Regis *et al.* 2001, Smith *et al.* 2005). Nine different toxins have been described in *Bt* strains El-(Bendary, 2006) These include an a -exotoxin of phospholipase C type, b -exotoxin (nucleotide thermostable, non selective toxic), g -exotoxin (toxic to sawflies), d -endotoxin (exploited commercially for pest control), louse factor exotoxin (only toxic against lice), mouse exotoxin (active to mice and lepidoptera), water soluble toxin Vip3A (vegetative insecticidal protein) and enterotoxin Delecluse *et al.*, 2000, WHO 1999).

These toxins bind to different receptors (phospholipids, phosphatidylcholine and sphingomyelin) in different insect species and with varying intensities, thus explaining species specificities (Adams *et al.* 1996, Baum *et al.* 1999). The potential presence of the b -exotoxin in certain *Bt* subspecies prompted the EPA to require acute toxicity testing of many *Bt* microbial pesticide formulations in rodents to confirm its absence. The presence of b -exotoxin in commercial *Bt* pesticide formulations has been prohibited by the EPA (McClintock *et al.* 1995) *Bt* strains containing mixtures of up to 6–8 different Cry proteins have been widely used as microbial pesticides since *Bt kurstaki* strain HD1 was originally commercialized in 1961. Cry protein-encoding genes were then visualized as an obvious and interesting choice to protect crops against pests. The first Cry gene was cloned and expressed in *E. coli* (Schnepf and Whiteley 1981) and later soon the first genetically modified *Bt*-protected tomato, tobacco and cotton plants were produced (Fischhoff *et al.* 1987, Perlak *et al.* 1993, Vaeck *et al.* 1987). Nowadays, *Bt* protected cotton, potato, and corn have been commercialized in the US and one or more of these products are marketed in Argentina, Mexico, China, Canada, South Africa, France, Australia, Spain, Ukraine and Portugal, (Betz *et al.* 2000) so that they represent about 1–2% of the global insecticide market (Baum *et al.* 1999)

These plants express one or more Cry proteins for the control of the Colorado potato beetle, tobacco budworm, cotton bollworm, pink bollworm, European corn borer, southwestern corn borer and corn earworm, among others. The total surface planted in the US for *Bt*-protected cotton, corn and potato exceeded 16 million acres in 1998, that is 17%, 18% and 4% of the total cotton, corn and potato acreage, respectively (Gianessi and Carpenter, 1999) This has represented a significant reduction (metric tons annually) in the use of insecticides to control these pests. More recently, a new insect-protected *Bt* corn variety has been developed that provides protection against corn rootworm, a pest that causes significant damage to the roots of corn grown in the Midwest and Eastern US (Miller 2002) In China, where there has been widespread adoption of *Bt* cotton in the last few years, the number of insecticide applications have been reduced from 20 to

7/ha/season, which means a decrease of organophosphate and organochlorine insecticides by up to 80%(Huang *et al.* 2002). In terms of cost, this represents a reduction of the insecticide cost by $762/ha/season.

In Europe, a number of *Bt*-derived products are used to control Lepidoptera pests in vegetables, tomatoes, top fruit, vines, olives and forestry. They include *Bt* serotype 3, *Bt* subsp. *azawai*, *Bt* subsp. *kurstaki* (*Btk*), d -endotoxin of *Bt* subsp. *kurstaki*, *Bt* subsp. *tenebrionis*, *Bt* subsp. *Israelensis* and *B. sphaericus*. (Butt *et al.* 1999) An illustrating example of the use of *Bt* in forestry was the control of an unexpected outbreak of the nun moth *Lymantria monacha*, which attacked more than 600,000 ha of forest in Poland in 1994. An emergency program, in which several public and private institutions were involved, was put into practice by applying a special forestry formulation of *Btk* from Novo Nordisk called Foray 48B. This material is suitable for spraying from the air at low volumes (4 l/ha). The results were excellent with *Btk* providing *ca.* 95% control of the pest while having a minimum impact on non-target and beneficial organisms. Since then, similar programs are being used in adjacent countries to control the pest. In North America, *Btk*-based products have become the first choice for controlling forest pest outbreaks, particularly for the gypsy moth *Lymantria dispar* (Butt *et al.* 1999).

With regard to toxicity, *Bt*-protected plants provide a safety degree which is unmatched by any other pest control product. First, Cry proteins are non toxic to vertebrate species (mice, rats, monkeys, humans) even at doses higher than 1_106 m g/kg body weight, while dosages acutely toxic to susceptible insects are about m g/kg body weight (McClintock *et al.* 1995, EPA 1998) Second, the acidic environment of the mammalian stomach does not favor solubilisation and activation of the Cry proteins. Third, these proteins are readily degraded very fast (often in some seconds), from 60–130 kDa to polypeptides less than 2 kDa that corresponds to peptides with 10 amino acids in length. The rapid degradation of these proteins by proteases in the mammalian gastrointestinal tract precludes their toxicity in mammals. Fourth, several studies in vertebrates have failed to find high affinity Cry protein binding sites on gut epithelial cell membranes (Hoffmann *et al.* 1988, Noteborn *et al.* 1993).

In addition, the Cry proteins produced in *Bt*-protected crops (Cry3Aa in *Bt* potato and Cry1Ab and Cry1Ac proteins in *Bt* cotton) have been shown to rapidly degrade in soil at rates comparable to the rate of degradation of Cry proteins in microbial *Bt* products (Palm *et al.* 1996, 1993) In summary, the benefits of using *Bt*-protected plants include: reduction of chemical insecticide treatments for the target pests, highly effectiveness in pest control, increase of crop yields, reduction of levels of fungal toxin, and additional pest control by preserving or enhancing populations of beneficial organisms.

While *Bt* is currently among the most widely used microbial insecticides, many other strains of *Bacillus* sp., including *Bacillus sphaericus* and *Bacillus subtilis*, as well as *Pseudomonas* sp. are also currently used as biopesticides (Sudakin 2003).

9.6.3 Viruses

Insects are attacked by many different types of virus, baculovirus being the most promising in insect control, particularly of Lepidoptera and Diptera (Cory and Hails, 1997) because of their specificity. In addition, insect baculovirus are safe to vertebrates, plants and non target organisms, and they are pathogenic, the death of the host is generally the outcome of a baculovirus infection. The major successes of microbial control with viruses have taken place in forestry, particularly for sawflies control in Europe and North America through the NPV of the European pine sawfly, *Neodiprion sertifer* (Lord 2005). Baculovirus control of pest insect populations was demonstrated in the 1940's but the first viral insecticide registered was that of *Helicoverpa* (*Heliothis*) *zea* in 1971 under the trade names Viron/H and later Elcar (Ignoffo 1973). This virus had been developed by the USDA for control of the tobacco budworm (*Heliothis virescens*), corn earworm (*Heliothis zea*) and *Heliothis armigera* on cotton, row crops, fruits and vegetables.

Since then, a number of baculovirus insecticides have been registered and produced commercially (Moscardi, 1999). They are all wildtype baculovirus, and like most microbial insecticides, have had limited success for various reasons. These include narrow spectrum of biological activity, slow mode of action (5–7 days after ingestion of NPVs and 7–14 days in the case of GV infections, see below) and photolability (particularly to solar radiation).

There have been different approaches directed to enhance the role of baculovirus as effective biopesticides. For instance, the effect of baculovirus may be enhanced by the synergistic action of specific chemical insecticides, such as the synthetic pyrethroids deltamethrin and permethrin (McCutchen and Flexner 1999) To improve potency and speed of action, several recombinant baculovirus have been developed (Bonning and Hammock 1996, Treacy 1999) Baculoviruses are classified into two genera: nuclear polyhedrovirus (NPV) and granulovirus (GV). The NPVs are further segregated into two groups based on the phylogenetic relationships of 20 different genes Herniou 2001). The baculovirus are characterized by double-stranded circular DNA genomes ranging from 88 to 153 kb. Baculoviruses produce two types of progeny: the budded virus (BVs) and the occluded virus (OVs).

The OVs of the NPV and GV are termed polyhedron (plural polyhedra) and granule, respectively. BVs are responsible for the systemic or cell-to-cell spread of the virus within an infected insect. OVs, in turn, are responsible for the larva-tolarva transmission of the virus (Inceoglu *et al.* 2006) During infection, the host larva is debilitated, resulting in reduction of development, feeding, and mobility and increasing exposure to predation (Young and Kring, 1991).

Post larval effects may include lower pupal and adult weights, as well as reduced reproductive capacity and longevity (Rothman and Myers 1996) Infected larvae usually climb to the upper parts of the plants, dying in 5–8 days although cessation of feeding may occur 2–4 days depending on biotic and abiotic factors. Diseased and dead larvae serve as inoculums for virus transmission, which may occur by rain and movement of arthropods on plants, or *via* predators and parasitoids. Fortunately, not many mammalian viruses are as successful and prevailing as baculoviruses in infecting insects (Inceoglu *et al.* 2006).

The BVs offer several unique advantages as a vector for the expression of a foreign gene within insect cells and insect larvae. Currently, *Autographa californica* NPV (AcNPV) is the choice for gene insertion or gene deletion research both in industry and academic institutions. Methods for the construction and use of recombinant BVs for the expression of heterologous genes have been described and are identical to those that were later used to generate recombinant Bv insecticides. Merrington *et al.* 1999, Richardson 1995) In this context, it should be noted the efforts conducted with AcNPV engineered to express an insect specific toxin derived from the Algerian scorpion *Androctonus australis* (Zlotkin *et al.* 1971) The recombinant AcNPV showed equal selectivity towards beneficial insects and pest species to that of the native AcNPV. In addition, the AcNPV-AaIT killed noctuid moth larvae in half the time required by the native AcNPV and several private companies have conducted promising field trials in cotton, tobacco and other crops (Copping and Menn 2000). Other recombinant baculovirus systems reported include: genes encoding the juvenile hormone esterase have been inserted in AcNPV, Hammock *et al.* 2001) genes insertions and genes deletions to express development-disrupting enzymes (Harrison and Bonning 2001) diuretic hormones (Popham Holly *et al.* 1998) specific toxins such as *Diguetia canities* and *Tegenaria agrestis* spider toxins (Popham Holly *et al.* 1998, Krapcho *et al.* 1995) scorpion toxins, Gershburg *et al.* 1998, Harrison and Bonning 2000, Regev 2003) and specific neurotoxin from the venom of *Leiurus quinquestriatus hebraeus* against *Trichoplusia ni* and other lepidoptera pests (Jinn *et al.* 2006).

9.6.4 Fungi

Another major class of microbial biopesticides includes fungi. Many fungi are pathogenic to the insect host, and these are referred to as entomopathogenic fungi. Many of the genera of entomopathogenic fungi currently under research either belong to the class Entomophthorales or Hyphoymycetes. Species within these classes can behave very differently, and thus, for example, insect host, infection levels, germination rates and optimum temperature of action can vary between species and isolates (Shaw *et al.* 2002, Sierotzki *et al.* 2000) Members of Hyphoymycetes are generally considered opportunistic pathogens infecting many species of several insect orders, and host death is commonly associated with toxin production overwhelming host defense mechanisms. In contrast, other groups of fungi are thought to have been evolved into higher parasitic forms. For example, infection and host death by Entomophthorales usually occurs by tissue colonization with little or no use of toxins (Shah and Pell, 2003).

In contrast to other microbials acting *via* ingestion by the insect host, entomopathogenic fungi attack insects by direct penetration through spiroacular openings in the cuticle. Fungal infections occur in arthropods but other non-host insects and/or species which are not pest of cultivated crops, such as spiders and ants, may also be infected. The fungal infection and transmission is produced by asexual fungal spores or conidia that are dispersed throughout the environment in which the insect host is present. The spores germinate in the insect's blood and germinating mycelia gradually kill the host (Copping and Menn 2000) Interactions between the entomopathogenic fungi and their arthropod hosts and the induced altered behaviors on the invertebrates resulting thereof has been recently reviewed Roy *et al.* (2006).

Although over 750 fungal species are known to infect insects, only a few have been commercialized in USA, Europe, Latin America, China and the former Soviet Union. A list of the commercially available fungi, the current producer and the target pests has been reported (Flexner and Belnavis 1998) The most notable mycoinsecticides currently in use are: *Beauveria bassiana*, *Beauveria brongniari*, *Metarhizium anisopliae*, and *Metarhizium flavoviride*. Among them, the most prominent is *B. bassiana*, from which several strains have been developed and marketed in the US, Europe, Russia, South America and probably also in China (Wraight and Carruthers 1999) The application of *B. bassiana* for the control of the pine moth *Dendrolimus* spp. in China probably represents the largest use of a biocontrol agent. The programs started in 1950 and it is calculated that at least one million hectares of pine forest are now involved in the treatments (Lord 2005) *B. bassiana* strain Bb-147 is registered on maize in Europe for control of the European corn borer *Ostrinia nubilalis* and the Asiatic corn borer *Ostrinia furnacalis*. *B. bassiana* strain GHA, in contrast, is registered in the US for control of the whitefly, thrips, aphids and mealybugs. *B. bassiana* strain ATCC 74040 is registered against many soft-bodied insects of the orders Homoptera, Heteroptera and Coleoptera. *B. brongniari* is registered on sugarcane and barley for control of white grubs and cockchafers *Melolontha melolontha*. *Melolontha anisopliae* is one of the first fungi used in biological control (the first account of its utilization dates from Russia in 1879 (McCoy *et al.* 1988)) because it has not adverse effects in mammalian toxicology tests, and presents no risk of infectious complications because the organism in incapable of growth at mammalian body temperatures. It is registered in the US for control of termites but is also produced for control of coleopteran and lepidopteran greenhouse pests. In addition, recent developments have disclosed that new formulations of *M. anisopliae* are being used in Africa, Australia and Brazil against locusts and grasshoppers (Lomer, *et al.* 2001) *Melolontha flavoviride* (*M. anisopliae* var. *acridum*) has been found effective against the brown locust *Locustana pardalina* in Africa (Price *et al.* 1997) *Locusta migratoria* in Madagascar (Delgado *et al.* 1997) and the Australian plague locust *Chortoicetes terminifera* and *L. migratoria* in Australia (Milner and Hunter 2001) With variable success, *M. flavoviride* has also been tested against the tree locust *Anacridium melanorhodon* in Sudan (Kooyman and Abdalla, 2001) the rice grasshopper *Hieroglyphus daganensis* in Benin, Mali and Senegal Lomer, *et al.* 1997) and the desert locust *Schistocerca gregaria* in Mauritania (Langewald *et al.* 1997). In the case of this highly mobile insect, an estimated plot size larger than 1000 ha would be necessary and such trials are expected to take place within the FAO Emergency Prevention Program. Very recently, *M. anisopliae* has been modified to express an insect-specific neurotoxin from the scorpion *A. australis*, under the control of a promoter that is active only in the presence of insect hemolymph (Wang and Leger 2007) This restricts expression of theneurotoxin to the time after the fungus has penetrated the cuticle.

The toxin dramatically increased pathogenicity and virulence against tobacco hornworm *Manduca sexta* and yellow fever mosquitoes *Aedes aegypti*. In comparison with the wild fungus, the modified fungus achieved the same mortality rates in tobacco hornworm at 22-fold lower doses and reduced survival times of infected mosquitoes by over 40% (Wang and Leger, 2007) Compared with most chemical insecticides, fungi are less toxic to mammals and have negligible environmental impacts. One obstacle in exploiting fungi for insect control, however, is that they may kill their host too slowly. Even highly virulent isolates take 2–5 days to kill an insect, and infected host can survive much longer, depending on dose and environmental conditions. Another constraints which may hamper a wider utilization of entomopathogenic fungi include dessication, UV light, host behavior, temperature, pathogenic strength, etc. (Lacey and Goettel, 1995) Many of these constraints are currently being addressed by recent advances in formulation technology (Flexner and Belnavis, 1998).

9.6.5 Nematodes

Although it was known for more than 60 years that certain nematodes infect and kill soil-inhabiting grubs, only until recently practical attention has been paid to entomopathogenic nematodes (EPNs) as biopesticide agents (Copping and Menn 2000) Nematodes are simple roundworms, colorless, unsegmented, and lacking appendages. They are not actually microbial organisms but are always included as part of the microbial control of insects. In fact, the non-feeding infective juveniles possess attributes of both insect parasitoids or predators and microbial pathogens. Like parasitoids/predators they have chemoreceptors and are motile; like pathogens they are highly virulent, killing their hosts quickly (Kaya and Gaugler 1993) In addition, in contrast to chemicals which should have to decay within few days, EPNs are persistent and recycle inside the host causing long-term and sustainable effects on pest populations. Peters 1996) They can be cultured easily *in vitro* and have a high reproductive potential. EPNs are environmentally friendly agents with no detrimental effects on non target organisms, mammals or plants (Ehlers 2001) In almost all countries EPNs are exempt from registration, which enables small and medium-sized companies to develop them as plant protection agents. Moreover, they can be stored for months and can be applied with conventional spraying equipment. On the other hand and like other biological control agents, nematodes are constrained by being living organisms that require specific conditions to be effective; they are fragile, subject to desiccation and sensitive to temperature changes and solar radiation (Copping and Menn 2000) They are ubiquitous and have been found in a wide range of ecologically diverse soil habitats including cultivated fields, forests, grasslands, deserts and even ocean beaches. Today, nematodes are mainly used where chemical insecticides fail, for instance in soil, galleries of boring insect pests or where resistance to insecticides has appeared (Ehlers 2001).

Of all nematodes studied, those from the genera *Steinernema* and *Heterorhabditis* have aroused the greatest interest and developed for insect control. These nematodes are characterized by the ability to carry specific pathogenic bacteria, *Photorhabdus* with *Heterorhabditidae* and *Xenorhabdus* with *Steinernematidae*, which are released into the insect hemocoel after penetration of the 3-stage infective juveniles (called "dauer juveniles", DJ). The DJ is a free-living, third juvenile stage that is well adapted to long-term survival in the soil. When a host is located, the nematodes penetrate into the insect body cavity, usually *via* natural body openings (mouth, anus, spiracles), and the DJ release the symbiont cells and the symbiotic bacteria (*Xenorhabdus* or *Photorhabdus*), which multiply rapidly and cause rapid insect death (within about 2 days after nematode invasion). The bacteria proliferate and produce suitable conditions for nematode reproduction. Feeding on the symbiont cells, juveniles develop into adults and produce offspring. When the nutrients are consumed, the offspring develop into DJ, which retain the symbiotic bacteria in the intestine. The life cycle is completed in a few days, and hundreds of thousands of new infective juveniles emerge in search of fresh host (Ehlers 2001, Gaugler 1997). Only a few strains of the symbiotic bacteria have been studied in detail from a genetically point of view, but they have gained considerable attention since their genes represent a possible alternative to *Bt* genes for expression in transgenic plants (Guo *et al.* 1999)

As noted, the EPNs are lethal to a notably broad range of insect pests in the laboratory, although in the field the host range is considerably more restricted. Important soil insects against which there are no effective nematodes commercially available include wireworms, grape phylloxera, fire ants or corn rootworms (Gaugler 1997) Of the nearly 30 steinernematid and heterorhabditid nematodes identified to date, eight species are commercially available (Flexner and Belnavis 1998) *Seiternema carpocapsae* is the most studied, versatile, and available of all EPNs and is particularly effective against lepidopterous larvae, including webworms, cutworms, armyworms, girdlers and wood borers.

Other EPNs marketed are Seiternema feltiae, Seiternema glaseri, Seiternema kushidai, Seiternema riobravis, Seiternema scapterisci, Heterorhabditis bacteriophora, and Heterorhabditis megidis (Gaugler 1997) the target insects being mushroom flies, fleas, fungus gnats, coleopterous larvae particularly scarabs, mole crickets, root weevils, white grubs, sciarid flies, slugs, etc. (Gaugler 2000) In some cases, the efficiency of the EPNs particularly against white grubs has been inconsistent and unsatisfactory (Klein 1993) Nematode efficacy and consistency may be improved by combining EPNs with insecticides, particularly with the nicotinoid insecticide imidaclòprid (Koppenhofer *et al.* 2003) Combination of both agents induced a synergistic increase in scarab mortality between 4 and 13 fold higher than in treatments with the nematodes alone and the insecticide did not interfere with nematode recycling (Koppenhofer *et al.* 2003).

9.6.6 Protozoa

Entomopathogenic protozoa are an extremely diverse group of organisms in relationship with insects ranging from commensal to pathogenic. They are generally host specific and slow acting, often producing chronic infections characterized by a general debilitation of the host (Lacey and Goettel 1995) The infections cause sluggishness and irregular or slowed growth finally resulting in reduced feeding, vigor, fecundity and longevity of the insect host. However, although they are undoubtedly important in natural biological regulation of insect populations, they do not possess the required attributes for a successful microbial insecticide(Flexner and Belnavis 1998)

Out of the approx. 15,000 protozoa described, around 1000 species, primarily microsporidia, are known to attack invertebrates, including many insect species such as grasshoppers and Heliothine moths (Copping and Menn 2000) Most microsporidia must be eaten to infect an insect, but there may also be natural transmissions, for instance by predators and parasitoids, within a pest population. The spore formed by the protozoan is the infectious stage in susceptible insect. They are ingested by the host and germinate in the midgut. Sporoplasm is then released invading target cells and causing massive infection to the host (Brooks 1988).

The most notable entomopathogenic protozoa belong to *Nosema* spp. and *Vairimorpha necatrix*. Among the *Nosema* spp., *Nosema locustae* is the only commercially available species of microsporidium, and marketed under several labels for control of grasshoppers and crickets. It was first described against the grasshopper *L. migratoria*, but it is now known that its potential host range involves at least more than 100 orthopteran species (Sokolova and Lange, 2002) *N. locustae* has been the subject of extensive molecular investigations regarding metabolism, evolution, and phylogenetics, and its complete genome is currently being sequenced, but details on the ultrastructure of intracellular stages were unknown until recently (Slamovits *et al.* 2004). It has been extensively tested in the field in the US, Canada, Argentina, Cape Verde, China and Mali (Lomer, *et al.* 2001). In Argentina, comprehensive studies have indicated a considerable decline in locust populations in treated areas (Lange and De Wysiecki, 1996) and in China surfaces of *ca.* 100,000 ha have been treated annually (Yan *et al.* 1986) However, the utility of *N. Locustae* as a grasshopper control agent remains questionable mainly because of the great difficulty underlying an adequate efficacy assessment in this highly mobile insect (Lacey and Goettel, 1995) Other microsporidia with far greater virulence against grasshoppers than that of *N. locustae* have been described, including *Nosema enthomophaga*, *Nosema cuneatum* and *Johenrea locustae*. However, their great virulence hampers development of an effective production system (Lomer, *et al.* 2001).

Nosema pyrausta is another beneficial microsporidium that reduces fecundity and longevity on adults of the European corn borer (Windels, *et al.* 1976) and also slow down larval development increasing mortality (Siegel *et al.* 1986) In the laboratory, *N. pyrausta*-infected females laid a significantly reduced number of eggs at low temperatures in comparison to that laid by non-infected insects (Bruck *et al.* 2001) These data have been claimed to be useful in predicting *O. nubilalis* populations (Bruck *et al.* 2001).

9.7 Semiochemicals

Chemical substances or mixtures that mediate interactions between organisms are named semiochemicals, and includes pheromones, kairomones, allomones, and other classes of behaviourally active compounds. Semiochemicals are considered to be safer and environmentally more acceptable than conventional pesticides because they occur naturally, are able to target the pest species only, elicit low acute toxicities to vertebrates, and are usually volatile chemicals that do not leave behind harmful residues (Punga, 1989) Semiochemicals may be used to monitor populations or in direct pest control strategies reducing populations by mass trapping, lure-and-kill or mating disruption.

9.7.1 Monitoring

Development of effective monitoring systems provides valuable information for coordination of the treatment schedule with pest phenology. Semiochemicals are used in traps to monitor changes in population levels allowing a better knowledge about the onset of adult emergence and the flight peak. Highly sensitive pheromone-based monitoring is crucial for detection of incipient infestations of introduced or exotic insects, such as the Mediterranean and Mexican fruit flies (Heath *et al.* 1996) wood boring and bark beetles (Brockerhoff *et al.* 2006) or for detection of pests with constantly expanding ranges, such as the gypsy moth (Sharov *et al.* 1997) the pink bollworm (Baker *et al.* 1990) or oriental beetles. (Alm *et al.* 1999).

The most widely used attractants in monitoring systems are sex pheromones to monitor (Cork *et al.* 2005, Wang *et al.* 2005) aggregation pheromones to monitor coleopteran species (Adachi *et al.* 2007, Allison *et al.* 2003) and host plant odors for dipteran species. 100. Although in the vast majority of cases a good correlation between catches and level of damage has been noticed (Faccoli and Stergulc 2004) sometimes the number of adults caught in traps was not correlated with the number of eggs laid and larvae found in plants (Godin and Boivin 1998).

In order to get a good prediction of the damage caused by the larvae of the next generation, several features should be considered.

Pheromone composition of the bait including dose and purity, both chemical and stereomeric, is the first parameter to bear in mind. The stereomeric purity of the lure is very important for an efficient monitoring in many species, as proved to be for the oak processionary moth (*Thaumetopoea processionea*) the Israeli pine bast scale *Matsucoccus josephi*) and the leafminer *Phyllonorycter blancardella*. In addition, it should also be considered the possible geographical variation in pheromone composition, as in the European pine sawfly *N. sertifer*, or in the apple leafminer moth, *Phyllonorycter ringoniella*. Variables like color, form and position of the traps may have also a substantial influence in the number of catches, and optimization needs to be performed for every species. Trap color may be also a significant variable in certain species, like stink bugs and the processionary moth *Thaumetopoea pityocampa* males, whereas other species do not show any preference for any specific color. The type of trap can alsobe essential for an optimum level of catches, as demonstrated for the Japanese beetle *Popillia japonica* or for the Nantucket pine tip moth *Rhyacionia frustana*. Position of the traps and the relative distance among them within the plot should also be considered (Bacca *et al.* 2006).

9.7.2 Mating disruption

Mating disruption is the most widely and successfully used control method for a variety of insects (El-Sayed and Suckling, 2005). It prevents mating and, hence, reduces the incidence of larvae in the next generation. This is normally done by releasing a large amount of pheromone or pheromone analogue in the treated area, and has been used against lepidopteran species) and other orders like Coleoptera, Hemiptera and Heteroptera (Kakizaki 2004).)

Some recent cases of mating disruption are shown in Table 9.2. Mating disruption has a number of advantages as pest management tactic. It is species-specific, has low environmental impact and is more sustainable than other broad spectrum techniques without evidence of resistance. In fact, it has proven to be one of the preferred control methods against insecticide-resistant populations. On the other hand, mating disruption has also important drawbacks. Due to its specificity, it is often more costly than the broad spectrum insecticides, particularly when more complex and/or unstable components are used. In addition, secondary pests can often emerge as important problems. Monitoring success of the experiments is often difficult and sometimes inconsistent, since possible reduction in trap catches in pheromone-treated blocks may not be corresponded with egg mass densities and damage produced by larvae of the next generation (Borchert and Walgenbach, 2000).

Table 9.2. Some recent cases of mating disruption experiments using sex pheromones or pheromone analogues

Common name	Species	Crop	Reference
Corn stalk borer	*Sesamia nonagrioides*	maize	Solé et *al.* 2008
Codling moth	*Cydia pomonella*	apple	Epstein et *al.* 2006
European corn borer	*Ostrinia nubilalis* *Tecia solanivora*	corn potato	Solé et *al.* 2008
Oriental fruit moth	*Grapholitha molesta*	apple orchard peach/pear orchard	Il'ichev et *al.* 2006
Jasmine moth	*Palpita unionalis*	olive groves	Hegazi et *al.* 2007
Cherry tree borer	*Synanthedon hector*	cherry orchard	Matsumoto et *al.* 2007
Light brown apple moth	*piphyas postvittana*	citrus	Mo et *al.* 2006
Olive pyralid moth	*Euzophera pinguis*	live groves	Ortiz et *al.* 2004
Clearwing moths	*Paranthrene robiniae*	poplar	Brown et *al.* 2006
Yellow stem borer	*Scirpophaga incertulas*	rice	Cork, 1998
Cone worm moths	*Dioryctria spp.*	pine seed orchard	DeBarr et *al.* 2000
Pink bollworm	*Pectinophora gossypiella*	Cotton field	Lykouressis et *al.* 2005

There has been considerable debate about the mechanisms underlying mating disruption (Miller *et al.* 2006) although there is general agreement now that more than one mechanism may be operational at the same time and that they may vary between species (Sanders 1997). Two possible mechanisms are recognized in the literature for mating disruption: competitive (competitive attraction false-plume following) and non-competitive (camouflage, desensitization, and sensory imbalance) (Miller *et al.* 2006).

Competitive attraction appears to be the prevalent mechanism in mating disruption when the pheromone is deployed in specific point sources. This is the suggested mechanism operating, for instance, in the mating disruption of the codling moth *Cydia pomonella* (Epstein *et al.* 2006). However, studies on *Choristoneura rosaceana* (Lepidoptera: Tortricidae) suggest that disruption of the matefinding behavior can be explained by a combination of mechanisms including adaptation of antennal receptors, camouflage of the female-produced plume, and false-trail following (Evenden *et al.* 2000).

The main parameters to be considered for a successful mating disruption experiment include: type of dispenser, dispenser release rate and blend composition. In addition to the classical dispensers, such as hollow-fibers, microencapsulated sprayables, laminates, polyethylene tubing, and aerosols, emulsified paraffin wax has also been recently introduced as a long-season dispenser that works efficiently against *Grapholitha molesta*, offering several advantages like low cost, self adhesion and biodegradation (Lame *et al.* 2007).) Also, new sprayable microencapsulated pheromone formulations have been recently effective against *G. molesta*) Another type of controlled-release dispensers of pheromone recently used are MSTRS (Metered Semiochemical Timed Release System) devices, which are very efficient in food stores cranberryfields and apple orchards. With respect to the dispenser release rate, this should be constant and long life to cover entirely the flight duration of the moth. It should be noted that increasing pheromone concentration does not always provide improved control of high populations. An optimization of the pheromone concentration is then necessary to allow reductionof the applied pheromone and therefore lowering the cost of the treatment. Blend composition is also important since the different components should be stable under environmental conditions, and optimized to have the least expensive formulation possible, for instance using only the major component. Other important factors for success imply to know the infestation level and the pheromone concentration in the treated areas. Low levels of infestation are clearly preferred for mating disruption since the lower number of insects flying the lower the probability to encounter each other for mating.

This has been confirmed in field experiments against the gypsy moth and codling moth. In addition, when the population density of the pest is very high, neither false trails nor plume masking significantly affect the ability of males to find the females and, furthermore, individuals from the surrounding zones may be attracted to the treated areas by the applied pheromone. To measure pheromone concentrations in the field three different techniques are commonly used: chemical analysis, field electroantennogram recordings and single sensillum recordings in the field. These methods, however, do not indicate the success of the experiment directly, which should be assessed by other methods.

Recently, there have been assayed successfully blends of attractive formulations targeting more that one species at the same time, like four different leaf roller species in apple orchards the codling moth *C. pomonella* and the Oriental fruit moth (*G. molesta*) and the pest vineyard (*Lobesia botrana*) Other ways to improve mating disruption techniques have been tried in the context of IPM programs. For instance, integration of mating disruption with deployment of transgenic apples appeared to offer prospects for delaying the evolution of resistance to *Bt* toxins in the lightbrown apple moth (*Epiphyas postvittana*) Also, programs using combination of mating disruption with insecticide treatments have been developed but in some cases the use of insecticides has not improved the level of damage (Tcheslavskaia *et al.* 2005).

Little work in pest control has been done using pheromone antagonists. These compounds, that alter the behaviour or physiology of the insect communication system may be pheromone components of closely-related species with a sufficiently similar structure to that of the natural pheromone to bind to the pheromone receptor sites, and therefore competing with the natural attractant) Successful control of the navel orangeworm (*Amyelois transitella*) and the pea moth (*Cydia nigricana*) has been reported using pheromone inhibitors. Very recently, a significant reduction of damage by the Mediterranean corn borer (MCB) *Sesamia nonagrioides* and the European corn borer *O. nubilalis* in maize fields has been noticed by utilization of a trifluoromethyl ketone analogue of the MCB pheromone (Solé *et al.* 2008) Effectiveness of the treatment was particularly high for the most damaging second generation of both pests.

9.7.3 Mass trapping

In mass trapping, a very high proportion of insects must be caught in traps baited with chemical lures before mating or oviposition to reduce the pest population. This reduction must then be translated into a reduction in plant damage. For success it is required that the lure be very attractive, over surpassing if possible the effect of the naturally occurring attractant. For Lepidoptera it is essential that males are trapped before mating and this is most likely to occur with insects that mate only once. For Coleoptera it is highly recommended that both sexes are caught (if trapping is based on aggregation pheromones) before eggs are laid or damage is inflicted by feeding adults.

Mass trapping techniques have been used successfully to control a wide range of insect pests, typically species in Lepidoptera, Coleoptera and Diptera orders. The mechanism of population reduction *via* trapping differs depending on the type of semiochemicals used. In Lepidoptera and some Coleoptera the female sex pheromone is the agent used to attract males. In other Coleoptera, for instance Scolytidae, aggregation pheromones are generally used, either alone or in combination with food attractants (Faleiro *et al.* 2003). In other cases, insects can be attracted only by volatiles from the host as the beetle *Hoplia communis*, which can be caught in great numbers by 2-

phenylethanol, a major volatile component of the host flowers (*Rosa* spp.) or just by synthetic food-based baits as the Mediterranean fruit fly *Ceratitis capitata* (McQuate *et al.* 2005)

Several studies have compared the efficiency of mass trapping with other methods of pest management. In some of them mating disruption was considered to be more effective than mass trapping in trials against the pink bollworm (*Pectinophora gossypiella*) Trematerra considered mass trapping better than mating disruption on small, hilly sites, whereas other found mating disruption too expensive (Mafra-Neto and Habib 1996).

There may be cases in which a combination of mass trapping and mating disruption would be effective, such as mass trapping females by using kairomones and using sex pheromone to disrupt males (El-Sayed *et al.* 2006).

One of the handicaps of this technique is the cost. A cost reduction should be taken into account for mass trapping success, particularly if it implies a reduction in the use of insecticides. In Costa Rica, for instance, control of the American palm weevil *Rhynchophorus palmarum* as vector of the red ring nematode *Bursaphelencus cocophilus*, the infecting agent of the red ring diseased oil palms, was successfully achieved using less than one trap per five hectares. Also, in Ethiopia, the implementation of an adaptive tsetse population management system along with the application of geostatistical methods to discover patches with increased fly densities allowed the authors to reduce the number of traps from 216 to 127, maintaining previously achieved levels of occurrences of the pest) On the other hand, the number of catches should also be as high as possible. For instance, in mass trapping experiments against *Monochamus clamator* and *Monochamusscutellatus* using pheromone components of scolytid bark beetles, traps baited also with ethanol, a - pinene and ipsenol captured twice as many beetles as traps baited with host volatiles alone. Another way to reduce costs is the control of more than one species at the same time. Thus, experiments in Costa Rica and Honduras allowed control of the Indian sugarcane weevil *Metamasius hemipterus* and the American palm weevil *R. palmarum* by using lures emitting a mixture of their male-produced aggregation pheromones (Alpizar *et al.* 2003).

9.7.4 Lure and kill

Lure and kill tactics combine lures with insecticides. Many preliminary studies carried out on dipteran and lepidopteran species have been performed with no conclusive results (El-Sayed *et al.* 2006). Charmillot and Hofer (1997) developed one attract and kill formulation against the codling moth, which was registered in Switzerland by

Novartis. More recently, a bait of (Z)-9-tricosene, the only commercially available pheromone for use in lure and kill approaches to housefly control, in combination with sugar/insecticide has been widely used in indoor livestock farms (Hanley *et al.* 2004) The formulation was, however, inefficient to provide adequate control of houseflies outdoors. (Williams *et al.* 2007) have claimed that pyrethroid bifentrin-treated lethal ovitraps can be used as a lure and kill device against the dengue vector *A. aegypti* and that they should be effective in the field for at least four weeks. Kairomonal attractants can also be used in lure and kill tactics, as shown by the attracticides developed against the navel orangeworm *A. transitella* in almonds (Phelan and Baker 1987).

9.8 Botanical insecticides

Herbs, animal products and inorganic materials have been for centuries the ingredients of drugs and pesticides. Long before knowing the structure of plant naturally occurring chemicals (botanicals), plants or derivatives thereof were extensively used in agriculture as insecticides. Pyrethrum from *Chrysanthemum cinerariifolium* Viz. (Compositae), rotenone from *Lonchocarpus nicou* or *Derris elliptica* (Leguminosae), and nicotine from *Nicotiana tabacum* (Solanaceae), are outstanding among other examples (Jacobson and Crosby 1971).

Use of botanical insecticides in commercial agriculture was dramatically reduced when synthetic insecticides were developed. DDT was used extensively to keep World War II soldiers free of head and body lice and also proved very effective against mosquito-transmitted diseases, such as malaria or yellow fever and Discovery and increasing development of synthetic insecticides followed (major classes: organochlorinated, organophosphate, carbamate) driven by their lower cost, effectiveness and longer lasting properties (Casida and Quistad, 1998).

But public concern for long-term risks for health or the environment, brought by toxicological problems (acute and chronic poisoning of applicators, farmworkers or consumers) and environmental contamination (wildlife destruction or disruption of pollination and natural pest control), resulted in severe regulatory restrictions or even banning (Isman, 2006). Pyrethroid, being one class based on a natural product model is an example of successful synthetic pesticide chemistry (Casida and Quistad 1995).

As a consequence, increased attention and interest have followed on botanical insecticides as natural pesticides for IPM strategies (Weinzierl 1998) However, "natural" plant-derived compounds may be as toxic (or more toxic) to humans and beneficial insects as many common "synthetic" insecticides are, and a "reduced risk" status is to be proved.

At present, current botanicals in commercial use for insect control fall into four major types (pyrethrum, rotenone, neem and essential oils) and a few more (ryania, nicotine and sabadilla) are of limited use (Isman, 2006). The expanding body of literature reporting new plant derivatives with prospective bioactivity against insect pests is also considered here.

9.8.1 Pyrethrins

Pyrethrum is an extract from dried flowers of *Tanacetum* (*Chrysanthemum*) *cinerariifolium* (Asteracae) or related species, and the active principles "pyrethrins" may reach a 3% content. Chemical components of the extract are sesquiterpenes, triterpenes and sterols, flavonoids, *n*-alkanes and fatty acids. The "pyrethrins" are meroterpenes (mixed biosynthesis: a terpene-derived unit is attached to a non-terpene moiety), esters of the chrysanthemic or pyrethric acid with ketocyclopentene alcohols (Fig. 1) : pyrethrolone (pyrethrins I and II), cinerolone (cinerins I and II) and jasmolone (jasmolin I and II). Often available at 25–50% concentration, pyrethrin I and II are present in greatest amounts. Pyrethrins show neurotoxic action by blocking voltage-gated sodium channels (a mechanism qualitatively similar to that of DDT), prolonging their opening and thereby causing a rapid knock-down effect and death (Isman, 2006).

Synthetic analogs (pyrethroids) have been developed to overcome low stability to UV light without compromising biodegradability, and to maintain insecticidal potency while minimizing fish toxicity. The structure of modern pyrethroids differs quite a lot from that of model pyrethrins as well as their molecular mode of action.

9.8.2 Nicotinoids

Nicotine is an alkaloid isolated from tobacco plants, mainly *Nicotiana tabacum* and *Nicotiana rustica* (Solanaceae), and it is the most toxic botanical insecticide, with an LD50 (rat oral) at 50 mg/kg, and extremely harmful to humans. As a fast-acting nerve toxin, it works as a contact poison (symptons similar to those caused by organophosphate or carbamate insecticides). Acting at the nicotinic acetylcholine receptor it leads to uncontrolled continuous firing of the neuroreceptor. Nicotine is most effective on soft-bodied insects and mites, including aphids, thrips, leafhoppers and spider mites (Isman, 2006).

Commercial tobacco contains other alkaloids such as nornicotine (found specially in *Nicotiana glutinosa* and *Nicotiana sylvestris*) and anabasine (*Nicotiana glauca* being itsmost important source) and a range of other "minor" related alkaloids (Fig. 1).

Related to nicotine in action and partly in structure, the highly insecticidal neonicotinoids originated from screening novel synthetic chemicals. This newest major class of neuroactive insecticides is increasingly used because of their outstanding potency and systemic action for crop protection against plant piercing-sucking pests, and being highly effective for flea control on cats and dogs (Tomizawa and Casida 2003).

9.8.3 Rotenoids

Rotenone is an isoflavonoid molecule, and the major constituent of insecticidal, acaricidal, and piscicidal cubé resin. It is commonly available as a dust containing 1–5% of active ingredients from rhizomes or roots of the tropical *Lonchocarpus*, *Derris* or *Tephrosia* (Leguminosae), or extracts (as resins) usually with up to 45% total rotenoids. The four major active ingredients of cubé resin are rotenone, deguelin, rotenolone and tephrosin (Fig. 1), totaling 77 wt%. From the resin 12 new compounds have been isolated and identified, out of 29 rotenoid constituents, and their biological activities assayed for all of them (Fang and Casida 1999). Fifty-two rotenoids and seven rotenoid glucosides were reported in the period 1855–1981) and fifty more new ones, and again seven rotenoid glucosides, were isolated in the period 1982–1999 (Paz Parente and Pereira da Silva 2001) A recent review on isoflavonoids covers the period 1997–2004.).

9.8.4 Neem and azadirachtin-related tetranortriterpenoids

Derived from *Azadirachta indica* (*Melia azadirachta*) (Meliaceae), neem-based insecticides contain azadirachtin, a complex tetranortriterpene, as the most important active principle (Fig. 2). Enthusiasm for neem was a direct result of the numerous and beneficial reported effects on insects: repellence, feeding and oviposition deterrence, interference with reproduction (sterility in adult female insects) or growth and development (ecdysteroid synthesis and release). An overwhelming research effort has been devoted to study also other trees of the family Meliaceae (*Melia*, *Toona*, *Trichilia*, *Khaya*, *Turrea*, *Aglaia* and many others) (Koul and Wahab 2004.). A large number of intact triterpenes (protolimonoids, as very likely biosynthetic precursors of the limonoids), and limonoids (degraded triterpenes tetranortriterpenoids; remainder side-chain as a furan ring) of different ring systems (intact; A-seco; B-seco; Cseco; D-seco) have also been isolated and characterized, along with azadirachtin and closely related compounds (such as the meliacarpins) where the furan ring has been modified.). Three International Neem Conferences and later on World Neem Conferences (WNC: Bangalore, India, 1993; Gatton College near Brisbane, Australia, 1996; University of British Columbia, Vancouver, Canada, 1999; Mumbai, India, 2002; Coimbatore, India, 2007) have been held and the proceedings from several are available. Schmutterer *et al.* 1981).

9.8.5 Annonaceous acetogenins

Pesticide activity of Annonaceae species was attributed to their content of isoquinoline alkaloids. However, the insecticidal activity was traced to a newly isolated type ofcompound, displaying a long-chain fatty acid structure (usually C-32 or C-34, such as for asimicin). Since then, well over four hundred of them have been reported). Their structures might be described by four partial units as ZHCL. L stands for a g -lactone head and H for a characteristic heterocyclic fragment, whereas C is a linking chain from L to H, and Z a terminal unit. Both C and Z may contain functional groups (hydroxyl as most usual, occasionally double or triple bond, epoxy ring or carbonyl).

One major group displays a 4-methyl-2-substituted butenolide (L1, as in asimicin) as g -lactone head. Less common moieties within this group are 4-hydroxy-L1, 4-(hydroxymethylene) butanolide, 3-hydroxy-butanolide, and 3-methoxy- 4-methylenebutenolide. The second major skeletal type displays a 2-(2-oxopropyl)-4-substituted butanolide (L2) with the corresponding butenolide as less common function. The most common heterocyclic fragments are built out of tetrahydrofuran rings [mono-THF, adjacent bis-THF (as inasimicin), non-adjacent bis-THF, adjacent tris-THF]. Less common heterocycles are tetrahydropyran or epoxy rings. These compounds are potent inhibitors of complex I (NADH:ubiquinone oxidoreductase) in mitochondrial electron transport systems, and slow-acting stomach poisons, a mode of action identical to that of rotenone.

9.8.6 Piperamides and isobutylamide-related compounds

Insecticidal activity of unsaturated aliphatic isobutylamides was first reported for an active principle of "pyrethri radix" drug (roots of *Anacyclus pyrethrum* DC, Compositae), and named initially as "pyrethrine". Isolated as pellitorine (Fig. 4), this last name was adopted to avoid confusion with the isolate from *Chrysanthemum cinerariifolium* Vis. (Compositae). It was recognized as a complex mixture and finally elucidated the main component as (2*E*,4*E*)-*N*-isobutyl-deca-2,4-dienamide. Other than in Asteraceae species (*Anacyclus*, *Achillea*, *Spilanthes*, *Echinacea*, *Acmella*, *Artemisia*, *Brachycome*)/ Compositae (*Chrysanthemum*), unsaturated aliphatic isobutylamides are common in Piperaceae (*Piper*, *Peperonia*, *Ottonia*), Rutaceae (*Zanthoxylum*, *Heliopsis*), Aristolochiaceae and Poaceae (*Ctenium*). The secondary metabolites isolated from Piper species were near to six hundred (classified under twelve categories) by 1997 (Parmar *et al.* 1997) and recently the phytochemistry, insecticidal activity and mode of action of *Piper* spp. (Piperaceae) has been reviewed (Scott *et al* 2008). In addition to isobutylamides, derivatives of pyrrolidine and piperidine, with different degrees of unsaturation and end-groups, are also common amongst the alkaloid/amide group. The compounds cause rapid knockdown and kill of flying insects, acting as voltage-dependent blockers of the sodium channel.

9.8.7 Essential oils

Plant essential oils (or their constituents) have been valued as insecticides, owing to their broad spectrum of activity. Direct toxicity, oviposition and feeding deterrence, repellency or attraction appear to result from interaction with the insect nervous system, either by acetylcholinesterase inhibition or antagonism of the octopamine receptors. The oils consist of complex mixtures (hydrocarbon or oxygenated mono and sesquiterpenes; aliphatics; aromatics; *etc*) with a few major constituents such as 1,8-cineole (rosemary, eucalyptus), menthol (mints), pulegone (pennyroyal), eugenol (clove), *etc.* (Fig. 5). Most come from highly aromatic Lamiaceae (basil, mints, salvia, lavender, sage, rosemary, thyme, *etc.*), Rutaceae (lemon, lime, amyris), Myrtaceae (myrtle, clove), *etc.*, plant species also known as flavorings and spices and many considered to have medicinal use (Dewick 1997, Pavela 2005).

Applications to crop protection (stored product pests), mosquito repellency (citronella oil), control of domestic pests (cockroaches, ants, fleas, *etc.*), *Varroa* mite control, as aphicides and acaricides (cinnamon oil) and urban insect control (eugenol-based products from basil or clove) have been reported (Jacobson and Crosby, 1971, Isman *et al.* 2007) A range of insecticide products may be also blends of plant oils, such as clove, peanut oil, thyme, lemon grass, cinnamon or pennyroyal as components. (Jefferies *et al.* 1992).

9.8.8 Minor plant insecticides

The powdered stemwood of *Ryania speciosa* (Flacourtiaceae) has been used as "ryania insecticide", the activity attributed to ryanodine, a pyrrole-2-carboxylic ester of a complex diterpene (ryanodol). Eleven ryanoids have been obtained from different preparations and their insecticidal activity, mammalian toxicity and potency at the Ca_2 release chanel complex determined (Jefferies *et al.* 1992) Related ryanodane diterpenes have been isolated from *Persea indica.* Fraga *et* 2001). The veratrum alkaloids derived from *Sabadilla* and *Veratrum* species are effective by blocking the sodium channel in insects, but they also have rather high mammalian toxicity.

They were used extensively in Europe and the USA between the late nineteenth and mid-twentieth centuries. Major "sabadilla" bioactive principles, cevadine and veratridine, are esters of a steroidal alkaloid named veracevine. (Ujváry and Casida, 1997, Li, *et al.* 2006)

Quassia is a weak insecticide derived from the wood of *Quassia amara* L (Simarubeaceae/Ailanthus) and related genera such as *Aeschrion excelsa* ("Jamaica quassia"). Aqueous extracts of Quassia chips were used as an insecticide from the late 18th to the mid 20th centuries. Quassinoids are highly oxygenated degraded triterpenes with a variety of interesting biological properties (Polonsky *et al.* 1999). Recent new quassinoid isolations have been reported from *Ailanthus, Brucea*, *Picrasma*, *Eurycoma*, and *Simaruba* species (Connolly and Hill, 2005, 2007).

9.9 Precautions while Using Biorationals

Biorationals can't be used the same way as traditional means because most of them are not cheap. Secondly, unlike the older synthetic insecticides, many biorationals have narrower spectra of activity, minimal to no residual activity, or short persistence. Many are active only after contacting the target pest. Therefore, it is critical to use them properly to make them cost and pest effective. It is imperative to use biortaionals within an integrated pest management (PM) program to ensure the correct selection of the most appropriate product according to the pest species, the developmental stage(s) being targeted, and optimal timing and placement of an application. Most biorationals are better suited to use as preventative rather than curative treatments. This requires basic scouting and monitoring components be in place. Your own experiences will prepare you are the type of pests that may be encountered on certain crops from year to year, which is where good record-keeping pays off. This information will help you formulate a plan.

Some products also need to be used on an "in-first basis. The biofungicides Root Shield and Plant Shield serve to illustrate this point well. These products need to be applied before diseases invade plant tissues. The fungus used in these preparations, Trichoderma harziunum T-22, helps block the entry of plant pathogens into the plant tissues by colonizing niches on the root, stem or leaf surface, which are preferred sites of infection for these pathognes. The fungus can also actively attack and kill invading pathogens.

Rhapsody, which is based on the bacterium Bacillus subtilise, has a similar mode of action. But it is a preventative, not curative, biocontrol product. Understanding how your control tools work and what they can do, or perhaps more importantly what they can't do, is critical to their successful use.

Early pest detection is vital, therefore a strategy to prevent populations getting out of control can be implemented as soon as possible. Consider the rapid potential for thirps' and aphids' population growth and spread if allowed to go unchecked. Aphids can spread approximately 120 square feet per day in a chrysanthemum crop. Correct identification of the pest is key to the selection of the best product for its control.

Nicotinoid, Marathon or other systemic insecticides, and Flagship work well against piercing and sucking in-sects like aphids and whiteflies, but no U-trips. Endeavor, which is also a systemic but with a completely different and new chemistry, works well against aphids, but not so well against whiteflies. The IGRs Distance and buprofezin work well against whiteflies, but not against aphids. Conserve has excellent efficacy against thrips, caterpillar pests, and leaf miners, but does not work against aphids and whiteflies.

Many of these new products have a much restricted spectrum of activity than their predecessors. It is no longer a case of one-stop shopping for all your insecticide needs. Product selections need to be made carefully, according to the target pest and your own experience with that pest. Learning from other growers, extension specialists, meetings, and trade magazine articles can also help guide this choice. Also, be sure to read the label for proper usage.

9.9.1 While Mixing Biorationals

Tank-mixing some products such as soaps, oils, or need-based products with other insecticides is beneficial against certain pests. But because environmental factors and varietal and plant growth stage may affect phytotoxic expression, phytotoxicity should be assessed before treating large numbers of plants.

For fungal products such as BotaniGard, Mycotech's reference data sheets show compatibility with most biortaional insecticides like Azatin, Avid, soaps, oils, and IGRs. There is also considerable compatibility with chemical fungicides, and tank-mixing is possible with biorational fungicides such as Phyton 27 and Cleary's 3336. Fungal use may be integrated with many other products, but applications need to be made several days before or after fungicides have been used.

Fungicides are also compatible with most predators and parasitoids, although the oils used in the emulsifiable formulations are initially repellent to some natural enemies such as whitefly parasitoids. Thus, formulation choice is important, depending on the crop, pest, and overall crop management program. The biofungicide Root Shield also has proven to be compatible with many fungicides, insecticides, and growth regulators, and to date appears to have no adverse effects on natural enemies.

Bacillus thuringiensis (*e.g.,* Gnatrol and Dipel) is compatible with most crop protection, and has few or no adverse effects against a range of predators and parasitoids. Nematodes, which are primarily used for fungus gnat control, are also very compatible with biorational pesticides.

9.10 Integration of Biorationals and Beneficial

Compared to traditional pesticides, biorationals are generally much "softer" on natural enemies. Soaps, horticultural oils, and IGRs may have some negative impact on beneficial when applied as foliar sprays that directly contact the natural enemy, but these effects are short-lived due to their limited residual activity.

Koppert's adult Macrolophus, Encarsia, and Diglyphus are susceptible to need products, and bees may be adversely affected. Neem products may be safe for Orius, Aphidius, and predatory mites. Similarly, there has been considerable debate about the safety of Conserve for natural enemies, which in itself highlights some of the difficulties associated with trying to assess compatibility.

In general, when spray residues are dry there is a high degree of compatibility with honey bees and bumble bees, Predatory bugs such as Orius, ladybeetles, and lacewings. Effects on predatory mites are less clear but appear to be more chronic, which can reduce longevity and fecundity, and adversely affect establishment.

Parasitic wasps are very sensitive and releases must be timed to prevent direct exposure to high levels on treaed plants. Avid is very harmful to some predators and parasitoids, al-though it affects some developmental stages more than others. Systemic insecticides such as Marathon and Endeavour are harmful to many predatory and parasitic species.

Koppert has compiled what is probably the best source of information on compatibility. This publication is in English and Dutch, with up-to-date in-formation available at www.koppert.nl.

9.11 Biorational Availability as a constraint

While alluding to many products in this article, the main purpose has been to highlight what it takes to use them effectively. Here is a list of some biorational products. Existing IGRs include Adept, Precision, Distance, Enstar II, Azatin XL, Neemazad, Ornazin, and Citation. Most have 12-hour REIs, but Azatin and Enstar II have four-hour REIs.

Horticultural oils and soaps, and microbially-derived preparations, such as Avid and Conserve, are biorationals.

The microbial biopesticides BotaniGard, and Bt products have a variety of trade names such as Dipel, Xentari, and Gnatrol to name a few.

The borderline biorationals are Marathon and Endeavor.

Registrations in the near future include Marathon look-alikes Acetamiprid and Flagship for aphid whiteflies, and mealybugs, and the IGR buprofezin (Applaud).

For mites, Floramine, Ovation, Hexygon, and Pylon have been newly registered, or appear to be on the verge of registration for greenhouse ornaments. They appear to be fairly compatible with most biological, although some precautions should be observed if predatory mites form part of an IPM program.

Messenger is another interesting new product with a different mode of action. This has just been registered for use on ornamentals and other crops, and works by provoking a plant's natural defense mechanisms in a manner analogous to the way in which vaccines elicit in immune response in humans. The full range of crops and diseases against which this material might promote a defense response remains to be determined, but certainly the early results are encouraging, and it is very appropriate for an IPM program.

Fungicides fitting the biorational, profile include Phyton 27, Cleary's 3336, and the biofungicides RootShield and PlantShield (*T. rzianum*), Rhapsody, and Serenade (*Bacillus subtilise*).

9.12 Biorationals and IPM

- Here are some considerations when incorporating biorataional into a spray program:
- Think about your total pest management program. Ideally, the program should be though through ahead of time.
- Early and accurate identification of pests (or diseases, and the predominant life stages is vital.
- Biorationals are often only active against certain pests or life stages.
- Some products are preventative, not curative, and work on an "in-first" basis.
- Thorough coverage of foliage is essential, with sprays targeted at the pest. Many biorationals must directly contact the pest to be effective.
- Biorationals are often only active against specific life-stages, therefore repeat sprays may be necessary.
- Do not overuse products with the same chemistry. Use spray rotations.
- Even though most of these products are considered soft on natural enemies, still consider the choice of product and the timing of sprays/releases to minimize any negative interactions.

9.13 Need of developing Biorationals

The use of chemical pesticides in agriculture has generated an abundance and quality of food and fiber unequalled in the history of mankind. Despite strong current advocacy of low tillage sustainable agricultural practices, the universal expectation of cheap, plentiful agricultural products and freedom from disease eliminates the possibility of returning to more pastoral, non interventional agricultural and public health protection is one of the cornerstones of modern life in developed countries.

With our current pesticides, the efficacy, ease of use, and incompatibility with other pest control strategies discouraged the active development of alternative methods of insect control until recently. This reliance on a single crop protection strategy has rendered many crops registered for use on a given crop and to the development of resistance in exacerbated by the widespread misuse of pesticides; continuous pressure due to spraying on a calendar schedule whether it was needed or not, the spraying of cocktails of several pesticides at once, and the reluctance to rotate pesticides to preserve their fields efficacy. These practices have accelerated different modes of action. It is widely recognized that continued development of chemical pesticides with the same modes of action or the same receptor target as pesticides in current use is a case of diminishing returns; the effective lifetime of toxin compounds, which are merely variations on current pesticides, is increasingly brief. The development of juvenile hormone (JH) analos, an ecdysone agonist or molt accelerating chemical and plants bioengineer with B.T. toxins is prime examples of new pesticides with novel modes of action. However, the JH analogs are only effective against those insects where adults are the problem, the ecdysone agonists are still unproven, and insects are all but certain to develop resistance to B.t. toxins, thus requiring a continuing process to develop novel approaches to insect control.

The goal is to develop new and effective pest management tactics based on the chemistry, biology, and biochemistry of insects and plants. As an added benefit, these tactics will generally be mutually compatible with and complementary to other environmentally sound methods of insect control, for example the use of biological control agents such as parasitic and predatory insects. This is of crucial importance because future pest control efforts will of necessary be characterized by the integration of several control strategies, rather than total reliance on a single strategy as is often the case with current programs based on insecticide use.

9.14 Conclusion

Natural products have had a great impact in crop protection over the years. It is remarkable the number of compounds that have reached the market place and remain there after a long period of time. A more tolerant regulatory policy from the developed countries will ease the registration of new natural products that together with advances in mass production technologies will benefit the commercialization of new bioactive control agents. On the other hand, utilization of the skeletons of natural products as base for the development of new biopesticides is one of the most successful applications so far of natural products discovery to crop protection. The commercial success of the synthetic pyrethroids is a classic example, with their share in today's market being over 1 billion US $. It is likely that discoveries will continue in the future to identify new areas of chemistry that justify the efforts directed in the synthesis of such natural products analogues.

Pheromones will continue to play a significant role in pest control although its economical impact in the market is limited. There is an ever-increasing interest by consumers in developed countries on the use of much more ecologically acceptable pesticides and more benign production systems. This trend towards a greater market acceptance of new alternatives is coupled with the increasing failures of broad-spectrum pest management tactics due to insecticide resistance or other undesirable side effects resulting thereof. This is the case, for example, of the Colorado potato beetle which is resistant to almost all available synthetic insecticides. Very recently, the male aggregation pheromone (*S*)-3,7-dimethyl-2-oxo-6- octen-1,3-diol, is being investigated as an attractant for control of this pest (Kuhar *et al.* 2006) Mating disruption has shown remarkable successes, particularly in cotton for the control of the pink bollworm, and in orchards, vineyards and olive groves for control of the olive fly, codling moth and grape berry moth.

However, the specificity of the compounds is a major constrain for crops suffering attack from two or more different insect species. Therefore, new blends of attractive formulations targeting more than one species at the same time have been assayed successfully, for example to control the codling moth, the Oriental fruit moth and the grape vine moth. With regard to living organisms, more work should be done to overcome the drawbacks inherent to the living systems, particularly to enhance the shelf life, the speed of kill, the biological spectrum and the field efficacy. However, great successes have been noticed in the last years, particularly with *Bt* transgenic crops. Molecular biology, not covered in this review, and specially molecular genetics of microorganisms and genetic engineering technology will help to identify the modes of action of many biocontrol agents and set up the basis for the development of new strategies for their subsequent improvement and use.

The biological and microbial insecticides (such as neem extracts and Bt) are generally compatible with organic agriculture and most of them have low toxicity to humans and natural enemies. But some agents that have been used in organic agriculture (e.g. rotenone, pyrethrins and ryania) may have toxic effects on non-target animals such as fish, bees and various beneficial arthropods. Hence, although regulatory authorities are familiar with conventional insecticides, safety assessment of organic insecticides and bioinsecticides requires different approaches (Plimmer 1993). Biological insecticides include living organisms such as viruses, bacteria, fungi, and nematodes.

The main drawbacks of biological insecticides are their instability under conditions of heat, UV radiation, or aridity. Another problem is the short shelf life as compared with conventional insecticides. They are also not sufficiently effective in controlling high population levels or heavy attacks of insect pests. Biological insecticides have slow and selective activity, and their strict management requirements render them economically unattractive to conventional agriculture. In addition, the use of biological insecticides is fairly expensive and thus restricted to organic agriculture.

In this chapter information has been provided on the newer groups of insecticides that have been developed and commercialized during the past two decades. In many of these new groups only a single product has been successfully employed in the field, e.g. spinosad, chlorfenapyr, and fipronil; in other groups, such as the neonicotinoid and benzoylphenyl urea, a series of insecticides have been developed with similar modes of action and probable cross-resistance.

There have been efforts to develop biorational insecticides with selective properties to act on biochemical sites present only in the target arthropods and not in non-target organisms. This approach has led to the development of insecticides such as the neonicotinoids (which affect nicotine acetylcholine receptors, with much higher affinity to those of insects than mammals), the BPUs and buprofezin (which inhibit chitin synthesis), and the JH mimics and ecdysone agonists (for controlling lepidopterans, coleopterans, scale insects and whiteflies). Compounds originating from natural products, which act on GABA, glutamate receptors and the chloride channels (such as the avermectins and the spinosyns) and at the mitochondrial level (chlorfenapyr) with much more association to the insects than mammals, have been developed and used successfully to combat diverse insect pests and spider mites. Additional new groups of insecticides include the phenylpyrazoles (fipronil), the pyridines (pymetrozine) and the oxadiazines (indoxacarb), which act on GABA receptors, on the nerves controlling the salivary pump and on the sodium channel, respectively. All the above, are considered biorational, or "low risk" insecticides, which have relatively low detrimental effect on the environment and its inhabitants, rendering them important components in IPM programs.

The newer insecticides are not immune to evolution of resistance in the insect pests, and since their registration, several cases of resistance to those insecticides have been reported; but the development of diverse modes of action has assisted in managing resistance to insect pests (Denholm *et al.* 1998b). As most of the biorational insecticides have shown effectiveness against different strains of resistant species, with no evidence of cross-resistance, they can play an important role in IRM strategies.

References

Abrol, D. P. and Shankar, U. 2014. Pesticides, Food Safety and Integrated Pest Management. In: *Integrated Pest Management* vol. 3, D. Pimentel, R. Peshin (eds.), Springer Science+Business Media Dordrecht 2014. Pp. 167-199.

Adachi I., Uchino K. and Mochizuki F. 2007. Development of a pyramidal trap for monitoring fruit-piercing stink bugs baited with *Plautia crossota stali* (Hemiptera: Pentatomidae) aggregation pheromone. *Appl. Entomol. Zool.* 42, 425–431.

Adams L. F., Liu C. L., McIntosh S. C. and Starnes S. L. 1996. Crop Protection Agents from Nature: Natural Products and Analogues," ed. by L. G. Copping, Royal Society of Chemistry, Cambridge, UK, pp. 360–388.

Ahmad Z. and Attique M. R. 1993. Control of pink bollworm with gossyplure in Punjab, Pakistan. Bulletin OILB/SROP 16:141–148.

Albajes R., Konstantopoulu M., Etchepare O., Eizaguirre M., Frerot B., Sans A., Krokos F., Ameline A. and Mazomenos B. 2002. Mating disruption of the the corn borer sesame noagrioides (Lepidoptera: Noctuidae) using sprayable formulations of pheromones. *Crop Prot.* 21, 217–225.

Albrecht CP, Sherman M 1987. Lethal and sublethal effect of avermectin B1 on three fruit fly species (Diptera: Tephritidae). J Econ Entomol 80: 344-347

Al-Deeb MA, Wilde GE, Zhu KY (2001) Effect of insecticides used in corn, sorghum, and alfalfa on the predator *Orius insidiosus* (Hemiptera: Anthocoridae). J Econ Entomol 94:1353-1360

Aldrich JR, Cantelo WW 1999. Suppression of Colorado potato beetle infestation by pheromone-mediated augmentation of the predatory spined soldier bug, *Podisus maculiventris* (Say) (Heteroptera: Pentatomidae) Agric Forest Entomol 1:209-217

Allison, J. D., W. D. Morewood, J. H. Borden, K. E. Hein, and I. M. Wilson. 2003. Differential bioactivity of *Ips* and Dendroctonus pheromone components for *Monochamus clamator* and *M. scutellatus* (Coleoptera: Cerambycidae). Environ. Entomol. 32: 23–30.

Alm S. R. and Dawson C. G. 2003. Evaluation of Two Prototype Traps and Existing Trap Designs for Captures of Japanese Beetles (Coleoptera: Scarabaeidae). *J. Econ. Entomol.* 96, 453–455.

Alm, S. R., M. G. Villani, and W. L. Roelofs. 1999. Oriental beetles (Coleoptera: Scarabaeidae) current distribution in the United States and optimization of monitoring traps. J. Econ. Entomol 92:931–935.

Alpizar D., Fallas M., Oehlschlager A. C., Gonzalez L. M., Chinchilla C. M. and Bulgarelli J. 2002. Pheromone mass trapping of the West Indian sugarcane weevil and the American palm weevil (Coleoptera: Curculionidae) in palmito palm. *Fla. Entomol.* 85, 426–430.

Andaloro JT, Wing KD, Green JH, Lang EB 2000. Steward™ dispersion and cotton leaf interactions: impact on cotton insect pests and safety to beneficial arthropods. *In:* Dugger CP, Richter DA (eds) Proc 2000 Beltwide Cotton Prod Conf, National Cotton Council of America, Memphis, pp 939-940

Anderbrant O., Lofqvist J., Hogberg H. E., Hedenstrom E., Baldassari N., Baronio P., Kolmakova G., Lyons B., Naito T., Odinokov V.*et al.* 2000. Selective elimination of aphid endosymbionts: effects of antibiotic dose and host genotype, and fitness consequences. *Entomol. Exp. Appl.* 95, 229–239.

Angeli G., Anfora G., Baldessari M., Germinara G. S., Rama F., Cristofaro A. De and Ioriatti C. 2007. Mating disruption of codling moth Cydia pomonella with high densities of Ecodian sex pheromone dispensers. *J. Appl. Entomol.* 131, 311–318.

Arena JP, Liu KK, Paress PS, Frezier EG, Cully DF, Mrozik H, Schaeffer J (1995) The mechanism of action of avermectin in *Caenorhabditis elegans*: correlation between activation of glutamate-sensitive chloride current, membrane binding and biological activity. J Parasitol 81: 286-294

Asaro C., Cameron R. S.,. Nowak J. T, Grosman D. M., Seckinger J. O. and Berisford C. W. 2004. Efficacy of Wing Versus Delta Traps for Predicting Infestation Levels of Four Generations of the Nantucket Pine Tip Moth (Lepidoptera: Tortricidae) in the Southern United States. Environm. Entomol. 33, 397–404.

Asjes CJ, Blom-Barnhoorn GL 2002. Control of aphid vector spread of lily symptomless virus and lily mottle virus by mineral oil/insecticide sprays in Lilium. Acta Hort 570:277-281

Athanassiou C. G., Kavallieratos N. G., Gakis S. F., Kyrtsa L. A., Mazomenos B. E. and Gravanis F. T. 2007. Influence of trap type, trap colour, and trapping location on the capture of the pine moth, Thaumetopoea pityocampa. *Entomol. Exp. Appl.* 122, 117–123.

Aukema B. H., Dahlsten D. L. and Raffa K. F. 2000. Exploiting Behavioral Disparities Among Predators and Prey to Selectively Remove Pests: Maximizing the Ratio of Bark Beetles to Predators Removed During Semiochemically Based Trap-Out. Environm. Entomol. 29, 651–660.

Bacca T. Lima ER. Picanco MC. Guedes RNC. Viana JHM. Optimum spacing of pheromone traps for monitoring the coffee leaf miner *Leucoptera coffeella.* Entomol Exp App. 2006; 119:39–45.

Bai D, Lummis SCR, Leicht W, Breer H, Sattelle DB 1991. Actions of imidacloprid and a related nitromethylene on cholinergic receptors of an identified insect motor neurone. Pestic Sci 33: 197-204

Baker T. C., Fadamiro H.Y. and Cosse A.A. 1998. Widely-spaced, high-emission-rate pheromone sources suppress mating of European corn borer females. *Proceedings 6th Australisian Applied Entomological Research Conference,* Brisbane, Australia, pp. 279–288

Baker T. C., Staten R. T. and Flint H. M. 1990. Behavior-Modifying Chemicals for Insect Management," ed. by R. L. Ridgway, R. M. Silverstein, M. N. Silverstein and M. N. Inscoe, Marcel Dekker Inc., New York, pp. 417–436.

Baum J. A., Johnson T. B. and Carlton: B. C. 1999. Biopesticides Use and Delivery," ed. by F. R. Hall and J. J. Menn, Humana Press, Totowa, New Jersey, pp. 189–209. Bengtsson M., Karg G., Kirsch P. A., Lofqvist J., Sauer A. and Witzgall P. 1994. Pheromone release by codling moth females and mating disruption dispensers *J. Chem. Ecol.* 20, 871–887.

Bermejo AFigadère., B. Zafra-Polo M.C., Barrachina I., Estornell E. and Cortes D.2005. Acetogenins from Annonaceae: recent progress in isolation, synthesis and mechanisms of action . *Nat. Prod. Rep.* 22, 269–303.

Betz F. S., Hammond B. G. and Fuchs R. L. 2000. *Regul.* Safety and advantages of *Bacillus thuringiensis*-protected plants to control insect pests. *Toxicol. Pharm.* 32, 156–173.

Biddinger DJ, Hull LA 1995. Effects of several types of insecticides on the mite predator, *Stethorus punctum* (Coleoptera: Coccinellidae), including insect growth regulators and abamectin. J Econ Entomol 88:358-366

Black BC, Hollingworth RM, Ahammadsahib KI, Kukel CD, Donovan S (1994) Insecticidal action and mitochondrial uncoupling activity of AC-303,630 and related halogenated pyrroles. Pestic Biochem Physiol 50: 115-128

Blackshaw R. P. and Vernon R. S. 2008. Spatial relationships between two Agriotes click-beetle species and wireworms in agricultural fields. *Agr. Forest Entomol.* 10, 1–11.

Bloem S., Hight S. D., Carpenter J. E. and Bloem K. A. 2005. Developing the most effective trap to monitor the geographical expansion of cactus moth Cactoblastis cactorum (Lepidoptera:Pyralidae). *Fla. Entomol.* 88, 300–306

Bloomquist JR (1993) Toxicology, mode of action, and target site-mediated resistance to insecticides acting on chloride channels. Comp Biochem Physiol 106C:301-314

Bloomquist JR (2001) GABA and glutamate receptors as biochemical sites for insecticide action. In: Ishaaya I (ed) Biochemical sites of insecticide action and resistance. Springer, Berlin Heidelberg New York, pp. 17-41

Bonning B. C. and Hammock B. D. 1996. Development of Recombinant Baculoviruses for Insect Control. *Annu. Rev. Entomol.* 41, 191–210.

Borchert D. M. and Walgenbach J. F. 2000. Comparison of pheromone-mediated mating disruption and conventional insecticides for management of tufted apple bud moth (Lepidoptera: Tortricidae).*J. Econ. Entomol.* 93, 769–776 (2000).

Bortolotti L, Porrini C, Sbrenna AM, Sbrenna G (2000) Ovicidal action of fenoxycarb on a predator, *Chrysoperla carnea* (Neuroptera: Chrysopidae). Appl Entomol Zool 35:265-270

Bosa C. F., Cotes A. M., Osorio P., Fukumoto T., Bengtsson M. and Witzgall P. 2006. *J. Econ. Entomol.* 99, 1245–1250.

Boyd ML, Boethel DJ (1998) Susceptibility of predaceous hemipteran species to selected insecticides on soybean in Louisiana. J Econ Entomol 91: 401-409

Breuer M., Kontzog H. G., Guerrero A., Camps F. and De Loof A. 2003.*J. Chem. Ecol.* 29, 2461–2468.

Brickle D S, Turnipseed SG, M. J. Sullivan MJ 2001. Efficacy of insecticides of different chemistries against *Helicoverpa zea* (Lepidoptera: Noctuidae) in transgenic *Bacillus thuringiensis* and conventional cotton. J Econ Entomol 94:86-92

Brockerhoff E. G., Jones D. C., Kimberley M. O., Suckling D. M. and Donaldson T. 2006. *Forest Ecol. Manag.* 228, 234–240.

Brooks F. M. 1988. Handbook of Natural Pesticides. Microbial Insecticides, Part A, Entomogenous Protozoa and Fungi; ed. By C. M. Ignoffo, CRC Press, Inc., Boca Raton, FL, pp. 1–149.

Broumas T., Haniotakis G., Liaropoulos C., Tomazou T. and Ragoussis N. 2002. The efficacy of improved form of the mass-trapping method for the control of olive fruit fly Bactocera oleae (Gmelin) (Diptera: Tephritidae) pilot scale feasibility studies. *J. Appl. Entomol.* 126, 217–223.

Brown J. J., Kittelson N. T., Hannon E. R. and Walsh D. B. 2006. An endemic population of western poplar clearwing moths (Lepidoptera: Sesiidae) invades a monoculture of hybrid poplar. *J. Econ. Entomol.* 99, 771–779.

Bruck D. J., Lewis L. C. and Gunnarson R. D. 2001. *J. Invertebr. Pathol.* 78, 210–214. Buchholz A, Nauen R (2002) Translocation and translaminar bioavailability of two neonicotinoid insecticides after foliar application to cabbage and cotton. Pest Manag Sci 58:10-16

Butt T. M., Harris J. G. and Powell K. A. 1999. Biopesticides Use and Delivery," ed. by F. R. Hall and J. J. Menn, Humana Press, Totowa, New Jersey, pp. 23–44.

Calkins C. O. and Faust R. J. 2003. Overview of areawide programs and the program for suppression of codling moth in the western USA directed by t5he united states department of agriculture-agriculture research service. *Pest Manag. Sci.* 59, 601–604.

Caprio M. A. and Suckling D. M.1995. Presidential address to the 47th conference 1994. *Proceedings 48th New Zealand Plant Protection Conference* pp. 52–58.

Cardé R. T. and Minks A. K. 1995. Control of moth pests by mating disruption: successes and constraints. *Annu. Rev. Entomol.* 40, 559–585. Carvalho M. O. and Mexia A. 2003. *Adv. Stored Prod. Prot.* 222–229.

Casida J. E. and Quistad G. B. 1998. Golden age of insecticide research: past, present or future. *Annu. Rev. Entomol.* 43, 1–16.

Casida JE, Quistad GB 1998. Golden age of insecticide research: past, present, or future. Annu Rev Entomol 43:1-16

Chandler LD, Pair SD, Harrison WE (1992) RH-5992, a new insect growth regulator active against corn earworm and fall armyworm (Lepidoptera: Noctuidae) J Econ Entomol 85: 1099-1103

Charles J. F., Nielsen-Leroux C. and Delecluse A. 1996. Bacillus sphaericus toxins: molecular biology and mode of action. *Annu. Rev. Entomol.* 41, 451–472. Charmillot P. J. and Hofer D. 1997. *IOBC/WPRS Bull.* 20, 139–140.

Chaton PF, Ravanel P, Tissut M, Meyran JC 2002. Toxicity and bioaccumulation of fipronil in the nontarget arthropodan fauna associated with subalpine mosquito breeding sites. Ecotox Environ Saf 52:8-12

Childers CC, Villanueva R, Aguilar H, Chewning R, Michaud JP 2001. Comparative residual toxicities of pesticides to the predator *Agistemus industani* (Acari: Stigmaeidae) on citrus in Florida. Exp Appl Acar 25:461-474

Cisneros J, Goulson D, Derwent LC, Penagos DI, Hernández O, Williams T 2002. Toxic Effects of Spinosad on Predatory Insects. Biol Cont 23: 156-163

Clarke, S. 2001. Review of the operational IPM programme for the southern pine beetle. *IPM Review, 6*, 293–301.

Cole L, Nicholson R, Casida JE 1993. Action of phenylpyrazole insecticides at the GABA-gated chloride channel. Pestic Biochem Physiol 46: 47-54

Coll J. 2005. New Method of DNA Isolation from Two Food Additives Suitable for Authentication in Polymerase Chain Reaction Assays *J. Agric. Food Chem.* 53, 3749–3750.

Connolly J. D. and Hill R. A. 2005. Triterpenoids *Nat. Prod. Rep.* 22, 487–503. Triterpenoids

Connolly J. D. and Hill R. A. 2007. Triterpenoids *Nat. Prod. Rep.* 24, 465–486.

Copping L. G. and Menn J. J. 2000. Biopesticides: a review of their action, applications and efficacy. *Pest Manag. Sci.* 56, 651–676.

Cork A. 1998. Proceedings 6th Australisian Applied Entomological Research Conference, 1998, Brisbane, Australia, pp. 304–313.

Cork, A., Alam, S.N., Rouf, F.M.A., and Talekar, N.S. 2005. Development of mass trapping technique for control of brinjal shoot and fruitborer, *Leucinodesorbonalis* Lepidoptera:Pyralidae). *Bull. Entomol. Res.* 95, 589–596.

Cory J. S. and Hails R. S. 1997. The ecology and biosafety of baculoviruses. *Curr. Opin. Biotech.* 8, 323–327.

Curtis C. E., Clark J. D., Carlson D. A. and Coffelt J. A. 1987. A pheromone mimic: disruption of mating communication in the navel orangeworm, *Amyelois transitella,* with Z,Z-1,12,14-heptadecatriene. *Entomol. Exp. Appl.* 44, 249–255.

De Cock A, de Clercq P, Tirry L, Degheele D 1996. Toxicity of diafenthiuron and imidacloprid to the predatory bug *Podisus maculiventris* (Heteroptera: Pentatomidae). Environ Entomol 25:476-480

De Cock A, Degheele D 1998. Buprofezin: a novel chitin synthesis inhibitor affecting specifically planthoppers, whiteflies and scale insects. In: Ishaaya I, Degheele D (eds) Insecticides with novel modes of action: mechanism and application. Springer, Berlin Heidelberg, New York, pp 74-91

De Lame F. M. and Gut L. J. 2006. Effect of Monitoring Trap and Mating Disruption Dispenser Application Heights on Captures of Male *Grapholita molesta* (Busck; Lepidoptera: Tortricidae) in Pheromone and Virgin Female-Baited Traps. *Environm. Entomol.* 35, 1058–1068.

De Lame F. M., Miller J. R., Atterholt C. A. and. Gut L. J. 2007. *Development and Evaluation of an Emulsified Paraffin Wax Dispenser for Season-Long Mating Disruption of Grapholita molesta in Commercial Peach Orchards. J. Econ. Entomol.* 100, 1316–1327.

DeBarr G. L., Hanula J. L., Niwa C. G. and Nord J. C. 2000. Synthetic pheromones disrupt male *Dioryctria* spp. moths in a loblolly pine seed orchard. *Can. Entomol.* 132, 345–351.

Delecluse A., Juarez-Perez V. and Berry C. 2000. Entomopathogenic bacteria: From Laboratory to Field Application," ed. By J. F. Charles, A. Delecluse and C. Nielsen-Leroux, Kluwer Academic Publishers, Dordecht, Netherlands, pp. 101–125, 2000.

Delgado F. X., Britton J. H., Lobo-Lima M. L., Razafindratiana E. and Swearingen W. 1997. "New Strategies in Locust Control," ed. by S. Krall, R. Peveling and D. Ba Diallo, Birkhauser, Basel, pp. 171–176.

Denholm I, Cahill M, Dennehy TJ, Horowitz AR 1998a. Challenges with managing insecticide resistance in agricultural pests, exemplified by the whitefly, *Bemisia tabaci*. Phil Trans Roy Soc Series B 353:1757-1767

Denholm I, Horowitz AR, Cahill M, Ishaaya I 1998b. Management of resistance to novel insecticides. *In:* Ishaaya I, Degheele D (eds) Insecticides with novel modes of action: mechanisms and application. Springer, Berlin, pp 260-282.

Dennehy TJ Williams L 1997. Management of resistance in *Bemisia* in Arizona cotton. Pestic Sci 51:398-406

Dewick P. M. 1997. "Medicinal natural products," John Wiley and Sons, Chichester, pp. 163–166.

Dhadialla TS, Calson GR, Le DP 1998. New insecticides with ecdysteroial and juvenile hormone activity. Annu Rev Entomol 43:545-569

Dorn J, Frischknecht ML, Martinez V, Zurflüh R, Fischer U 1981. A novel nonneurotoxic insecticide with a broad activity. Z Pflanzenkr Pflanzenschutz 88: 269-275

Driesche RG-van, Hoddle MS, Lyon S, Sanderson JP, van-Driesche RG, Lenteren JCvan (1999) Use of insect growth regulators to reduce rates of *Eretmocerus eremicus* needed for biological control of whiteflies on poinsettia. Bull OILBSROP 22:61-64

Duffie WD, Sullivan MJ, S.G. Turnipseed SG 1998. Predator mortality in cotton from different insecticide classes. *In:* Dugger CP, Richter DA (eds) Proc 1998 Beltwide Cotton Prod Conf, National Cotton Council of America, Memphis, pp 1111-1112

Dunkelblum E., Harel M., Assael F., Mori K. and Mendel Z. 2000. *J. Chem. Ecol.* 26, 1649–1657.

Dybas RA 1989. Abamectin use in crop protection. *In:* Campbell WC (ed) Ivermectin and Abamectin, Springer, Berlin Heidelberg New York, pp 287-310.

E. M. Hegazi, M. A. Konstantopoulou, P. Milonas, A. Herz, B. E. Mazomenos, W. E. Khafagi, A. Zaitun, S. M. Abdel-Rahman, I. Helal and S. El-Kemny: *Crop Prot.* 26, 837–844 2007.

Ehlers R. U. 2001.Mass production of entomopathogenic nematodes for plant protection. *Appl. Microbial. Biotechnol.* 56, 623–633.

El-Bendary M. A. 2006. Bacillus thuringiensis and Bacillus sphaericus biopesticides production. *J. Basic Microbiol.* 46, 158–170.

Elbert A, Nauen R, Leicht W 1998. Imidacloprid, a novel chloronicotinyl insecticide: Biological activity and agricultural importance. In: Ishaaya I, Degheele D (eds) Insecticides with novel modes of action: mechanism and application. Springer, Berlin Heidelberg New York, pp 50-73

Elbert A, Overbeck H, Iwaya K, Tsuboi S 1990. Imidacloprid, a novel systemic nitromethylene analogue insecticide for crop protection. Proc 1990 Brighton Crop Prot Conf – Pests and diseases, pp 21-28

Ellsworth PC (1998) Whitefly management in Arizona: looking at whole systems. *In:* Dugger CP, Richter DA (eds) Proc 1998 Beltwide Cotton Prod Conf, National Cotton Council of America, Memphis, pp. 65–68

Ellsworth PC, Martinez-Carrillo JL 2001. IPM for *Bemisia tabaci*: a case study from North America. Crop Prot 20:853-869

El-Sayed A. M. and Suckling D. M. 2005. Behavioural Observations of Mating Disruption in Three Lepidopteran Pests. *Behaviour* 142, 717–729.

El-Sayed A. M., Suckling D. M., Wearing C. H. and Byers J. A. 2006. Potential of Mass Trapping for Long-Term Pest Management and Eradication of Invasive Species. *J. Econ. Entomol.* 99, 1550–1564.

El-Sayed, A.M., Wainman, L.I., Santangelo, E.M., Unelius, C.R., and Trimble, R.M. 2005b. Sex pheromone of a North American population of the spotted tentiform leafminer, Phyllonorycter blancardella. Entomol. Exp. Appl. 116:143-148.

Elzen GW 2001. Lethal and sublethal effects of insecticide residues on *Orius insidiosus* (Hemiptera: Anthocoridae) and *Geocoris punctipes* (Hemiptera: Lygaeidae). J Econ Entomol 94:55-59

EPA: Registration Elegibility Decision (RED) *Bacillus thuringiensis*, EPA 738–R–98-004, 1998.

Epstein D. L., Stelinski L. L., Reed T. P., Miller J. R. and Gut L. J. 2006. Higher densities of distributed pheromome sources provide disruption of codling moth (Lepidoptera:Tortricidae)superior to that of lower densities of clumped sources. *J. Econ. Entomol.* 99, 1327–1333.

Evenden M. L. and McClaughlin J. R.2005. *Male Oriental Fruit Moth Response to a Combined Pheromone-Based Attracticide Formulation Targeting Both Oriental Fruit Moth and Codling Moth (Lepidoptera: Tortricidae). J. Econ. Entomol.* 98, 317–325.

Evenden M. L., Judd G. J. R. and Borden J. H. 1999. *Mating Disruption of Two Sympatric, Orchard-Inhabiting Tortricids, Choristoneura rosaceana and Pandemis limitata (Lepidoptera: Tortricidae), with Pheromone Components of Both Species' Blends. J. Econ. Entomol.* 92, 380–390.

Evenden M. L., Judd G. J. R. and Borden J. H. 2000. *Disruption of pheromone communication in three sympatric* leafroller (Lepidoptera: Tortricidae) pests of apple in British Columbia. *J. Insect Behav.* 13, 499–510.

Faccoli, M.; Stergulc, F. 2004. *Ips typographus* (L.) pheromone trapping in south Alps: Spring catches determine damage thresholds. *J. Appl. Entomol.*, *128*, 307–311.

Fadamiro H. Y. and Baker T. C. 2002. Pheromone puffs suppress mating *Plodia interpunctella* and *Sitotroga cerealella* in and infested corn store *Entomol. Exp. Appl.* 102, 239–251.

Fadamiro, H.Y., Cosse, A.A., Dittl T. and Baker, T.C. 1998. Suppression of mating by blackheaded fireworm, Rhopobota naevana in Wisconsin craneberry marshes by using MSTRS TM DEVICES . J.Agr. Entomol. 15:377-386.

Faleiro J. R., Rangnekar P. A. and Satarkar V. R. 2003. Age and fecundity of red palm weevils *Rhynchphorus ferrugineus* (Olivier) (Coleoptera: Rhynchophoridae) captured by pheromone traps in coconut plantations of India. *Crop Prot.* 22, 999–1002.

Fang N. and Casida J. E. 1999. Cube resin insecticide: identification and biological activity of 21 rotenoid constituents. *J. Agric. Food Chem.* 47, 2130–2136.

Feely WF, Crouch LS, Arison BH, Van den Heuvel WJA, Colwell LF, Wislocki PG (1992). Photodegradation of 4"-(epimethylamino)-4"-deoxyavermectin B1a thin films on glass. J Agric Food Chem 40: 691-696

Ferguson JS, Koenig JP, White SM, Dunbar DM, Lawson DS 1999. Evaluation of Fulfill(R) 50wg (pymetrozine) for cotton aphid control in 1998 field trials. In: Dugger CP, Richter DA (eds) Proc 1999 Beltwide Cotton Prod Conf, National Cotton Council of America, Memphis, pp 1016-1019

Figuls M, Castañé C, Gabarra R 1999. Residual toxicity of some insecticides on the predatory Bugs *Dicyphus tamaninii* and *Macrolophus caliginosus*. BioControl 44:89-98

Fischhoff D. A., Bowdish K. S., Perlak F. J., Marrone P. G.,. McCormick S. M, Nidermeyer J. G., Dean D., Kusano-Kretzmer K., Mayer C. J., Rochester D. E. *et al.* 1987. Insect tolerant tomato plants.*Bio/Technol.* 5,807–813.

Fisher MH, Mrozik H (1989) Chemistry. In: Campbell WC (ed) Ivermectin and abamectin. Springer, Berlin Heidelberg, New York, pp 1-23

Flexner J. L. and Belnavis D. L. 1998. "Biological and Biotechnological Control of Insect Pests," *ed. by* J. E. Rechcigl and N. A. Rechcigl, Lewis Pub., Boca Raton, FL, pp. 35–62,1998.

Flückiger CR, Kristinsson H, Senn R, Rindlisbacher A, Buholzer H, Voss G (1992 a) CGA 215'944 - a novel agent to control aphids and whiteflies. Brighton Crop Prot Conf - Pests and Diseases, Vol 1, pp 43-50

Flückiger CR, Senn R, Buholzer H 1992b. CGA 215'944 - opportunities for use in vegetables. Proc 1992 Brighton Crop Prot Conf - Pests and Diseases, Vol 3, pp 1187-1192

Foster SP, Denholm I, Thompson R 2002. Bioassay and field-simulator studies of the efficacy of pymetrozine against peach-potato aphids, *Myzus persicae* (Hemiptera: Aphididae), possessing different mechanisms of insecticide resistance. Pest Manag Sci 58:805-810

Fraga B. M., Terrero D., Gutiérrez C. and González- Coloma A. 2001. Minor diterpenes from Persea indica: their antifeedant activity. *Phytochemistry* 56, 315–320.

French NM II, White CA, Treacy KM 1996. Pirate insecticide-miticide for control of cotton insects in mid-south cotton. In: Dugger CP, Richter DA (eds) Proc 1996. Beltwide Cotton Prod Conf, National Cotton Council of America, Memphis, pp 865-870

Funderburk J, Stavisky J, Olson S 2000. Predation of *Frankliniella occidentalis* (Thysanoptera: Thripidae) in field peppers by *Orius insidiosus* (Hemiptera: Anthocoridae). Environ Entomol 29: 376-382

Fuog D, Fergusson SJ, Flückiger C 1998. Pymetrozine: a novel insecticide affecting aphids and whiteflies. In: Ishaaya I, Degheele D (eds) Insecticides with novel modes of action: mechanism and application. Springer, Berlin Heidelberg, New York, pp 40-49

Gaugler R. 1997. Biological Control: A guide to Natural Enemies in North America, ed. by C. Weeden, T. Shelton and J. Hoffmann, *http://www.nysaes.cornell.edu/ent/biocontrol/pathogens/ nematodes.html.,* 1997.

Gaugler R., Grewal P., Kaya H. K. and Smith-Fiola D. 2000. Quality assessment of commericially produced entomopathogemnic nematodes. *Biol. Control* 17, 100–109.

Georghiou GP (1983) Management of resistance in arthropods. In: Georghiou GP, Saito T (eds) Pest resistance to pesticides. Plenum Press, New York, pp 769-792

Gunning R 2002. *Heliothis* insecticide resistance continues to climb. Aust Cotton Grower 23 (3):71-72

Gershburg E., Stockholm D., Froy O., Rashi S., Gurevitz M. and Chejanovsky N. 1998.Baculovirurs–mediated expression of a scorpion depressant toxin, improves the insecticidal efficacy achieved with excitatory toxins. *FEBS Lett.* 422, 132–136.

Gianessi L. P. and Carpenter J. E. 1999. Agricultural Biotechnology: Insect Control Benefits,.

Godin C. and Boivin G. 1998. Seasonal Occurrence of Lepidopterours pest of cruciferous crops in South-Western Quebec in relation to degree-day accumulation. *Can. Entomol.* 130, 173–185.

Goodell, G. E. 1984. Challenges to integrated pest management research and extension in the Third World: Do we really want IPM to work? *Bulletin of the Entomological Society of America, 30*, 18–26.

Gordon D., Zahavi T., Anshelevich L., Harel M., Ovadia S., Dunkelblum E. and Harari A. R. 2005. *Mating Disruption of Lobesia botrana (Lepidoptera: Tortricidae): Effect of Pheromone Formulations and Concentrations. J. Econ. Entomol.* 98, 135–142.

Gronning E. K., Borchert D. M., Pfeiffer D. G., Felland C. M., Walgenbach J. F., Hull L. A. and Killian J. C. 2000. *Effect of Specific and Generic Sex Attractant Blends on Pheromone Trap Captures of Four Leafroller Species in Mid-Atlantic Apple Orchards. J. Econ. Entomol.* 93, 157–164.

Gunning RV, Devonshire AL 2002. Negative cross-resistance between indoxacarb and pyrethroids in Australian *Helicoverpa armigera*: a tool for resistance management. Resist Pest Manag Newsletter 11:(2)52

Guo L., Fatig R. O. III, Orr G. L., Schafer B. W., Strtickland J. A., Sukhapinda K., Woodsworth A. T. and Petell J. K. 1999. Purification and characterization of toxin A and toxin B. *J. Biol. Chem.* 274, 9836–9842.

Hammes G. G, Sherrod D, Mitchell WH 1998. A five year summary of beneficial arthropod studies with "Steward" (DPX MP 062). *In:* Dugger CP, Richter DA (eds) Proc Beltwide Cot Conf, Nat Cot Council, Memphis TN, pp 940

Hammock B. D., Bonning B. C., Posse R. D., Hanzlik T. N. and Maeda S. 1990. Expression and effecls of Ihe juvenile hormone eslerase in a baculovirus veclor, *Nature* 344, 458–461.

Hamon N, Shaw R, Yang H 1996. Worldwide development of fipronil insecticide. *In:* Dugger CP, Richter DA (eds) Proc 1996 Beltwide Cotton Prod Conf, National Cotton Council of America, Memphis, pp 759-765

Hanley M. E., Dunn D. W., Abolins S. R. and Goulson D. 2004. *Evaluation of (Z)-9-tricosene baited targets for control of the housefly (Musca domestica) in outdoor situations J. Appl. Entomol.* 128, 478 482.

Hara AH (2000) Finding alternative ways to control alien pests – part 2: New insecticides introduced to fight old pests. Hawaii Landscape 4 (1) 5

Harari A. R., Zahavi T., Gordon D., Anshelevich L., Harel M., Ovadia S. and Dunkelblum E.: *Pest management programmes in vineyards using male mating disruption. Pest Manag. Sci.* 63, 769–775 2007.

Hardee DD, Adams LC, Elzen GW 2001. Monitoring for changes in tolerance and resistance to insecticides in bollworm/tobacco budworm in Mississippi, 1996-1999. Southwest Entomol 26: 365-372

Harder HH, Riley SL, McCann SF, Irving SN 1996. DPX-MP062: a novel broad spectrum, environmentally soft, insect control compound. Proc 1996 Brighton Crop Prot Conf – Pests and diseases, vol 2, pp 449-454

Harder HH, Riley SL, McCann SF, Sherrod DW 1997. DPX-MP062: a novel broad spectrum, environmentally soft, insect control compound. *In:* Dugger CP, Richter DA (eds) Proc 1997 Beltwide Cotton Prod Conf, National Cotton Council of America, Memphis, pp 48-50

Harrison R. L. and Bonning B. C. 2000. *Use of scorpion* neurotoxins to improve the insecticidal activity of Rachiplusia ou multicapsid nucleopolyhedrovirus. *Biol. Control* 17, 191–201.

Harrison R. L. and Bonning B. C. 2001. *Use of proteases* to improve the insecticidal activity of baculoviruses. *Biol. Control* 20, 199–209.

Haseeb M, Amano H, Nemoto H 2000. Pesticidal effects on mortality and parasitism rates of *Diadegma semiclausum*, a parasitoid of the diamondback moth. BioControl 45: 165-178

Hathaway D. O., Moffitt H. R. and George D. A. 1985. Codling moth (Lepid.: Tortricidae): Disruption of sexual communication with an antipheromone ((E,E)-8,10-dodecadien-l-ol acetate). *J. Ent. Soc. B.C.* 82, 18–22 1985.

Heath R. R., Epsky N. D., Dueben B. D. and Meyer W. L. 1996. *Liste des especes de la* famille *des* Cheliferidae qui habitant l'A1gérie et le Maroc. *Fla. Entomol.* 79, 144–153.

Herniou E. A., Luque T., Chen X., Vlak J. M., Cory J. S. and O'Reilly D. 2001. Use of Whole Genome Sequence Data To Infer Baculovirus Phylogeny. *J. Virol.* 75, 8117–8126

Hillier N. K., Dixon P. L. and Larson D. J. 2004. *Trap Captures of Male Grapholita libertina (Lepidoptera: Tortricidae) Moths: Relationship to Larval Numbers and Damage in Wild Lingonberry. Environ. Entomol.* 33, 405–417.

Hines RL, Hutchison WD 2001. Evaluation of action thresholds and spinosad for lepidopteran pest management in Minnesota cabbage. J Econ Entomol 94:190-196

Hoffmann C., Luthy P., Hutter R. and Pliska V. 1988. Binding of the delta endotoxin from Bacillus thuringiensis to brush-border membrane vesicles of the cabbage butterfly (Pieris brassicae). *Eur. J. Biochem.* 173, 85–91.

Hofte H. and Whitely H. R. 1989. Insecticidal crystal proteins of Bacillus thuringiensis. *Microbiol. Rev.* 53, 242–255

Holloway JW, Forrester NW, Leonard BR 1999. New insecticide chemistry for cotton IPM. *In:* Dugger CP, Richter DA (eds) Proc 1999. Beltwide Cotton Prod Conf, National Cotton Council of America, Memphis, pp.1086-1089

Hong S. C. and Williamson R. C. 2004. *Comparison of Sticky Wing and Cone Pheromone Traps for Monitoring Seasonal Abundance of Black Cutworm Adults and Larvae on Golf Courses. J. Econ. Entomol.* 97, 1666–1670.

Horowitz AR, Forer G, Ishaaya I 1994. Managing resistance in *Bemisia tabaci* in Israel with emphasis on cotton. Pestic Sci 42:113-122

Horowitz AR, Ishaaya I 1994. Managing resistance to insect growth regulators in the sweetpotato whitefly (Homoptera: Aleyrodidae). J Econ Entomol 87: 866-871

Horowitz AR, Ishaaya I 2004. Biorational insecticides – mechanisms, selectivity and importance in pest management programs. *In:* Horowitz AR, Ishaaya I (Eds.) *Insect Pest Management – Field and Protected Crops.* Springer, Berlin, Heidelberg, New York, pp. 1–28

Horowitz AR, Kontsedalov S, Denholm I, Ishaaya I 2002. Dynamics of insecticide resistance in *Bemisia tabaci* - a case study with an insect growth regulator. Pest Manag Sci 58:1096-1100

Horowitz AR, Mendelson Z, Cahill M, Denholm I, Ishaaya I 1999. Managing resistance to the insect growth regulator pyriproxyfen in *Bemisia tabaci.* Pestic Sci 55:272-276

Horowitz AR, Mendelson Z, Ishaaya I 1997. Effect of abamectin mixed with mineral oil on the sweet potato whitefly (Homoptera: Aleyrodidae). J. Econ Entomol 90:349-353

Horowitz AR, Mendelson Z, Weintraub PG, Ishaaya I 1998. Comparative toxicity of foliar and systemic application of acetamiprid and imidacloprid against the cotton whitefly, *Bemisia tabaci* (Hemiptera: Aleurodidae). Bull Entomol Res 88: 437-442

http://www.epa.gov/agriculture/pesticide.html, 2007.

http://www.epa.gov/oppbppd1/biopesticides/ingredients/index. htm, 2003.

http://www.pesticidesafety.uiuc.edu/newsletter/html/v17n304. pdf

Huang J., Rozelle S., Pray C. and Wang Q. 2002. Plant Biotechnology in China. *Science* 295, 674–677 (2002).

Hunt DA, Treacy MF 1998. Pyrrole insecticides: a new class of agriculturally important insecticides functioning as uncouplers of oxidative phosphorylation. *In:* Ishaaya I, Degheele D (eds) Insecticides with Novel Modes of Action, Mechanism, and Application, Springer Verlag, Berlin, pp 138-151

I. Ujváry and J. E. Casida: Partial synthesis of 3-O-vanilloylveracevine, an insecticidal alkaloid from Schoenocaulon officinale. *Phytochemistry* 44, 1257–1260 (1997).

Ignoffo C. M. 1973. Development of a viral insecticide: concept to commercialization. *Exp. Parasitol.* 33, 380–406.

Il'ichev A. L., Stelinski L. L., Williams D. G. and Gut L. J. 2006. *Sprayable Microencapsulated Sex Pheromone Formulation for Mating Disruption of Oriental Fruit Moth (Lepidoptera: Tortricidae) in Australian Peach and Pear Orchards. J. Econ. Entomol.* 99, 2048–2054.

Il'ichev AL, Williams DG, Gut LJ 2007. Dual pheromone dispenser for combined control of codling moth, Cydia pomonella L. and oriental fruit moth Grapholita molesta (Busck) (Lep., Tortricidae) in pears. J. Appl. Entomol. 131, 368 -376.

Imai T., Maekawa M., Tsuchiyaand S. Fujimori T.1988. Field attraction of Hoplia communis to 2-phenylethanol, a major volatile component from host flowers, Rosa spp *J. Chem. Ecol.* 24, 1491–1497.

Inceoglu A. B., Kamita S. G. and Hammock B. D. 2006. Genetically modified baculoviruses: a historical overview and future outlook. *Adv. Virus Res.* 68, 323–360.

Ishaaya I 1990. Benzoylphenyl ureas and other selective control agents – mechanism and application. *In:* Casida JE (ed) Pesticides and Alternatives, Elsevier Science Publisher BV, Amsterdam, pp 363-376

Ishaaya I 2001. Biochemical processes related to insecticide action: an overview. *In:* Ishaaya I (ed) Biochemical Sites of Insecticide action and Resistance. Springer, Berlin Heidelberg New York, pp 1-16

Ishaaya I 2003. Introduction: Biorational insecticides – mechanism and application. Arch Insect Biochem Physiol 54: 144

Ishaaya I, Casida JE 1974. Dietary TH 6040 alters cuticle composition and enzyme activity of house fly larval cuticle. Pestic Biochem Physiol 4:484-490

Ishaaya I, De Cock A, Degheele D 1994. Pyriproxyfen, a potent suppressor of egg hatch and adult formation of the greenhouse whitefly (Homoptera: Aleurodidae). J Econ Entomol 87:1185-1189

Ishaaya I, Horowitz AR 1992. Novel phenoxy hormone analog (pyriproxyfen) suppresses embryogenesis and adult emergence of sweetpotato whitefly (Homoptera: Aleyrodidae). J Econ Entomol 85: 2113-2117

Ishaaya I, Horowitz AR 1995. *Pyriproxyfen, a novel insect growth regulator for controlling whiteflies: mechanism and resistance management.* Pestic Sci 43: 227-232

Ishaaya I, Horowitz AR 1998. Insecticides with novel modes of action: an overview. *In:* Ishaaya I, Degheele D (eds) Insecticides with Novel Modes of Action, Mechanism and Application. Springer-Verlag, Berlin, Heidelberg, New York, pp 1-24

Ishaaya I, Horowitz AR, Tirry L, Barazani A 2002a. Novaluron (Rimon), a novel IGR – mechanism, selectivity and importance in IPM programs. Med Landbouww Rijksuniv Gent 67:617-626

Ishaaya I, Kontsedalov S, Horowitz AR 2002b. Emamectin, a novel insecticide for controlling field crop pests. Pest Manag Sci 58:1091-1095

Ishaaya I, Kontsedalov S, Horowitz AR 2005. Biorational insecticides: mechanism and crossresistance. Arch Insect Biochem Physiol 58: 192–199

Ishaaya I, Kontsedalov S, Mazirov D, Horowitz AR 2001. Biorational agents - mechanism and importance in IPM and IRM programs for controlling agricultural pests. Med Landbouww Rijksuniv Gent 66:363-374

Ishaaya I, Mendelson Z, Melamed-Madjar V 1988. Effect of buprofezin on embryogenesis and progeny formation of sweetpotato whitefly (Homoptera: Aleyrodidae). J Econ Entomol 81:781-784

Ishaaya I, Yablonski S, Horowitz AR 1995. Comparative toxicity of two ecdysteroid agonists, RH-2485 and RH-5992, on susceptible and pyrethroid-resistant strains of the Egyptian cotton leafworm, *Spodoptera littoralis.* Phytoparasitica 23:139-145

Ishaaya I, Yablonski S, Mendelson Z, Mansour Y, Horowitz AR. 1996. Novaluron (MCW 275), a novel benzoylphenyl urea, suppressing developing stages of lepidopteran, whitefly and leafminer pests. Proc 1996 Brighton Crop Prot Conf – Pests and Diseases, pp 1013-1020

Isman M. B. 2006. Botanical insecticides, deterrents, and repellents in modern agriculture and increasingly regulated world. *Annu. Rev. Entomol.* 51, 45–66 (2006).

Isman M. B., Machial C. M., Miresmailli S. and Bainard L. D.: 2007. "Pesticide chemistry," ed. by H. Ohkawa, H. Miyagawa and P. W. Lee, Wiley-VCH.

Itaya W 1987. Insect juvenile hormone analogue as an insect growth regulator. Sumitomo Pyrethroid World 8: 2-4

Izawa Y, Uchida M, Sugimoto T, Asai T 1985. Inhibition of chitin biosynthesis by buprofezin analogs in relation to their activity controlling *Nilaparvata lugens* Stål. Pestic Biochem Physiol 24: 343-347

Jacobson M. and Crosby D. G. (eds.). 1971. "Naturally Occurring Insecticides," Marcel Dekker Inc., New York,

Jamses DG (2002) Selectivity of the acaricide, bifenazate, and aphicide, pymetrozine, to spider mite predators in Washington hops. Int J Acar 28:175-179

Jansma JE, Keulen H van, Zadoks JC 1993. Crop protection in the year 2000: a comparison of current policies towards agrochemical usage in four West European countries. Crop Prot 12:483-489

Jansson RK, Dybas RA 1998. Avermectins: Biochemical mode of action, biological activity and agricultural importance. In: Ishaaya I and Degheele D (eds)

Insecticides with Novel Modes of Action – Mechanisms and Application, Springer, Berlin Heidelberg New York, pp 153-170

Jefferies P. R., Toia R. F., Brannigan B., Pessah I. and Casida J. E. 1992. Ryania insecticide: analysis and biological activity of 10 natural ryanoids. *J. Agric. Food Chem.* 40, 142–146

Jinn T. R., Tu W. C., Lu C. I. and Tzen J. T. C. 2006. Enhancing insecticidal efficacy of baculovirus by early expressing an insect neurotoxin, IghIT2, in infected *Trichoplusiani* larvae. *Appl. Microbiol. Biot.* 72, 1247–1253.

Kakizaki M. 2004. The sex pheromone components for mating disruption of the rice leaf bug, *Trigonotylus caelestialium* (Heteroptera: Miridae)," *Appl. Entomol. Zool.* 39, 221–228, 2004.

Kaya H. K. and Gaugler R. 1993. Entomopathogenic nematodes. *Annu. Rev. Entomol.* 38, 181–206.

Kayser H, Kaufmann L, Schürmann F 1994. Pymetrozine (CGA 215'944): a novel compound for aphid and whitefly control. An overview of its mode of action. Proc 1994 Brighton Crop Prot Conf – Pests and diseases, vol 2, pp 737-742

Kerns D. L. 2000. Mating disruption of beet armyworm (Lepidoptera: Noctuidae) in vegetables by a synthetic pheromone. *Crop Prot.* 19, 327–334.

Khetan S. K. 2001. Microbial pest control," ed. by S. K. Khetan, Marcel Dekker, New York, pp. 3–42.

Kirst HA, Michel KH, Mynderse JS, Chao EH, Yao RC, Nakatsukasa WM, Boeck LD, Occlowitz J, Paschel JW, Deeter JB, Thompson GD 1992. Discovery, isolation and structure elucidation of a family of structurally unique fermentation-derived tetracyclic macrolides. *In:* Baker DR, Fenyes JG, Steffens JJ (eds) Synthesis and chemistry of agrochemicals III. Ame Chem Soc, Washington DC, pp 214-225

Klein M. G. 1993. Nematodes and the Biological Control of Insect Pests, ed. by R. Bedding, R. Akhurst and H. Kaya, CSIRO, East Melbourne, Australia, pp. 49–58, 1993.

Klis SFL, Vijverberg HPM, van den Berken J (1991) Phenylpyrazoles, a new class of pesticides: electrophysiological investigation into basis effects. Pestic Biochem Physiol 39: 210-218

Koehler PG, Patterson RJ 1991. Incorporation of pyriproxyfen in German cockroach (Dictyoptera: Blattellidae) management program. J Econ Entomol 84: 917-921

Kooyman C. and Abdalla O. M. 1998. *Application of Metarhizium* flavoviride (Deutermycotina: Hypomycetes) spores against the tree locust, Anacridium melanorhodon (Orthoptera: Acridadae) in Sudan. *Biocontrol Sci. Technol.* 8, 215–219.

Koppenhofer A. M., Cowles R. S., Cowles E. A., Fuzy E. M. and Kaya H. K. 2003. Effects of neonicotinoid insecticide synergists on entomopathogenic nematodes fitness. *Entomol. Exp. Appl.* 106, 7–18.

Koppenhofer A. M., Polavarapu S., Fuzy E. M., Zhang A. J., Ketner K. and Larsen T. 2005. Mating disruption of oriental beetle (Coleopteran: Scarabaediae) in turf grass using encapsulated formulations of sex pheromone components. *Environm. Entomol.* 34, 1408–1417.

Koppenhofer AM, Cowles RS, Cowles EA, Fuzy EM, Baumgartner L 2002. Comparison of neonicotinoid insecticides as synergists for entomopathogenic nematodes. Biol Cont 24:90-97

Koul O. and Dhaliwal G. S. (eds.) 2002. Microbial biopesticides, Taylor and Francis, London.

Koul O. and Wahab S. 2004. *"Neem: Today and in the New Millennium,"* Springer, Netherlands, 2004.

Kovanci O. B., Schal C., Walgenbach J. F. and Kennedy G. G. 2005. Comparison of mating disruption with pesticides for management of oriental fruit moth (Lepidoptera : Tortricidae) in North Carolina apple orchards. *J. Econ. Entomol.* 98, 1248–1258

Kovanci O. B., Walgenbach J. F., Kennedy G. G. and Schal C. 2005. *Phytoparasitica* 33, 334–342.

Krapcho K. J., Kral R. M., Vanwagenen B. C., Eppler K. G. and Morgan T. K. 1995. Characterization of voltage sensitive sodium channel gene coding sequences from insecticide susceptible and knockown resistant housefly strains. *Insect Biochem. Mol. Biol.* 25, 991–1000.

Kuhar T. P., Mori K. and Dickens J. C. 2006. Potential of a synthetic aggregation pheromone for integrated pest management of Colorado potato beetle *Agr. Forest Entomol.*, 77–81 (2006

Lacey L. A. and Goettel M. S. 1995. *Current developments* in microbial control of insect pests and prospects for the early 21st century. *Entomophaga* 40, 3–27.

Lange C. E. and. De Wysiecki M. L 1996. The *fate of Nosema* locustae (Microsporida: Nosematidae) in Argentine grasshoppers (Orthoptera: Acrididae). *Biol. Control* 7, 24–29.

Langewald J., Kooyman C., Douro-Kpindou O.-K., Lomer C. J., Dahmoud A. O. and Mohamed H. O. 1997. *Field treatment of Desert* Locust (Schistocerca gregaria Forskål) hoppers in Mauritania using an oil formulation of the entomopathogenic fungus Metarhizium flavoviride *Biocontrol Sci. Technol.* 7, 603–611.

Langley P (1990) Control of the tsetse fly using a juvenile hormone mimic, pyriproxyfen. Sumitomo Pyrethroid World 15: 2-5

Lasota JA, Dybas RA 1991. Avermectin, a novel class of compound: implication for use in arthropod pest control. Annu Rev Entomol 36:96-117

Le DP, Thirugnanam M, Lidert Z, Carlson GR, Bryan JB 1996. RH-2485: a new selective insecticide for caterpillar control. Proc 1996 Brighton Crop Prot Conf – Pests and diseases, vol 2, pp 481-486

Leonard PK 2000. Chlorfenapyr, a novel IPM compatible resistance management tool for fruit production. Acta Hort 525: 257-275

Li H-J., Jiang Y. and Li P. 2006. Chemistry, bioactivity and geographical diversity of steroidal alkaloids from the Liliaceae family. *Nat. Prod. Rep.* 23, 735–752.

Liu B, Sengonca C 2002. Ivestigations on side-effects of the mixed biocide GCSCBtA on different predators of *Plutella xylostella* (L.) (Lep., Plutellidae) in southeastern China. Anz Schadling 75:57-61

Liu MY, Casida JE 1993. High affinity binding of [3H]-imidacloprid in the insect acetylcholine receptor. Pestic Biochem Physiol 46: 40-46

Lomer C. J., Bateman R. P., Johnson D. L., Langewald J. and Thomas M. 2001. Biological control of locusts and grasshoppers. *Annu. Rev. Entomol.* 46, 667–702.

Lomer C. J., Thomas M. B., Godonou I., Shah P. A., Douro-Kpindou O. K. and Langewald J 1997. Control of grasshoppers, particularly Hieroglyphus daganensis, in northern Benin using Metarhizium flavoviride *Mem. Entomol. Soc. Can.* 171, 301–311.

Lord J. C. 2005. From Metchnikoff to Monsanto and beyond: the path of microbial control. *J. Invertebr. Pathol.* 89, 19–29.

Ludwig S, Oetting R 2001. Effect of spinosad on *Orius insidiosus* (Hemiptera: Anthocoridae) when used for *Frankliniella occidentalis* (Thysanoptera: Thripidae) control on greenhouse pot chrysanthemums. Florida Entomol 84:311-313

Lykouressis D., Perdikis D., Samartzis D., Fantinou A. and Toutouzas S. 2005. Management of the pink bollworm Pectinophora gossypiella (Saunders) (Lepidoptera: Gelechiidae) by mating disruption in cotton fields. *Crop Prot.* 24, 177–183

MacConnell JG, Demchak RJ, Preiser FA, Dybas RA 1989. Relative stability, toxicity, and penetrability of abamectin and its 8, 9-oxide. J Agric Food Chem 37:1498-1501

Maeda S. 1989. Use of proteases to improve the insecticidal activity of baculoviruses. *Biochem. Biophys. Res. Comm.* 165, 1177–1183.

Mafra-Neto A. and Habib M. 1996. Evidence that mass trapping suppresses pink bollworm in cotton fields. *Entomol. Exp. Appl.* 81, 315–323.

Manceva S. D., Pusztai-Carey M. and Butko P. 2004. Effect of pH and ionic strength on the cytolytic toxin Cyt1A: a fluorescence spectroscopy study. *Biochem. Biophys. Acta* 1699, 123–130.

Marquini F, Guedes RNC, Picanço MC, Regazzi AJ 2002. Response of arthropods associated with the canopy of common beans subjected to imidacloprid spraying. J Appl Entomol 126:550-556

Masner P, Angst M, Dorn S 1987. Fenoxycarb, an insect growth regulator with juvenile hormone activity: a candidate for *Heliothis virescens* (F.) control on cotton. Pestic Sci 18: 89-94

Matsumoto K., K. Nakamuta and T. Nakashima 2007. Mating disruption controls the cherry tree borer, Synanthedon hector (Butler) (Lepidoptera: Sesiidae), in a steep orchard of cherry trees. Journal of Forest Research 12:34-37.

McClintock J. T., Schaffer C. R. and Sjoblad R. D. 1995. A comparative review of the mammalian toxicity of *Bacillus thuringiensis-based pesticides Pest. Sci.* 45, 95–105.

McCoy C. W., Samson R. A. and Boucias D. G. 1988. Handbook of Natural Pesticides," ed. by C. M. Ignoffo and N. B. Mandava, CRC Press, Boca Raton, FL, pp. 151–236,

McCutchen W. F. and Flexner L. 1999. Biopesticides Use and Delivery, ed. by D. R. Hall and J. J. Menn, Humana Press, Totowa, N.J., pp. 341–355, 1999.

McNair C., Gries G. and Sidney M. 1999. Feeding behaviour of the first instar *Christoneura fumiferana* and *Christoneura pinus* (Lepidoptera: Tortricidae). *Can. Entomol.* 131, 97–105.

McNally P, Mullins W 1996. The role of Provado in western cotton IPM programs. In: Dugger C P, Richter DA (eds) Proc 1996 Beltwide Cotton Prod Conf, National Cotton Council of America, Memphis, pp 859-862

McPherson RM, Jones DC, Bertrand PF, Csinos AS 2002. Impact of thrips (Thysanoptera: Thripidae) management practices on suppression of tomato spotted wilt virus and aphid (Homoptera: Aphididae) control in flue-cured tobacco. J Entomol Sci 37: 143-153

McQuate G. T., Sylva C. D. and Jang E. B. 2005. *Mediterranean fruit fly* (Dipt., Tephritidae) suppression in persimmon through bait sprays in adjacent *coffee* plantings. *J. Appl. Entomol.* 129, 110–117.

Medina P, Budia F, Tirry L, Smagghe G, Vinuela E 2001. Compatibility of spinosad, tebufenozide and azadirachtin with eggs and pupae of the predator *Chrysoperla carnea* (Stephens) under laboratory conditions. Biocontrol Sci Tech 11:597-610

Medina P, Smagghe G, Budia F, Estal P del, Tirry L, Vinuela E, del Estal P 2002. Significance of penetration, excretion, and transovarial uptake to toxicity of three insect growth regulators in predatory lacewing adults. Arch Insect Biochem Physiol 51:91-101

Mellin TN, Busch RD, Wang CC 1983. Postsynaptic inhibitions of invertebrate neuromuscular transmission by avermectin B1a. Neuropharmacology 22: 89-96

Menn J. J. and Hall F. R 1999. "Biopesticides Use and Delivery," *ed. by* F. R. Hall and J. J. Menn, Humana Press, Totowa, New Jersey, pp. 1–10

Merrington C. L., King L. A. and Posse R. D. 1999. Protein Expression: A practical approach, ed. by S. J. Higgins and B. D. Hames, Oxford University Press, pp. 101–127, 1999.

Michaelides P, Irving SN 1998. Cotton insect pest control with indoxacarb: a novel insecticide. Proc World Cotton Research Conf-2, Athens, Greece 1998, pp 773-776

Miller D. 2002. The Sustainable Learning Community: One University's Journey to the Future. *Prog. Farmer* 117, 22–24.

Miller J. R., Gut L. J., de Lame F. M. and Stelinski L. L. 2006. *Differentiation of competitive non-competitive* mechanisms mediating disruption of moth sexual com- munication by point sources of sex pheromone (part I): Theory. *J. Chem. Ecol.* 32, 2089–2114.

Miller J. R., Gut L. J., de Lame F. M. and Stelinski L. L. 2006. Differentiation of competitive vs. non-competitive mechanisms mediating disruption of moth sexual com- munication by point sources of sex pheromone (part 2): Case studies. *J. Chem. Ecol.* 32, 2115–2143.

Miller TP, Treacy MF, Gard IE, Lovell JB, Wright DP Jr, Adder RW, Kamki VM (1990) AC303,603 – summary of 1988-89 field trials results. Proc 1990 Brighton Crop Prot Conf – Pests and diseases

Milner R. J. and Hunter D. M. 2001. *Recent developments in the use of* fungi as biopesticides against locusts and grasshoppers in Australia. *J. Orthop. Res.* 10, 271–276.

Mizell RF III, Schiffhauer DE, Taylor JL (1986) Mortality of *Tetranychus urticae* Koch (Acari: Tetranychidae) from abamectin residues: effects of host plant, light, and surfactants. J Entomol Sci 21:329-337

Mo J. H., Glover M., Munro S. and Beattie G. A. C. 2006. *Evaluation of mating disruption for control of lightbrown apple moth* (Lepidoptera: Tortricidae) in citrus *J. Econ. Entomol.* 99, 421–426

Mori K, Gotoh T 2001. Effects of pesticides on the spider mite predators, *Scolothrips takahashii* (Thysanoptera: Thripidae) and *Stethorus japonicus* (Coleoptera: Coccinellidae). Int J Acar 27: 299-302

Moscardi F. 1999. Assessment of the application of baculoviruses for control of Lepidoptera. *Annu. Rev. Entomol.* 44, 257–289.

Moulton JK, Pepper DA, Dennehy TJ 2000. Beet armyworm (*Spodoptera exigua*) resistance to spinosad. Pest Manag Sci 56: 842-848

Moulton JK, Pepper DA, Jansson RK, Dennehy TJ 2002. Pro-active management of beet armyworm (Lepidoptera: Noctuidae) resistance to tebufenozide and methoxyfenozide: baseline monitoring, risk assessment, and isolation of resistance. J Econ Entomol 95:414-424

Mulder R, Gijswijk MT (1973) The laboratory evaluation of two promising new insecticides which interfere with cuticle deposition. Pestic Sci 4: 737-745

Nakatani M. 2000. "Bioactive compounds from natural sources," *ed. by* C. Tringali, Taylor and Francis, London, pp 527–554.

Naranjo SE, P. C. Ellsworth PC, Chu CC, T. J. Henneberry TJ 2002. Conservation of predatory arthropods in cotton: role of action thresholds for *Bemisia tabaci* (Homoptera: Aleyrodidae). J Econ Entomol 95:682-691

Nauen R, Ebbinghaus-Kintscher U, Elbert A, Jeschke P, Tietjen K 2001. Acetylcholine receptors as sites for developing neonicotinoid insecticides. *In:* Ishaaya I (ed) Biochemical Sites of Insecticide Action and Resistance. Springer- Verlag, Berlin, Heidelberg, New York, pp 77-105

Neves PMOJ, Hirose E, Tchujo PT, Moino Jr A 2001. Compatibility of entomopathogenic fungi with neonicotinoid insecticides. Neotrop Ent 30:263-268

Noteborn H. P. J. M., Bienenmann-Ploum M. E., Van den Berg J. H. J., Alink G. A., Zolla L.and Kuiper H. A. 1993. Food safety of transgenic tomatoes expressing the insecticidal crystal protein Cry 1a(b) from *Bacillus thuringiensis* and the marker enzyme APH (3')11 *Med. Fac. Landbouww. Univ. Gent.* 58/4b, 1851–1858 1993.

Oberlander H, Smagghe G 2001. Imaginal discs and tissue cultures as targets for insecticide action. *In:* Ishaaya I (ed) Biochemical sites of insecticide action and resistance. Springer, Berlin Heidelberg New York, pp 133-150

Oehlschlager A. C., Chinchilla C., Castillo G. and Gonzalez L. 2002. *Control of Red Ring* Disease by Mass Trapping of Rhynchophorus palmarum (Coleoptera: Curculionidae), *Fla. Entomol.* 85, 507–513.

Ortiz A., Quesada A. and Sanchez A. 2004. Allelopathic interactions in crop-weed mixtures: applications for weed management. *J. Chem. Ecol.* 30, 991–1000.

Palli SR, Ladd TR, Sohi SS, Cook BJ, Retnakaran A 1996. Cloning and developmental expression of Choristoneura hormone receptor 3, an ecdysoneinducible gene and a member of the steroid hormone receptor superfamily. Insect Biochem Mol Biol 26:485-499

Palm C. J., Schaller D. L., Donegan K. K. and Seidler R. J. 1996. Persistence in soil of transgenic plant produced *Bacillus thuringiensis* var. kurstaki delta-endotoxin. *Can. J. Microbiol.* 42, 1258–1262.

Palm C. J., Seidler R. J., Donegan K. K. and Harris D. 2003. Transgenic plant pesticides:fate and persistence in soil. *Plant Physiol. Suppl.* 102, 166 (1993).

Papadopoulos N. T., Katsoyannos B. I., Kouloussis N. A., Hendrichs J., Carey J. R. and Heath R. R. 2001. *Early detection* and population monitoring of Ceratitis capitata (Diptera: Tephritidae) in a mixed-fruit orchard in northern Greece. *J. Econ. Entomol.* 94, 971–978.

Parmar V. S., Jain S. C., Bisht K. S., Jain R., Taneja P., Jha A., Tyagi O. D., Prasad A. K., Wengel J., Olsen C. E.and. Boll P. M. 1997. Phytochemistry of the genus piper. *Phytochemistry* 46, 597–673 (1997).

Pavela R 2005. Insecticidal activity of some essential oils against larvae of *Spodoptera littoralis. Fitoterapia* 76, 691–696.

Paz Parente J. and Pereira da Silva B. 2001. "Recent Research Developments in Phytochemistry," ed. by S. G. Pandalai, Research Signpost, Trivandrum, Vol. 5, pp. 153–167.

Peleg BA 1988. Effect of new phenoxy juvenile hormone analog on California red scale (Homoptera: Diaspididae), Florida wax scale (Homoptera: Coccidae) and the ectoparasite *Aphytis holoxanthus* DeBache (Hymenoptera: Aphelinidae) J Econ Entomol 81: 88-92

Perlak F. J., Stone T. B., Muskopf Y. M., Petersen L. J., Parker G. B., McPherson S. A., Wyman J., Love S., Reed G., Biever D.and Fischhoff D. A. 1993. Genetically improved potatoes: protection from damage by Colorado beetle. *Plant Mol. Biol.* 22, 313–321.

Peters, A. 1996. The *natural host range* of Steinernema and Heterorhabditis spp. and their impact in insect populations. *Biocontrol Sci. Technol.* 6, 389–402 ().

Phelan P. L. and Baker T. C. 1987. An *attracticide for control* of Amyelois transitella (Lepidoptera: Pyralidae) in almonds. *J. Econ. Entomol.* 80, 779–783.

Pietrantonio PV, Benedict JH 1999. Effect of new cotton insecticide chemistries, tebufenozide, spinosad and chlorfenapyr, on *Orius insidious* and two *Cotesia* species. Southwest Entomol 24: 21-29

Plettner E. 2002. *Insect pheromone olfaction:* new targets for the design of species selective pest control agents. *Curr. Med. Chem.* 9, 1075–1085.

Plimmer JR (1993) Regulatory problems associated with natural products and biopesticides. Pestic Sci 39:103-108

Pluschkell U, Horowitz AR, Weintraub PG, Ishaaya I 1998. DPX-MP062 - a potent compound for controlling the Egyptian cotton leafworm, *Spodoptera littoralis.* Pestic Sci 54:85-90

Polavarapu S., Wicki M., Vogel K., Lonergan G. and Nielsen K. 2002. *Disruption of sexual communication* for oriental beetles (Coleoptera: Scarabaeidae) with a microencapsulated formulation of sex pheromone components in blueberries and ornamental nurseries, *Environm. Entomol.* 31, 1268–1275.

Polonsky J., Bhatnagar S. C., Griffiths D. C., Pickett J. A. and Woodcock C. M. 1989. Activity of quassinoids as antifeedants against aphids. *J. Chem. Ecol.* 15, 993–998

Popham Holly J. R., Prikhod'ko G. G., Felcetto T. J., Ostlind D. A., Warmke J. W., Cohen C. J. and Miller L. K.: Popham H.J.R., Prikhod'ko G.G., Felcetto T.J., Ostlind D.A., Warmke J.W., Cohen C.J., Miller L.K. 1988. Effect of deltamethrin treatment on lepidopteran larvae infected with baculoviruses expressing insect-selective toxins μ-Aga-IV, As II, or Sh 1. Biol. Control. 12:79–87. *Biol. Control* 12, 79–87.

Price R. E., Bateman R. P., Brown H. D., Butler E. T. and Müller E. J. 1997. *Aerial spray trials against Brown locust* (Locustana pardalina, Walker) *nymphs in South Africa using oilbased formulationsof Metarhizium flavoviride.* Crop Prot. 16, 345–351.

Punga N.: 1989. Insect Pheromones in Plant Protection," *ed. by* A. R. Jutsum and R. F. S. Gordon, Wiley, Chichester, UK, pp. 295–302.

Reddy G. V. P. and Guerrero A. 2001. *Optimum timing of insecticide* applications against diamondback moth Plutella xylostellain cole crops using threshold catches in sex pheromone traps. *Pest Manag. Sci.* 57, 90–94.

Regev A., Rivkin H., Inceoglu A. B., Gershburg E., Hammock B. D., Gurevitz M. and Chejanovsky N. 2003. *FEBS Lett.* 537, 106–110 .

Regis L., Silva-Filha M. H., Nielsen-Leroux C. and Charles J. F. 2001. Bacteriological larvicides of dipteran disease vectors. *Trends Parasitol.* 17, 377–380.

Renou M. and Guerrero A. 2000. *Annu. Rev. Entomol.* 48, 605–630.

Retnakaran A, Granett J, Ennis T 1985. Insect growth regulators. In: Kerkut GA, Gilbert LI (eds) Comprehensive insect physiology, biochemistry and pharmacology, Vol. 12, Pergamon Press, Oxford New York, pp 529-601

Richardson C. D. 1995. Baculovirus expression protocols, ed. By C. D. Richardson, Humana Press, Totowa, N.J., pp. 161–177, 1995.

Rola, A. C., & Pingali, P. L. 1993. *Pesticides, Rice Productivity, and Farmers' Health.* Manila: International Rice Research Institute. (Washington, D.C.: Word Resources Institute).

Rothman L. D. and Myers J. H. 1996. Debilitating effects of viral diseases on host Lepidoptera. *J. Invertebr. Pathol.* 67, 1–10.

Roy H. E., Steinkraus D. C., Eilenberg J., Hajek A. E. and Pell J. 2006. Implications of predator foraging on aphid pathogen dynamics *Annu. Rev. Entomol.* 51, 331–357.

Ruberson JR, Roberts PM, Abney MR 2002. Effects of novel pesticides on *Trichogramma pretiosum. In:* Dugger CP, Richter DA (eds) Proc 2002 Beltwide Cotton Prod Conf, National Cotton Council of America, Memphis.

Ruberson JR, Tillman PG 1999. Effect of selected insecticides on natural enemies in cotton: laboratory studies. *In:* Dugger CP, Richter DA (eds) Proc 1999 Beltwide Cotton Prod Conf, National Cotton Council of America, Memphis, pp 1210-1213

Sackmann P, Rabinovich M, Corley JC 2001. *Successful removal of German yellowjackets* (Hymenoptera: Vespidae) by toxic baiting. J Econ Entomol 94:811-816

Salgado VL 1998. Studies on the mode of action of Spinosad: insect symptoms and physiological correlates. Pestic Biochem Physiol 60: 91-92

Sanders C. J. 1997. Insect Pheromone Research: New Directions, ed. by R. T. Cardé and A. K. Minks, Chapman and Hall, New York, pp. 333–346.

Scharf ME, Siegfried BD 1999. Toxicity and neurophysiological effects of fipronil and fipronil sulfone on the western corn rootworm (Coleoptera: Chrysomelidae). Arch Insect Biochem Physiol 40: 150-156

Schlyter F., Zhang Q. H., Liu G.-T. and Ji L.-Z. 2001. *Int. Pest Manag. Rev.* 6, 185–196. Schmutterer H. (ed.). 1995. "The Neem Tree (*Azadirachta indica.* A. Juss.) and other Meliaceous plants," VCH, Weinheim.

Schmutterer H. and Ascher K. R. S. (eds.). 1984. Natural Pesticides from "The Neem Tree (*Azadirachta indica A. Juss.)* and Other Tropical Plants," Deutsche Gesellschaft für Technische Zusammenarbeit (GTZ), Eschborn, Germany.

Schmutterer H. and Ascher K. R. S. (eds.). 1987. "Natural Pesticides from the Neem Tree (Azadirachta indica A. Juss.) and Other Tropical Plants," Deutsche Gesellschaft für Technische Zusammenarbeit (GTZ), Eschborn, Germany

Schnepf H. E. and Whiteley H. R. 1981. Cloning and expression of the Bacillus thuringiensis crystal protein gene in *Escherichia coli. Proc. Natl. Acad. Sci. USA* 78, 2893–2897.

Schoonejans T, Staaij M van der 2001. Spinosad, a new tool for insect control in vegetables cultivated in greenhouses. Med Landbouww Rijksuniv Gent 66:375- 386

Sciarretta A., Girma M., Tikubet G., Belayehun L., Ballo S. and Baumgartner J.2005. *J. Med. Entomol.* 42, 1006–1019.

Scott I. M., Jensen H. R., Philogéne B. J. R. and Arnason J. T. 2008. A review of Piper spp. (Piperaceae) phytochemistry, insecticidal activity and mode of action. *Phytochem. Rev.* 7, 65–75.

Sechser B, Reber B, Bourgeois F 2002. Pymetrozine: selectivity spectrum to beneficial arthropods and fitness for integrated pest management. Anz Schadling 75:72-77

Shah P. A. and Pell J. K. 2003. Entomopathogenic fungi as biological control agents. *Appl. Microbiol. Biot.* 61, 413–423.

Sharov A. A., Liebhold A. M. and Roberts E. A. 1997. *Methods for monitoring* the spread of gypsy moth (Lepidoptera: Lymantriidae) populations in the Appalachian Mountains. *J. Econ. Entomol.*90, 1259–1266.

Shaw K. E., Davidson G., Clark S. J., Ball B. V., Pell J. K., Chandler D. and. Sunderland K. D 2002. *Laboratory bioassays to assess* the pathogenicity of mitosporic fungi to Varroa destructor (Acari: Mesostigmata), an ectoparasitic mite of the honeybee, Apis mellifera. *Biol. Control* 24, 266–276.

Shipp JL, Wang K, Ferguson G 2000. Residual toxicity of avermectin b1 and pyridaben to eight commercially produced beneficial arthropod species used for control of greenhouse pests. Biol Cont 17:125-131

Shoop W, Mrozik H, Fisher M 1995. Structure and activity of avermectins and milbemycins in animal health. Vet Parasitol 59:139-156

Siegel J. P., Maddox J. V. and Ruesink W. G. 1986. *Lethal and sublethal effects* of Nosema pyrausta on the European corn borer (Ostrinia nubilalis) in central Illinois. *J. Invertebr. Pathol.* 48, 167–173.

Sierotzki H., Camastral F., Shah P. A., Aebi M. and Tuor 2000. *Biological characteristics of* selected Erynia neoaphidis isolates. *Mycol. Res.* 104, 213–219

Singh R. P., Chari M. S., Raheja A. K. and Kraus W. (eds.). 1996. Neem and environment Oxford and IBH Publishing Co, New Delhi,

Slamovits C., Williams B. A. P. and Keeling P. J. 2004. *Transfer of Nosema locustae* (Microsporidia) to Antonospora locustae n. comb. Based on molecular and ultrastructural data. *J. Eukaryot. Microbiol.* 51, 207–213.

Smagghe G, Degheele D 1992. Effect of RH-5849, the first nonsteroidal ecdysteroid agonist, on larvae of *Spodoptera littoralis* (Boisd.) (Lepidoptera: Noctuidae). Arch Insect Biochem Physiol 21: 119-128

Smagghe G, Degheele D 1998. Ecdysone agonists: mechanisms and biological activity. *In:* Ishaaya I, Degheele D (eds) Insecticides with novel modes of action: mechanism and application. Springer, Berlin Heidelberg, New York, pp 25-39

Smith A. W., Camara-Artigas A., Brune D. C. and Allen J. P. 2005. Implications of high-molecular-weight oligomers of the binary toxin from Bacillus sphaericus *J. Invertebr. Pathol.* 88, 27–33.

Sokolova J. and Lange C. E. 2002. An Ultrastructural Study of Nosema locustae Canning (Microsporidia) from Three Species of Acrididae (Orthoptera) *Acta Protozool.* 41, 229–237.

Solé J., Sans A., Riba M., Rosa E., Bosch M. P., Barrot M., Palencia J., Castellà J. and Guerrero A. 2008. Reduction of damage by the Mediterranean corn borer, Sesamia nonagrioides , and the European corn borer, Ostrinia nubilalis , in maize fields by a trifluoromethyl ketone pheromone analog. *Entomol. Exp. Appl.* 126, 28–39.

Soroker V., Blumberg, D. Haberman A., Hamburger- Richard M., Reneh S., Talebaev, S.. Anshelevich L and Harari A. R. 2005. *Phytoparasitica* 33, 97–106.

Sparks TC, Crouse GD, Durst G 2001. Natural products as insecticides: the biology, biochemistry and quantitative structure-activity relationships of spinosyns and spinosoids. Pest Manage Sci 57: 896-905

Sparks TC, Kirst HA, Mynderse JS, Thompson GD, Turner JR, Jantz OK, Hertlein MB, Larson LL, Baker PJ, Broughton MC, Busacca JD, Lawrence C. Creemer LC, Huber ML, Martin JW, Nakatsukasa WM, Paschal JW, Worden TV 1996. Chemistry and biology of the spinosyns: components of spinosad (Tracer®), the first entry into DowElanco's naturalyte class of insect control products. In: Dugger P, Richter D (eds) Proc 1996 Beltwide Cotton Prod Conf, National Cotton Council of America, Memphis, pp 692-696

Stansly, P. A., T.-X. Liu, D. J. Schuster and D. E. Dean. 1996. Role of Biorational Insecticides in Management of *Bemisia*, pp. 605-615, *In* D. Gerling and R. T. Mayer [eds.], *Bemisia* 1995: Taxonomy, Biology, Damage, Control and Management. Intercept Ltd., Andover, Hants, United Kingdom.

Stelinski L. L., McKenzie D., Gut L. J., Isaacs R. and Brunner J. 2007. Comparison of Female Attractiveness and Male Response Among Populations of Choristoneura rosaceana (Harris) in *Western and Eastern U.S. Apple Orchards*. Environm. Entomol. 36, 1032–1039.

Stelinski, L.L., L. J. Gut, M. Haas, P. McGhee, and D. Epstein. 2007. Evaluation of aerosol devices for simultaneous disruption of sex pheromone communication in *Cydia pomonella* and *Grapholita molesta* (Lepidoptera: Tortricidae). *Journal of Pest Science.* 80: 225-233.

Stern VM, Smith RF, van den Bosch R, Hagen KS 1959. The integration of chemical and biological control of the spotted alfalfa aphid. The integrated control concept. Hilgardia 29(2):81–101

Suckling D. M. and Karg G. 1998. Biological and Biotechnological Control of Insect Pests, ed. by J. E. Rechcigl and N. A. Rechcigl, Lewis Publishers, Boca Raton, FL, pp. 63–99.

Suckling D. M., Daly J. M., Chen X. and Karg G. 2007. Field electroantennogram and trap assessments of aerosol pheromone dispensers for disrupting mating in *Epiphyas postvittana. Pest Manag. Sci.* 63, 202–209.

Sudakin D. L. 2003. Biopesticides. *Toxicol. Rev.* 22, 83–90.

Sullivan MJ, Turnipseed SG, Robinson D 1999. Insecticidal efficacy against a complex of fall and beet armyworms and soybean looper in South Carolina cotton. In: Dugger CP, Richter DA (eds) Proc 1999 Beltwide Cotton Prod Conf, National Cotton Council of America, Memphis, pp 1034-1036

Takahashi H. Mitsui J, Takausa N, Matsud M, Yoneda H, Suszuki J, Ishimitsi K, Kishimoto T 1992. NI-25, a new type of systemic and broad spectrum insecticide. Proc 1992 Brighton Crop Prot Conf – Pests and diseases, Vol 1, pp 88-96

Tanaka K, Shozo E, Hikaru K 2000. Toxicity of insecticides to predators of rice planthoppers: spiders, the mirid bug and the dryinid wasp. Appl. Entomol. Zool 35:177-187

Tcheslavskaia K., Brewster C., Thorpe K., Sharov A., Leonard D. and Roberts A.2005. Effects of intentional gaps in spray coverage on the efficacy of gypsy moth mating disruption. *J. Appl. Entomol.* 129, 475–480.

Thompson G, Hutchins S 1999. Spinosad. Pesticide Outlook 10:78-81

Tillman PG, Hammes GG, Sacher M, Connair M, Brady EA, Wing KD 2002. Toxicity of a formulation of the insecticide indoxacarb to the tarnished plant bug, *Lygus lineolaris* (Hemiptera: Miridae), and the big-eyed bug, *Geocoris punctipes* (Hemiptera: Lygaeidae). Pest Manag Sci 58:92-100

Tillman PG, Mulrooney JE 2000. Effect of selected insecticides on the natural enemies *Coleomegilla maculata* and *Hippodamia convergens* (Coleoptera: Coccinellidae), *Geocoris punctipes* (Hemiptera: Lygaeidae), and *Bracon mmellitor, Cardiochiles nigriceps, and Cotesia marginiventris* (Hymenoptera: Braconidae) in cotton. J Econ Entomol 93:1638-1643

Tillman PG, Mulrooney JE 2001. Effect of malathion on beneficial insects. Southwest Entomol 24:13-21

Tillman PG, Mulrooney JE, Mitchell W 1998. Susceptibility of selected beneficial insects to DPX-MP062. *In:* Dugger CP, Richter DA (eds) Proc 1998 Beltwide Cotton Prod Conf, National Cotton Council of America, Memphis, pp 1112-1114

Tomizawa M, Casida JE 2003. Selective toxicity of neonicotinoids attributable to specificity of insect and mammalian nicotinic receptors. Annu Rev Entomol 48:339-364

Tomizawa M, Yamamoto I 1992. Binding of nicotinoids and the related compounds to the insect nicotinic acetylcholine receptor. J Pestic Sci 17: 231

Tomizawa M. and Casida J. E. 2003. Selective toxicity of neonicotinoids attributable to specificity of insect and mammalian nicotinic receptors. *Annu. Rev. Entomol.* 48, 339–364.

Treacy M. F. 1999. Biopesticides Use and Delivery," ed. by D. R. Hall and J. J. Menn, Humana Press, Totowa, N.J., pp. 321–340, 1999.

Trematerrra , P. 1993. On the possibility of mass-trapping *Synanthedon myopaeformis* Bkh. (Lepidoptera: Sesiidae). *J.Appl. Entomol.* 115: 476-483.

Trimble R. M. 2007. *Susceptibility of Oriental Fruit Moth (Lepidoptera: Tortricidae) Larvae to Selected Reduced-Risk Insecticides. J. Econ. Entomol.* 100, 1815–1820 (2007).

Trimble R. M., Pree D. J. and Carter N. J. 2001. Integrated control of oriental fruit moth (Lepidoptera : Tortricidae) in peach orchards using insecticide and mating disruption. *J. Econ. Entomol.* 94, 476–485.

Uchida M, Asai T, Sugimoto T 1985. Inhibition of cuticle deposition and chitin biosynthesis by a new insect growth regulator buprofezin in *Nilaparvata lugens* Stål. Agric Biol Chem 49: 1233-1234

US Government 1996. An act to amend the federal insecticide, fungicide, and rodenticide act and the federal food, drug, and cosmetic act, and for other purposes. Document AE 2.110:104-170. US Government Printing Office, Washington, DC

Vaeck M., Reybnaerts A., Hofte J., Jansens S., DeBeuckeleer M., Dean C., Zabeau M., van Montagu M. and Leemans J. 1987. A novel mechanism of insect resistance engineered into tobacco. *Nature* 328, 33–37.

van Beck N. A. M. and Hugues P. R. 2003. *Effect of signal sequence* and promoter on the speed of action of a genetically modified Autographa californica nucleopolyhedrovirus expressing the scorpion toxin LqhIT2. *Biol. Control* 27, 53–64.

Van de Veire M, Sterk G, van der Staaij M, Ramakers PMJ, Tirry L 2002. Sequential testing scheme for the assessment of the side-effects of plant protection products on the predatory bug *Orius laevigatus*. BioControl 47:101-113

Veitch N. C. 2007. Isoflavonoids of the leguminosae. *Nat. Prod. Rep.* 24, 417–464.

Vickers R. A.1998. *Proceedings 6th Australisian Applied Entomological Research Conference,* Brisbane, Australia, pp. 409–415.

Walton V. M., Daane K. M., Bentley W. J., Millar J. G., Larsen T. E. and Malakar-Kuenen R. 2006. Pheromone-based mating disruption of Planococcus ficus (Hemiptera: Pseudococcidae) in California vineyards. *J. Econ. Entomol.* 99, 1280–1290.

Wang C. and Leger R. J. S. 2007. *A scorpion neurotoxin increases the potency of a fungal insecticide.* Nature Biotechnol. 25, 1455–1456.

Wang Y. M., Ge F., Liu X. H., Feng F. and Wang L. J. 2005. Auto-infection system for the sweet potato weevil, Cylas formicarius ... (Strand) (Lepidoptera: Lymantriidae) with synthetic sex phero- mone in south China. *Int. J. Pest Manag.* 51, 289–295.

Warrior P. 2000. Living systems as natural crop-protection agents† *Pest Manag. Sci.* 56, 681 687.

Weinzierl R. A. 1998. Biological and biotechnological control of insect pests," ed. by J. E. Rechcigl and N. A. Rechcigl, Lewis Pubs., Boca Raton, pp. 101–121.

WHO: Microbial pest control agent *Bacillus thuringiensis*, 1999.

Williams C. R., Ritchie S. A., Long S. A., Dennison N. and Russell R. C. 2007. Transmission of lecucocytozoon smithi (Sporozoa: Leucocytozoidae) by black flies (Diptera:Simulidae) in New York,USA. *J. Med. Entomol.* 44, 256–262.

Williams L III, Price LD, Manrique V (2003) Toxicity of field-weathered insecticide residues to *Anaphes iole* (Hymenoptera: Mymaridae), an egg parasitoid of *Lygus lineolaris* (Heteroptera: Miridae), and implications for inundative biological control in cotton. Biol Cont 26: 217-223

Wilps H, Levchenko V, Vernigor A 2002. Field trials with Imidacloprid (Confidor) on the Italian locust *Calliptamus italicus* in Kazakhstan. J Appl Entomol 126:436-443

Windels M. B., Chiang H. C. and Furgala B. 1976. *Effects of Nosema pyrausta* on pupa and adult stages of the European corn borer. *J. Invertebr. Pathol.* 27, 239–242.

Wing KD 1988. RH-5849, a nonsteroidal ecdysone agonist: effect on *Drosophila* cell line. Science, WA DC 241: 467-469

Wing KD, Sacher M, Kagaya Y, Tsurubuchi Y, Mulderig L, Connair M, Schnee M 2000. Bioactivation and mode of action of the oxadiazine indoxacarb in insects. Crop Prot 19:537-545

Wing KD, Slawecki RA, Carlson GR 1988. RH-5849, a nonsteroidal ecdysone agonist: effect on larval Lepidoptera. Science, WA DC 241: 470-472

Wislocki PG, Grosso LS, Dybas RA 1989. Environmental aspects of abamectin use in crop protection. In: Campbell WC (ed) Ivermectin and abamectin. New York, Springer-Verlag, pp 182-200

Witzgall, P., Krisch, P., & Cork, A. 2010. Sex pheromones and their impact on pest management. *Journal of Chemical Ecology*, *36*, 80–100.

Workman PJ, Martin NA, Zydenbos SM 2002. Towards integrated pest management of *Thrips tabaci* in onions. New Zealand Plant Prot 55:188-192

Wraight S. P. and Carruthers R. I. 1999.: Biopesticides Use and Delivery," ed. by D. R. Hall and J. J. Menn, Humana Press, Totowa, N.J., pp. 233–269.

Wright DJ, Loy A, Green ASJ and Daybas RA, 1985. The translaminar activity of abamectin (MK-936) against mites and aphids. Meded. Fac Landbouwwet Rijksuniv. Gent 50:633-637

Yamamoto I, Yabuta G, Tomizawa M, Saito T, Miyamoto T, Kagabu S 1995. Molecular mechanism of selective toxicity of nicotinoids and neonicotinoids. J Pestic Sci 20:33-40

Yan Y., Wang G., Yu X., Li S. and Zhang L. 1996. *Proceedings Conf. Technol. Transf. Biol. Control Res. Pract.*, Montpellier, France.

Yanase D, Andoh A (1989) Porphyrin synthesis involvement in dephenyl ether-like mode of action of TNPP-ethyl, a novel phenylpyrazole herbicide. Pestic Biochem Physiol 35: 70-79

Young S. and Kring T. 1991. Selection of healthy and nuclear polyhedrosis virus (NPV) infected Anticarsia gemmatalis. *Entomophaga* 36, 265–273.

Zalom FG, Lanini WT, Miyao G, Davis RM 2001. A continuum of integrated pest management practices in processing tomatoes. Acta Hort 542:55-62

Zehnder G., Gurr G. M., Kuhne S., Wade M. R., Wratten S. D. and Wyss E. 2007. Arthropod Pest Management in Organic Crops. *Annu. Rev. Entomol.* 52, 57–80.

Zlotkin E., Rochat H., Kopeyan C., Miranda F. and Lissitzky S. 1971. Purification and properties of the insect toxin from the venom of the scorpion *Androctonus australis* Hectar. Boichimie 53:1073-1078.

BIO-RATIONAL AND BIOLOGICAL MANAGEMENT OF MITE PESTS

10

Amit Kumar Singh and Uma Shankar

10.0 INTRODUCTION

The phytophagous mites belong to four important families viz., Tetranychidae (spider mites), Tenuipalpidae (false spider mites), Eriophyidae (gall mites) and Tarsonemidae (broad/yellow mites). In vegetable crops, spider mite damage alone causes 10 to 15 per cent loss in yield (Anon., 1991). Out of 37 mite species known to feed upon vegetable crops, six species viz., *Tetranychus cinnabarinus* (Boisduval) (=urticae Koch), *T. neocaledonicus* Andre, *T. ludeni* Zacher, *T. macfarlanei* Baker and Pritchard, *Aceria lycopersici* (Wolff.) and *Polyphagotarsonemus latus* (Banks) are serious pests mostly of vegetables like brinjal, okra, cucurbit, chilli, potato etc. in major parts of the country (Gupta, 1991).

Integrated Pest Management (IPM) is a decision-making process that supports a balanced approach to managing crop production systems. The goal is effective, economical and environmentally sound suppression of pests. The concept of IPM evolved in response to problems caused by an over-reliance on chemical pesticides. Some of these problems are development of pesticide resistance, elimination of natural enemies of pests, outbreaks of formerly suppressed mites, hazards to non-target species, and environmental contamination. The objective of IPM is not to eradicate a mite pest but to reduce its abundance to such levels that it no longer pose an economic threat to crops. IPM is not an 'organic' or non-pesticide approach to pest control (Pal *et al.*, 1989). Organic producers also use IPM and certain approved pesticides to protect their crops. One of the main goals of IPM is to promote the use of effective, less toxic pesticides only if and when necessary.

Identification is essential to select the best mite management practice or acaricide for the mite/crop combination. It is not worthy to use a registered pesticide on a crop or against a pest not listed on the pesticide level. Knowing what natural enemies are present and active in the crop will aid in selecting an acaricide that poses the least risk to them.

1. Monitoring mite populations gives information needed to decide if and when to apply a pesticide for optimal effectiveness. Knowing what and when beneficial organisms are present is useful to time pesticide application to minimize impact on the beneficial organisms.
2. Making control decisions based on pest or damage monitoring data, potential damage, cost of control methods, value of production, impact of other pests, beneficial organisms and the environment. Decide if the cost of the pesticide application is justified in terms of the value of the commodity protected. When making decisions consider the Economic Injury Level (EIL) and of Economic Threshold Level (ETL) where available. The EIL is the pest population level when the loss caused by the pest is equal to the cost of control measures. Both the ETL and EIL values can change from year to year depending on the crop value and control costs.
3. Use a combination of behavioural, biological, chemical, cultural and mechanical methods to reduce pest populations to acceptable levels.
   - *Behavioural control* applies to mites. It takes advantage of insect responses to different sticky traps in case of dispersed mite through wind.
   - *Biological control* uses natural enemies to control pests. Once established, biocontrol agents can often keep their hosts (pests) from becoming problems. If pesticide use is necessary, select products with no or little impact on biocontrol agents. This preserves their presence or allows the establishment of introduced biocontrol agents. The key to biological pest control is effective combinations of different control methods in an IPM program, which can suppress damage from all pests and diseases to below the economic threshold level. This often involves the use of chemical pesticides that are compatible with natural enemies. There is also potential for the selection of strains and species of natural enemies that are resistant to chemical treatments.
   - Although scientists may work on the control of a single pest, farmers in reality are faced with a complex or "guild" of pest

species. Biological control today is looking at these complexes of pest species, and also combinations of plant pests and pathogens. Reduced pesticide use is emphasized, and the protection of natural enemies, in order to allow the natural balance between pests and their predators and parasites to be restored and maintained. Cost and returns will be basic to acceptance by growers of the biological approach to pest control. Low costs essentially depend on effective methods of mass production, and a reasonably long storage life. Quality control is an important aspect of mass rearing.

- *Cultural control* relates to production practices that discourage the introduction, establishment and development of pest populations. It minimizes the need for pesticides. Examples include growing mite tolerant crop varieties (Misra *et al.*, 1990), crop rotation, optimizing plant growth through proper use of plant nutrient, sanitation and clean cultivation. Weeds serve as alternative host of a number of phytophagous (plant feeding mites) (Sarkar *et al.* 2004). Weeds like *Cyanodon dactylon, Ludwigia perviflora, Physalis minima* and *Sonches arvensis* should be taken care off.
- *Mechanical control* uses barriers or devices such as insect nets in seed beds, removing heavily infested foliage at a regular interval. These practices can reduce the need for pesticides.
- *Chemical control* uses pesticides to control pests. Often pesticides are used only when all other available control options have not kept pest populations below the ETL (Rai *et al.* 1996). New less toxic, more target-specific products (often referred to as reduced risk pesticides or bio-rational) are replacing older broad-spectrum synthetic pesticides in order to reduce risks to food, environmental and human safety. However, these products on an average cost more, and require more precise application timing and frequency.

Indiscriminate use of pesticides not only aggravated mite problem, but, often mites develop resistance to a wide range of pesticides and that property often shows cross resistance to chemically related and non-related compounds. Rarely any morphological difference appears in the resistant population. Appearance of resistance is more common with carbonate and 'OP' type compounds and less so with other acaricides.

The principles involved in chemical control of mites are basically the same as applicable in case of insects. Several factors should be borne in mind before going for chemical control of mite, e.g. the extent and nature of damage, presence of natural enemies of mites (Rai *et al.* 1991), weather condition, nature of pest alongwith its population density etc. Acaricides are somewhat specific in their action. Acaricides not only will by direct contact and penetration through the integument but also act as stomach poison and as fumigant. Therefore before adopting chemical control strategy, some factors like, its mode of action, its translocation properties, if any, its' residual property, the crop on which it is to be sprayed, the stage of the mite to be controlled, climatic condition prevailing before and upto spraying, population migration in relation to reinfestation etc. should be considered to assess the efficiency of the acaricide.

Apart from the timing of application of chemicals, the following possible causes may provide unsatisfactory results and promote biorational and biological management of phytophagous mite :

a. The application of unsuitable acaricide.
b. Insufficient coverage of acaricide with the habitat of plant feeding mites.
c. Development of resistance in mites against pesticides.
   The nature of chemical control of plant feeding mite is extremely dynamic because of several element. The susceptibility of various developmental stages of the species to pesticides. The effect of pesticides on the performance of host plant and yield, removal of competitive mite and suppressive action on natural enemies of prime importance.

Acaricidal selectivity

On the basis of the above parameters, acaricides can be broadly classified into three groups :

a. Primarily toxic to egg.
b. Highly toxic to mobile stages.
c. Toxic to both egg and mobile stages.

Acaricides showing toxicity to only egg stages usually are not highly effective to achieve complete control by simple application. Therefore repeated application is essential to combat mixture of all developmental stages. In case of high mite density, a relatively high density of non-egg stages is present, hence, the time required for complete control may be too long and damage on the foliage may increase.

Acaricides toxicity to mobile stages is more effective for commercial control, because the proportion of these stages in the mite population is generally high and the feeding damage is inhibited immediately. Larval, nymphal and adult stages more sufficiently to come into contact with toxic residue, that remains of the foliage.

Acaricides belonging to both ovicidal and adulticidal property may achieve good control by a simple application. But, the limiting factor here may be the growth of the host plant resulting in leaves free from residue and inadequate coverage of the host plant.

10.1 Plant growth and mite activity

The toxicity of acaricides to mite and then host plant may vary during growing season. Rapid expansion of young leaves may limit the effectiveness and efficiency of acaricides.

Distribution of mite population on host plant

Most of the phytophagous mites are photo "–ve" at low population densities. The proportion of population may increase on the upper side at higher densities. It is difficult to achieve a smooth coverage of the underside of foliage by conventional spray equipment.

Sources of re-infestation

Two important sources of reinfestation may occur – (a) reproduction of mites escaped control on the hot plant, and, (b) mites from secondary host plant.

The development of resistance in mites against acaricides is related to the frequency of treatments and to the nature of acaricides used. If an acaricide gives good control and has no side effect, the tendency to continue the use of the acaricides is self evident. The choice between acaricides from different chemical group is limited, thus restricting the possibilities for alterations in the materials used.

The major factor favoured by the short generation cycle of mites, is the rapid development of resistance towards originally effective acaricides, particularly 'OP' type. It is presumed that, lot of research and data generations have been made on the efficacy, mode of action of different chemical groups as effective acaricides, in comparison to insecticides. But in recent times, in anticipating the problems of mites several compounds of diversified chemical nature has come out as mite control agents. It is certain, however, that not only acaricides belonging to various types show differences in their mode of action, in addition, the action of a certain compound can be different at the various stages of development of mites.

In view of the tremendous problems by mites in agricultural crops and phasing out of number of good acaricides on the ground of environmental hazards, scientists are in continuous search of new molecules effective as acaricides. It is therefore, pertinent to look into the recent development in respect to mite control agents.

Pesticides will always be an important part of IPM programs. IPM promotes careful use of pesticides based on an informed decision-making process. This process involves preventing pests and combining control practices based on current information on pest and beneficial presence, seasonal abundance, economic impact, and evaluation of program performance.

Reduced risk pesticides are either synthesized chemicals or chemicals derived from animals, plants, fungi or bacteria (called bio-pesticides). Examples of synthesized reduced risk pesticides are plant and insect growth regulators and chemicals that kill the pest but not the natural enemies (Dhar *et al.*, 2000). The Pest Management Regulatory Agency (PMRA) of Health Canada defines reduced risk pesticides as pesticides "may reasonably be expected to accomplish one or more of the following–

i. Reduced the risks of pesticides to human health.
ii. Reduce the risks of pesticides to non-target organisms.
iii. Reduce the potential for contamination of groundwater, surface water or other valued environmental resources.
iv. Broaden the adoption of integrated pest management strategies, or make such strategies more available or more effective."

Biological enemies of mites and their status and role in vegetable mite management.

The reproductive capacity of mite pests is sufficient to destroy or seriously reduce crop yield in the absence of mortality factors. Several biological enemies are known to have the power to reduce or regulate the population of injurious plant feeding mites.

Out of the various biological agents the predatory mites and insects are more important while the spiders and disease producing pathogens have got minor value.

Acarine Predators

Acarine members belonging to Phytoseiidae, Stigmaeiidae, Bdellidae, Cunaxidae, Chyletide, Tydeidae are predaceous on many injurious plant mites. Out of these predators phytoseiids are recognized as the most valuable and effective on plant mites in many diverse crop systems. Ghai (1974) reported the presence of 64 species of Phytoseiid belonging to 12 genera in India. Gupta (1983) recorded 29 important species of phytoseiids predaceous on phytophagous mites in the country. Out of these *Amblyseius longispinosus, A. fallacis, A. ovalis, A. tetranychivorus, A. channabasavanni, A. largoensis, A. multidentatus, Indoseiulus ricini* and *Typholodromus rickeri* were found capable of suppressing the prey population (Gupta, 1983; Hariyappa, 1985; Borah, 1987).

Phytoseiids are very efficient predators since they have shorter developmental, period than their prey, good searching capacity, ability to survive on lower prey densities and better adaptability to the types of habitats where their preferred prey lives. These predators show a great diversity in feeding habits; thriving on, in addition to prey mites, other preferred plant or plant-derived foods such as pollen, nectare etc. Even some phytoseiids can use eggs and immature stages of certain insects like whitefly, thrips, various fungi as alternate sources of food. Therefore they have better chances of survival than those solely dependent on mite pray. *A. ovalis*, a dominant mite predator of out country was found to have shorter development period and higher reproduction potential. Predatory mites other than the phytosiids as mentioned earlier have limited role in plant mite control. Though some of the species belonging to these families are important bio-control agent in other parts of the world, their role in our country is yet to be exploited.

Arachnid predators (Spiders)

Spiders are almost ubiquitous and have long been known as predators of insect. The spiders belonging to other families Angiopidae, Threidiidae, and Linyphyidae are reported to be efficient predators on mites. *Chaeracautum* sp. of the family Clubionidae was observed to feed on *Polyphagotarsonemus latus* (Borah, 1987). However, not much is known on arachnid predators feeding on plant mites.

Insectan Predators:

Coleoptera

Coccinellids and staphylinids among the coleopteran are the important and efficient predators on mites. Insects of the genus *Stethorus* (Fam : Coccinellidae) are effective only on mites, but they may utilize other foods for survival. Many species are relatively small and remarkably well adapted to live on an search for prey in the habitats of plant feeding mites. In India *S. pauperculus* on *Tetranychus neocaledonychus* (Manjunath, 1983), *Eutetranychus orientalis*, *P. latus* (Hariyappa, 1985) and *S. tetranychivorus* on *Tetranychus* spp. are reported. *S. pauperculus* was found to consume, during the total life period, an overall total of 251.56 eggs and nymphs of *T. macfarlanei* (Jose *et al.,* 1989). Borah (1987) observed higher preference of *Stethorus* sp. to quiescent larvae and other slow moving stages like larva and adult females of *P. latus.* From the perusal of literature it is evident that *S. pauperculus* is an important coccinellid predator on plant feeding mites in India.

Larvae and adults of the beetles belonging to the genus *Oligota* (Fam: Staphylinidae) are recognized predators of injurious mites. *O. flaviceps, O. oviformes* and *O. pymaea* are some of the examples of staphylinid predator. They can consumed different stages of mites. Each larvae can consume 20 mites per day while an adult about 10 mites per day only. *O. oviformes* was reported to feed on *T. neocaledonychus* (Manjunath, 1989) and on *P. latus (*Borah, 1987).

Natural enemies are the base on which any IPM programme is built. The role of natural enemies in mite pest management is regulatory and this bears the advantages like safe, economic and permanency.

Rasmy (1989) suggested to use predaceous mites as bio-control agent in integrated management of mite pests in the following three ways :

1. IPM programmes exploiting the naturally occurring predacious mites,
2. By release of preparatory mites, and
3. By utilizing pesticides resistant strains of predaceous mites.

During the last several decades, the control of pests, predominantly by the use of synthetic chemicals, though broad spectrum pesticides offer powerful incentives in the form of excellent control, increased yield and high economic returns, they have serious drawbacks such as development of resistance to pesticides, resurgence of pests outbreak of secondary pests, harmful effects on human health and environment and presence of undesirable residues. Biological control tends to be long-term and delayed, whereas chemical control tends to be sudden and immediate. Successful programs of biological control always involve information for farmers, pointing out the advantages (*e.g.* cost and net returns, safety and persistence). In Malaysia, Vietnam and Thailand, government programs combine the production of natural enemies with extension information for farmers. This is an effective way of promoting biological control among the small-scale farmers of Asia.

In addition to information, it is the cost of biological control, which will determine its appeal to farmers. If biological control agents are purchased as commercial inputs, their cost must be related to the value of the crop. Biological control must not only be technically feasible, but give farmers a net profit.

In turn, the cost will depend on efficient methods of mass rearing. Research to find the optimal conditions take a long time and regional exchange of such information would reduce the cost of biological control agents. Quality control is a major issue in mass rearing. Storage is another problem. Since biological control agents are living organisms, they tend to have a short shelf life, so they must be mass produced shortly before they are used. This adds to the cost of production and distribution.

Some scientists see chemical control as the cornerstone, on which all biological control strategies must be built. Others see biological control as the best kind of pest control, which should determine how and whether chemicals are used. However, the conservation approach has been found to be remarkably effective. Simply reducing the level of pesticide sprays gives natural enemies the opportunity to increase their numbers and control pests. Sometimes this is enough in itself to control important insect pests. However, it takes time. The question is how long the delay will last, and how long the grower is willing or able to wait.

References

Borah, D.C. 1987. Bio-ecology of *Polyphagotarsonemus latus* (Banks); (Acari: Tarsonemidae) and *Scirtothrips dorsalis* Hood (Thysanoptera: Thripidae) infesting chilli and their natural enemies. Approved Ph.D thesis, Univ. Agric. Sci. Dharwad, pp. 330.

Dhar, T., Dey, P.K. and Sarkar, P.K. 2000. Influence of abiotic factors on population build up of red spider mite, *Tetranychus urticae* on okra vis-à-vis evaluation of some new pesticides for their control. *Pestology*, Vol. 24 (9): 34-37.

Ghai, S. 1974. Predatory mites of India. *In:* (Lecture Note) All India Summer Institute in Acarology held at UAS, Bangalore. June 17- July 13, pp. 55-68.

Gupta, S.K. 1983. The family phytoseiidae in India. pp. 16. *In:* Abstract of papers, 2nd All India Symp. Acarol. Pune, May 26-28, Acarol. Soc. India, UAS, Bangalore.

Hariappa, A.S. 1985. Studies on the natural enemies of Chilli *Polyphagotarsonemus latus* (Banks) and predator prey interaction. Approved M. Sc (Ag.) thesis, Univ. Agric. Sci., Bangalore, pp. 95.

Ishaaya, I. 2001. Biochemical processes related to insecticide action : an overview. In: Ishaaya, I. (Ed.). *Biochemical sites of insecticides action and resistance.* Springer Berlin Heidelberg New York, pp. 1-16.

Jose, V.T., Shah, A.H. and Patel, C.B. 1989. Feeding potentiality of some important predators of the spider mite, *Tetranychus macferlani* a pest of Cotton. *In:* G.P. Channa Basavnna and C.A. Viraktamth (Eds.) *Progress in Acarology*, Vol.2, Oxford and IBH Pub., Co., pp. 357-360.

Manjunath, M. 1989. Bio-ecology and control of spider mite *Tetranychus neocaledonicus* Andre (Acari: Tetranychidae) on vegetable. M. Sc (Agril.) thesis. Univ. Agric. Sc. Bangalore, pp. 188.

Misra, K.K., Sarkar, P.K., Das, T.K. and Somchoudhury, A.K. 1990. Incidence of *Tetranychus cinnabarinus* on some selected accessions of brinjal with special reference to physical basis of resistance. *Indian Agric.*, 34(3): 177-185.

Pal, P.K., Somhoudhury, A.K., Sarkar, P.K., Mukherjee, A.B. and Lohar. N. 1989. Approaches to integrated control of *Tetranychus cinnabarinus* on egg plant. *In:* G.P. Channa Basavnna and C.A. Viraktamth (Eds.) *Progress in Acarology*. Vol.2. Oxford & IBH publishing Co. Pvt. Ltd., New Delhi.

Rasmy, A.H. 1989. Prospects of biological control in integrded mngement of mite pests. *In:* G.P. ChannaBasavnna and C.A. Viraktamth (Eds.) *Progress in Acarology*, Vol. 2, Oxford and IBH Pub., Co., pp. 337-341.

Rai, A.B., Sejalia, A.S., Patel, C.B. and Shah, A.H. 1991. Studies on Okra mite, *Tetranychus macferlanei* and it's chemical control. In: *Modern Acarology*, (Eds. F. Dusbabek and V. Bukva), 580, Vol. I; Academia, Prague and SPB Academic Publishing by the Hague.

Rai, A.B., Bhanderi, G.R. and Patel, C.B. 1996. Effectiveness of neem based product (Azadiractin) against *Tetranychus macferlanei* (Acarina: Tetranychidae) on Okra and their safety to predatory mites. *Neem and Environment*, Vol. 2 : 1127 p. (Eds. Singh, R.P.; Chari, M.S.; Raheja, A.K. and Krus, W). Oxford and IBH Publication Pvt. Ltd, New Delhi.

Sarkar, P.K., Somchoudhury, A. K. and Ghosh, R.K. 2004. Rection of wild tperi, *Physlis minima* L. against Yellow mite *Polyophagotarsonemus latus* (Banks) Acari: Tarsonemidae) in entisole of India. Abs. (No. S6MT03P00). *4th Int. Weed Sci. Cong.*, Durban, Republic of South Africa, June 20-24, 2004. pp. 20.

BREEDING OILSEED CROPS FOR INSECT PESTS RESISTANCE

11

S. K. Gupta

11.0 INTRODUCTION

In practical agriculture, resistance represents the ability of a certain variety to produce larger crop of good quality than an ordinary variety at the same level of insect population. During the 19th Century, the French wine industry survived a major set-back largely due to the use of resistant root stocks imported from America and which still maintain their resistance levels against the beetle pest (Painter 1951). Another important example of resistant variety is the U-4 variety of cotton possessing high level of resistance to the jassid, *Emposasca* spp. (Parnell, 1925). The excessive use of pesticide on oilseed crops needs to be minimized owing to several reasons: (i) most of the pesticides are lipophilic and are likely to be carried in the oil (ii) in a few cases like Brassica, leaves and inflorescence are consumed as saag (pot herb), (iii) the oilseed crops provide the raw material for the vanaspati industry, cosmetics, etc., (iv) many of the oilseeds and oils are directly consumed without further processing, and (v) the oil cakes are used as cattle feed. Thus there is an urgent need to exploit the non-pesticidal methods of pest control, among which the use of resistant varieties assumes a great significance. The details of each oilseed crops are given below:

11.1 Rapeseed-mustard

Among the oleiferous crops, rapeseed and mustard occupy an important place for their value as a source of oil and cake. The yield of these oilseeds is being limited by several factors but ravages caused by the insect-pests are quite considerable. The ill effects of chemical control of insect-pests have necessitated the search for suitable alternatives.

Among these, role of resistant varieties in integrated pest management (IPM) of insect pests has been recognized as very useful and practical preposition. Under the prevailing situations, the resistant varieties comprise the foremost strategies of pest management because of their well documented advantages. In India in the recent past, extensive work has been done on various aspects of HPR to the key pest (*Lipaphis erysimi*) but the work on the other major insect-pests of rapeseed and mustard has been negligible.

11.1.1 Insect Pest Complex

The number of insect pests associated with the rapeseed and mustard crops has increased from 24 (Rai, 1976) to about 40 (Bakhetia and Sekhon, 1989) owing to changed cropping patterns and introduction of new crops like *Brassica napus* and B. *carinata*. The mustard aphid, *Lipaphis erysimi* (Kaltenbach) is the key pest responsible for bulk of yield losses. The other major and minor insect pests in order of their severity are listed in Table 11.1.

Table 11.1. List of important insects pests of rapeseed-mustard

S.No.	Name of the pest	Nature of damage
	Key Pest	
1	Mustard aphid *Lipaphis erysimi* (Kalt)	Sucks sap from leaf, inflorescence ad pods
	Major pests	
2	The saw-fly *Athalia Lugens proxima* (Klug)	Feeds on leaves of young seedlings
3	Painted bug *Bagradan hilaris* (Burm.)	Sucks sap from leaves of seedlings, and also from pods and seeds
4	Pea leaf miner *Chromatomyia horticola* (Goureau)	Mines the leaves
5	Bihar hairy caterpillar *Spilosoma oblique walker*	Defoliates the vegetative parts
	Minor Pests	
6	Cabbage butterfly *Pieris brassicae* (Linn.)	Defoliates the foliage
7	Flea beetle *Phylotreta cruciferae* (Goeze)	Makes holes in young leaves
8	Green peach aphid *Myzus persicae* (Sulzer)	Sucks sap from the leaves flowers and pods
9	Diamond back-moth *Plutella xylostella* (Linn.)	Feeds on leaves and cuts holes
	New Pests	
10	The larger moth *Crocidolomia binotallus* (Zeller)	Damages the inflorescence

S.No.	Name of the pest	Nature of damage
11	Cabbage top borer *Hellula undalis* Fab.	Damages the inflorescence
12	Cabbages aphid *Brevicoryne brassicae* (Linn.)	Sucks sap from the foliage and inflorescence
13	The white fly *Bemisia tabaci* (Linn.)	Sucks sap from the leaves
14	Cabbages semi-looper *Plusia orichalcea* (Fab)	Defoliates the leaves etc.
15	Turnip moth *Agrotis segetum Dennis & Schiff*	Cuts the germinating Seedlings
16	Ants *Monomorium* sp	Takes away the seeds before germination
17	Pod fly *Dasineura brassicae* (Winnertz)	Damages the flowers and pods

Source: Kalia and Gupta (1997): Recent advances in oilseed brassicas.

11.1.1.1 Mustard Aphid

The mustard aphid, being a cosmopolitan, attains more importance on *Brassica* crops in which it can cause yield losses upto 96 per cent under varied agro-ecological situations (Baketetia, 1989; Landin, 1982). The continuous feeding of the mother aphids and their colonies drains out cell sap resulting plants with yellow, curled, crinkled and finally failed to grow.

11.1.1.2 Screening techniques

Various screening criteria have been employed by different authors for evaluation of cruciferous genotypes in order to get the résistance sources. This aspect has been extensively reviewed recently by Bakhetia and Singh (1992) (Table 11.2).

Table 11.2. Resistance source identification criteria used in Brassica species against the mustard aphid.

Technique	Resistance modality	Remarks
Seedling Survival	Non-preference, antibiosis, tolerance	useful in preliminary screening in screen house
Aphid Injury	Resistance, tolerance	Suitable for specific screening on a smaller scale in Screen-house and field too
Aphid Population	Non-preference, antibiosis, tolerance	A far more laborious method suited best for screening on small scale in the field or screen house

Technique	Resistance modality	Remarks
Fecundity and development	Antibiosis	Screening of select germplasm preferably in laboratory or under caged conditions
Grain yield	Tolerance	Best suited for large scale screening in the field or under Specially designed cages
Honey-dew Secretion	Host-food suitability, resistance	Most laborious and time consuming best suited for laboratory testing of a few selected ones only

Source: Kalia and Gupta (1997): Recent advances in oilseed brassicas.

11.2 Seedling survival

This technique was developed by Bakhetia and Bindra (1977) which advocates using the optimum level of aphid population per seedling or plant under the screen house conditions. This technique of resistance screening against the mustard aphid is widely accepted because of its higher efficiency and ease in application at seedling stage, thus has become an integral part of mass screening programme.

11.3 Aphid injury

Aphid damage symptoms expressed as injury grades (0-4) were proposed by Pathak (1961) and Teotia and Lal (1970) while Rajan (1961) and Bakhetia and Sandhu (1973) used a different set of injury grades formed on the basis of injury symptom (Table 11.3).

Table 11.3. Scoring methods used by different workers for aphid resistance screening in rapeseed-mustard (Injury symptoms as explained in different reports)

Injury Grade	Pathak 1961), Teatia and Lal (1970)	Rajan (1961)	Bakhetia and Sandhu (1973)
0	Normal	Plants unaffected by aphids	Free from aphid infestation
1	Yellowing in a few leaves	Apparently healthy plants with a few young ones in the flower	Normal growth no yellowing or curling of leaves, a few aphids without injury Symptoms, flowering and pod setting Almost normal
2	Yellowing and curling of some leaves, medium growth	Plant severely infested, inflorescene and pods full of aphids, no visible symptoms of damage	Average growth, flowering and pod setting; curling and yellowing of few leaves

Injury Grade	Pathak 1961), Teatia and Lal (1970)	Rajan (1961)	Bakhetia and Sandhu (1973)
3	Poor (stunted) Growth and flowers	Plants completely damaged and in different stages of decay	Curling and yellowing of some branches below average growth, poor flowering with very little pod setting
4	Stunting, plants defoliated and	–	Very poor growth, heavy curling of leaves, stunting of plants, a little or no flowering and pods formation. Heavy aphid colonies
5	–	–	Severe stunting of plants, curling and yellowing of almost all leaves. No flowering and pod formation. Plants full of aphids.

Source: Kalia and Gupta (1997): Recent advances in oilseed brassicas.

Aphid population

Pathak (1961) screened the Brassica germplasm at the flowering stage on the basis of the number of aphids/plant part under laboratory conditions which are given in the following grades:

11.4 Identified aphid resistant strains/varieties

There are numerous reports on testing of cruciferous germplasm for resistance to aphid and further resulted in identification of good number of lines showing promise against the mustard aphid under varied climatic conditions (Table 11.4).

Table 11.4. List of lines/strains of cruciferous oilseeds reported as possessing resistance/tolerance against the mustard aphid

Species	Strains/lines	Modality of	Reported by Resistance
B.alba	Lethbridge 22 A Metapollaska	R	Sachan and Singh (1983)
B. campestris	B 787 Pusa Kalyani	R T	Prasad (1983) Singh *et al.* (1982)
B. hirta	ACC 751516	R	Bakhetia (1985)
B. juncea	Laha 101 T 6342	HR	Lal (1969), Teotia and Lal (1970)
		R	Bakhetia & Sandhu (1973) Asthana and Das (1978) Brar and Sandhu (1978) Chander (1993)
	RLM 198	R	Bakhetia (1975), Bakhetia & Bindra (1977), Brar and Sandhu (1978)

	RK 2, RK 9, RH 785	R/T	Bakhetia *et al.* (1982)
	Rai-75 -1	R	Singh *et al.* (1982)
	T 16	R	Chand and Singh (1977)
	IB 680	R	Prasad (1983)
	RC 1425 (non-waxy) B 85 (Glossymutant White flower)	T	Yadava *et al.* (1985), Bakhetia (1985), Chatterjee and Sen-Gupta (1987)
	IM 76-103 , RLC	HR	Ran *et al.* (1989)
	1036, RLC 1037, RJ LM 198, PR 43, RW-white glossy, B 85, RLM 198	R	Rohilla (1990)
	T 6342, RH 7361, RH 7847, T 6342, RE 5	R	Kumar *et al.* (1990)
	RW-2-2, RW-32-2, B 85	R	Chander (1993)
B. napus	PI 171538, GSL 1501, GS 47,	R	Jarvis (1970)
	GS 140, GS 825,	R	Rohilla (1990)
	GSL 8851 & 8965	R	Chander (1993)
C. abyssinica	PI 247310	R	Jarvis (1969)
Eruca sativa	ITSA	HR/T	Bakhetia & Bindra (1977), Narang (1982), Dilawari & Dhaliwal (1988)
	T 27	HR	Chander (1993)

HR = Highly resistant; T = Tolerant; R = Resistant; LS = Low Susceptibility; S = Susceptible; HS = highly susceptible Modified after Bakhetia and Singh (1993)

Source: Kalia and Gupta (1997): Recent advances in oilseed brassicas.

11.5 Morphological traits versus resistance.

The morphological traits of plants imparting relative resistance to aphid injuries are not clearly understood (Table 11.5). More hairy plants of E. *sativa* were not preferred by the aphid for settlement, movement and oviposition and thus acted as a barrier (Rai and Sehgal, 1975; Narang, 1982). Chander (1993) recorded that T27 showed minimum aphid population in the field because of its antibiosis and not due to the presence of hairy structures around the inflorescence. It is concluded that possibly the role of biochemical constituents is of greater significance in the relative resistance of rapeseed-mustard varieties.

Table 11.5. Association of morphological characters of plants with the aphid resistance

Characters	Species	Reaction	Author(s)
Smaller and hardy inflorescence with loose packing of buds	*B. juncea*	Resistance	Rai and Sehgal (1975)
Darkers leaves	*B. juncea* *B. campestris* *(Brown sarson)*	Resistance	Pathak (1961)
Non-wariness (Glossy)	*B. rapa,* *B. napus*	Resistance	Srinivasachar and Malik (1972), Yadava *et al.* (1985) Angadi *et al.* (1987) Bakhetia *et al.* (1992) Narang (1982)
Profuse branching and wider angle of orientation	*B. juncea* *B. campestris*	Negative correlation with aphid population	
Hairiness Yellow or greenish-yellow leaves	*E.sativa* *Brassica* spp	Anti-xenosis Attraction	Narang (1982) Dilawari (1985)
White petal mutant	*B. juncea*	Anti-xenosis and negative correlation with population	Bakhetia *et al.*(1992)
Leaf colour, leaf thickness, plant height, inflorescence (type and colour), leaf size, leaf hairness	*Brassica* spp *E. sativa*	No. correlation with aphid population in the field	Chander (1993)

Source: Kalia and Gupta (1997): *Recent advances in oilseed brassicas.*

11.6 Biochemical constituents versus resistance

Many workers have tried to attribute the aphid resistance phenomenon to the biochemical composition of the plants (Table 11.6 & 11.7). Total sugars and sulphur contents in Brassica species have been found to be co-related negatively with aphid reproduction and population build-up (Bakhetia *et al.* 1982). Higher concentrations of total sugars and iron are known to impart resistance while nitrogen, zinc, copper and manganese seem unrelated (Narang, 1982). It has also been found that total protein, free amino acids, reducing sugars and total sugars were higher in susceptible strains while higher flavanoid contents in *B. napus* imparted resistance to the aphids (Malik, 1981).

In the olfactometer studies, higher and lower release rates of allylisothiocyanate proved as repellent and stimulant, respectively (Dilawari 1985; Dilawari and Atwal 1989b). Similarly, Dilawari and Atwal (1987) had found that the number of probes increased and feed intake drastically reduced on the artificial media containing higher levels of glucosinolates.

Table 11.6. Biochemical constituents versus aphid resistance in cruciferous species

S. No.	Biochemical constituents(s)	Biological effect	Plant species	Author(s)
1	Glucosinolates and sinigrin	Toxicant/growth inhibition/deterr ent	*Brassica* spp.	Anand (1976) Malik (1981) Rani (1981) Kumar *et al.* (1990) Chander (1993)
2	Total sugars, flavanoids, phenols, ascorbic acid K,S, Fe	Growth inhibition and toxicant	*Brassica* spp.	Rani (1981) Narang (1982) Gill and Bakhetia (1985) Chander (1993)
3	Proteins, free amino acids, ash N,P	Stimulants	*Brassica* spp.	Malik (1981, 1988) Gill and Bakhetia (1985)
4	Free amino acids, N,Cu, Zn, Mn	No relation	*E.sativa* *Brassica* spp.	Kundu and Pant (1967), Narang (1982) Weibull and Melin (1990) Chander (1993)
5	Gluconapin, Allyl isothiocyanate, volatile isothiocyanate	Stimulant, repellent, attractant	*Brassica* spp. *E. sativa*	Malik (1981) Dilawari (1985) Chander (1993)

Source: Kalia and Gupta (1997): Recent advances in oilseed brassicas.

Table 11.7. Morphological, anatomical and biochemical factors of resistance in rapeseed-mustard against the leaf miner

S.No.	Character/chemical	Species	Response/effect	References
	MORPHOLOGICAL			
1	Plant apparency wider angles	*Brassica* spp.	Easy host finding	Singh (1985)
2	Thinner and non-hairy leaves	*Brassica* spp.	Host selection	Singh *et al.* (1986)
3	Yellow colour	*Brassica* spp.	Attraction	Singh *et al.* (1986)
	ANATOMICAL			
1	Thick cuticle thick epidermis closely packed nesophyll tissue	*Brassica* spp.	Reduced larval feeding and adult oviposition	Singh *et al.* (1986)
2	Tissue proliferation	B. juncea	Destruction of eggs and larvae	Singh *et al.* (1986)

	BIOCHEMICAL			
1	Carotenoid pigments	*B. juncea*	Visual cues attraction	Singh and Bakhetia (1983)
2	Glucosinolates, volatile ITC	*Brassica* spp.	Oviposition stimulant/ deterrent larval feeding and growth inhibitor, olfactory attractant etc.	Singh (1985) Singh *et al.* (1986)
3	Phenolics and synaptic acids	*Brassica* spp.	Feeding inhibition oviposition deterrent toxicant	
4	Absence of histidine and arginine	*Brassica* spp.	Oviposition inhibition	
5	Progoitrin	*Brassica* spp.	Growth and feeding inhibition	

Source: Kalia and Gupta (1997): Recent advances in oilseed brassicas.

11.7 Resistance to groundnut pests

It has been reported that the bunch varieties had higher population of, and showed more injury owing to, leaf-webber, aphid and grey weevil, whereas the spreading and semi-spreading types harboured lower populations (Bakhetia, 1969-1971, 1977, 1978; Brar and Sandhu, 1975). This difference may be due to more of foliage and more of succulence of the bunch types which provide conditions favourable for providing with better shelter and preferred food.

11.8 Resistance to sesame leaf-roller

It has been reported that the branching pattern and maturity of the plants are positively correlated with their resistance to leaf-roller. Profuse branching and early maturing varieties provide conditions which are more conducive to pest development. The late maturing varieties NP 29, C 1036 and Chanda 3 have been reported to the less susceptible to the pest.

11.9 Breeding for resistance to insect-pests

A breeding programme for resistance to insects depends on their behavior and the nature of the crop involved as well as on the information available at the beginning. Painter (1951) suggested the following steps for such a programme:

i. Discover possible sources of resistance to insects among varieties and strains locally available.

ii. Search for new germplasm that may carry resistance.
iii. New selections out of older varieties.
iv. Plant introductions out of older varieties.
v. Varieties of related species of plants, if necessary.
vi. Determination of some of the basic properties of the plants responsible for resistance; at least a separation of the three classes: preference, antibiosis and tolerance.
vii. Hybridization to combine genes for resistance with desirable agronomic characters.
viii. Study of genetics of resistance to insects where possible.
ix. The testing of resistance in advanced-generation hybrids.
x. The study of resistance of released varieties in plots and on farms to evaluate as an insect control method.

In case where selection for resistance to a particular pest is difficult, perhaps because assessment is complex or environmental effects mask the genetic expression of resistance. In such cases, it may be possible to use indirect selection to achieve the objectives by selecting for early recognizable characters which is highly correlated with resistance or pleiotropic effects. Either way they can provide convenient method for the selecting insect resistance strains.

11.10 Conclusion

The more important oilseed crops comprise groundnut, rapeseed-mustard and sunflower which occupy a pride place in the present-day agriculture. These are valued for their edible oils for human consumption and the cakes for animals. In India, the former two crops contribute largely to the oilseed production while sunflower is gaining importance. The important cruciferous oilseeds grown on commercial scale world over include Brassica *campestris*, B. *napus*, B. *juncea* and B. *carinata*. Resistant varieties are very useful in pest management. The varieties reported to be resistant to a particular pest (or pests) should be introduced directly for cultivation in certain areas. Such resistant varieties should be used extensively in a hybridization programme aimed at incorporating pest resistance into the well adapted varieties of the region. Similarly, the species known to carry pest resistance may be used in inter-specific or intergeneric crosses through the genetic and cytogenetic techniques known for chromosomal manipulations. Mutation breeding is also an important tool for developing new varieties with desired characters of pest and disease resistance.

Among the alternative approaches, use of host plant resistance (HPR) in IPM programme has been given considerable importance in the past two decades, which has resulted in the development and standardization of suitable techniques for germplasm screening in the screen house and fields as well.

Despite these endless scientific endeavours, it is unfortunate that resistant/tolerant parents have not been evolved which could be possibly used in conventional breeding programmes to develop new resistant varieties against the mustard aphid. However, the identification of resistance imparting gene would solve the problem through more inter-generic transfer of genes from Brassica species. The more recent advancements in biotechnology have explored the possibility of transferring the foreign genes like NPT II, DHFR (mouse), napin and NOS into B. *napus* and B. *oleracea* plants directly or through *Agrobacterium tumefaciens* mediated genes transfers.

References

Anand, I.J. 1976. Oil quality vis-à-vis aphid resistance in Brassicae. Paper presented in the seminar-cum-workshop on oilseed research in India. PKV Nagpur, April, 5-9, 1976.

Asthana, K.S., Dass, S.R. 1978. Varietal evaluation of rapeseed-mustard to aphids (L. erysimi). Proc. Bihar Academy Agric. Sci. 24: 184-186.

Bakhetia, D.R.C., 1989. Current status of breeding for aphid resistance in rapeseed-mustard. In Bakhetia D.R.C., Kurl, S.P. and Rajendran, T.P. (Eds). The mustard aphid. Aphidological Society, India. MMPG College, Modinagar.

Bakhetia, D.R.C., Bindra, O.S. 1977. Screening technique for aphid resistance in Brassica crops. SABRAO J. 9: 91-107.

Bakhetia, D.R.C., Sandhu, R.S. 1971. Preliminary studies on the resistance of some Brassica species to the leaf miner, *Phytomyza horticola* (Meigen) infestation. Proc. 58th Indian Sci. Congress Part II, p. 753 (Abstract).

Bakhetia, D.R.C., Sandhu, R.S. 1977. Susceptibility of some Brassica species and Crambe abyssinica to the leaf miner, *Phytomyza horticola* (Meigen). Crop Improv. 4: 221-223.

Bakhetia, D.R.C., Sekhon, B.S. 1989. Insect pests and their management in rapeseed-mustard J. Oilseeds Res. 6: 269-299.

Bakhetia, D.R.C., Sekhon, B.S., Chatterjee, S.D. 1992. Management of mustard aphid: Review of the work from 1979-80 to 1989-90. Summing up concluding seminar on Indo-Swedish Collaborative Research Programme on Rapeseed-mustard Improvement. Organized by ICAR, Swedish Embassy and Swedish University of Agricultural Sciences, Uppasala, Sweden.

Bakhetia, D.R.C., Singh P. 1982. Host plant resistance in oilseeds, pp: 208-248. *In:* Dhaliwal, G.S. and Dilawari, V.K. (Eds) Adv. Host. Plant Resistance to Insects. Kalyani Publishers, New Delhi.

Bakhetia, D.R.C., Sushma Rani, Ahuja, K.L. 1982. Effect of suphur nutrition of some Brassica plants on their resistance response to the mustard aphid, L. erysimi (Kalt). Calicut Univ. Res. J. (Special Conference Number), May 3-5, 1982, p. 36.

Brar, K.S. Rataul, H.S. Labana, K.S. 1975. Differential reaction of mustard aphid Lipaphis erysimi (Kalt) to different rapeseed and mustard varieties under natural and artificial infestation. J. Res. Punjab Agric. Univ. 13: 14-18.

Chander, H., 1993. Identification and characterization of sources of resistance in some crucifers against the mustard aphid, Lipaphis erysimi (Kaltenbach), Ph.D. Dissertation, Punjab Agricultural University, Ludhiana 140 pp.

Dilawari, V.K. 1985. Role of glucosinolates in host selection, development and multiplication of mustard aphid, L. erysimi (Kalt). Ph.D. Dissertation, Punjab Agric. Univ. Ludhiana, 111 p.

Dilawari, V.K., Atwal, A.S. 1987. Effects of cruciferous glocosinolates on probing pattern and feed uptake by mustard aphid, L. erysimi (Kalt). Proc. Indian Acad. Sci. (Animal Sciences) 96: 695-703.

Dilawari, V.K., Atwal, A.S. 1989a. Response of mustard aphid, L. erysimi (Kalt) to visual stimulus –colour. J. Res. Punjab Agric. Univ. 26: 231-236.

Dilawari, V.K., Atwal, A.S. 1989b. Response of mustard aphid, L. erysimi (Kalt) to visual stimulus –colour. J. Res. Punjab Agric. Univ. 2: 103-108.

Dilawari, V.K.,Dhaliwal, G.S. 1988. Population build-up of mustard aphid, Lipaphis erysimi (Kaltenbach) on cruciferous cultivars. J. insect. Sci. 1: 149-153.

Downey, R.K. 1987. Breeding Brassica oilseeds crops. IRDC: MR 153e: 321-325.

Gill, R.S., Bakhetia, D.R.C. 1985. Resistance of some Brassica *napus* and *B.campestris* strains to Lipaphis erysimi (Kalt.). J. Oilseeds Res. 2: 227-239.

Jarvis, J.L. (1969). Differential response of introductions of Crambe to the turnip aphid. J.Econ. . Ent. 63: 1498-1502.

Kumar, V., Singh, D., Singh, H., Singh, H. 1990. Combining ability analysis for resistance parameters to mustard aphid (Lipaphis erysimi (Kalt.). Indian J. Agric. Sci. 57: 227-279.

Kundu, G.G., Pant, N.C. 1967. Studies on the Lipaphis erysimi (Kaltenbach) with special reference to insect plant relationship. I. Susceptibility of different varieties of Brassica and Eruca sativa to the mustard aphid infestation. Indian. J. Ent. 29: 241-251.

Landin, J. 1982. Life history, population dynamics and dispersal in the mustard aphid Lipaphis erysimi: A literature study, 18pp, Swedish University of Agricultural Sciences, Research Information Centre, Uppasala, Sweden.

Lindley, G. 1831. A guide to the orchard and kitchen garden. Longmans Press, London, 601 pp.

Malik, R.S. 1981. Morphological, anatomical and biochemical bases of mustard aphid, Lipaphis erysimi (Kalt.) resistance in cruciferous species. Sver. Utsad. Tidsk. 91; 25-35.

Narang, D.D. 1982. Studies on the basis of resistance in Brassica *campestris* var. yellow sarson, B. juncea and Eruca sativa to mustard aphid, Lipaphis erysimi (Kaltenbach). Ph. D. Dissertation, Punjab Agricultural University, Ludhiana 103 p. Painter, R.H. 1951. Insect Resistance in Crop Plants. The University press of Kansas, Lawrence, pp. 520.

Parnell, F.R. 1925. Breeding jassid resistant cottons. Emp. Cotton Growers Rev. 2: 330-336.

Pathak, M.D. 1961. Preliminary notes on the differential response of yellow and brown sarson and rai to mustard aphid (Lipaphis erysimi Kaltnbach). Indian Oilseeds J. 5:39-43.

Prasad, S.K. 1983. Varietal susceptibility of rapeseed and mustard cultivars to the aphid (L. erysimi Kalt.). Indian J. Ent. 45; 501-503.

Rai, B.K. 1976. Pests of Oilseed Crops in India and their Control, ICAR, New Delhi, pp. 121.

Rai, B., Sehgal, V.K. 1975. Field resistance of Brassica germplasm to mustard aphid, Lipaphis erysimi (Kalt.). Sci. & Cult. 41: 444-445.

Rajan, S.S. 1961. Aphid resistance of tetraploid toria. Indian Oilseeds J. 5: 251-255.

Ram, S., Gupta M.P., Maurya, R.P. 1989. Mustard varieties resistant to aphid (Lipaphis erysimi Kalt.) in India. Trop. Pest. Managm. 35: 150-153.

Rani, S. 1981. Role of sulphur nutrition in Brassica species for aphid resistance, M.Sc. Thesis, Punjab Agricultural University, Ludhiana, 74 pp.

Sachan, G.C., Singh C.P., 1983. Relative susceptibility of some exotic Brassica and sinapsis species to mustard sawfly, aphid and flea bettle. Annual Report Res. Dir. Ext. Stn. G.B. Pant Univ. Agric. Tech. Pantnagar.

Singh, P., Bakhetia, D.R.C. 1983. Comparative development and host selection mechanism of the pea leaf miner, Phytomyza horticola Goureau on rapeseed and mustard. Proc. 6th Int. Rapeseed Conf. 17-19 May 1983, Paris, pp. 1135-1141.

Singh P., Bakhetia D.R.C., Sidhu H.S., Setia R.C. 1986. Plant apparency and physical defences in Brassica species as index for resistance against the leaf miner, Chromatomyia horticola Goureau. In: Namopeth B. and Subhadvabandhu S. (Eds). New Frontiers in Breeding Researches; Kasetsart University, Bangkok, Thailand, pp. 755-772.

Singh, P., Mavi, G.S., Bakhetia, D.R.C., Sidhu, H.S. 1985. Identification of leaf miner, Chromatomyia horticola Goureau punctures in Brassica strains. Annals Biol. 1:226-228.

Singh, R.N., Dass, R., Saran, G., Singh, R.K. 1982. Differential response of mustard varieties to Lipaphis erysimi (Kalt.) Indian J. Ent. 44: 408.

Teotia, T.P.S., Lal., O.P. 1970. Differential response of different varieties and strains of Brassicae to aphid *Lipaphis erysimi* (Kalt.). Labdev J. Sci. Technol. 8B: 219-226. Weibull, J., Melin, G. 1990. Free amino acid content of phloem sap from Brassica plants in relation to performance of Lipaphis erysimi. Ann. Biol. 116: 417-423.

Yadava, A.K. Singh, H., Yadava, T.P. 1985. Inheritance of non-waxy trait in Indian mustard and its reaction to aphids. J. Oilseeds Res. 2: 339-342.

UTILIZATION OF ENTOMOPATHOGENIC NEMATODES IN INSECT PEST MANAGEMENT

12

Virender Kaul and Uma Shankar

12.0 INTRODUCTION

Entomopathogenic nematodes of the genera *Steinernema* and *Heterorhabditis* (Nematoda: Rhabditida) have emerged as excellent insect biological control agents. They are obligate parasites of insects and kill their hosts with the aid of bacteria. The potential of EPN as a safer alternative to insect control has been proved and are being utilized for control of several pests in many parts of the world. The nematode-bacterium complex kills insects rapidly. This rapid mortality permits the nematodes to exploit a range of hosts that spans nearly all insect orders, a spectrum of activity well beyond that of any other microbial control agent.

The losses due to insect pests and diseases are one of the major constraints in agricultural production in India. Growing awareness about public safety, soil and water pollution, insecticide resistance, effects on non-target organisms and degradation of pesticides have increased the pressure to shift from chemical intensive management to safer alternative control strategies. Entomopathogens as biological control agents are receiving a great deal of attention for they provide environmentally safe insect control. The use of most entomopathogens has focused on aboveground pests. Several entomopathogens (e.g., *Bacillus popilliae*, *B. thuringiensis* subsp. *japonensis* Buibui strain, *Beauveria* spp., *Metarhizium* spp.) are available for the control of soil dwelling pests, but to date the most successful have been the Etomopathogenic nematodes. Several species have been used as bioinsecticides against soil pests.

EPN of the genera *Steinernema* and *Heterorhabditis* (Nematoda: Rhabditida) have emerged as excellent insect biological control agents. They are obligate parasites of insects and kill their hosts with the aid of bacteria. The infective juvenile nematode (IJ), the only free living stage, enters the host *via* natural openings or insect cuticle and release bacteria responsible for killing. Nematodes complete 1- 3 generations within and IJ's exit to find new hosts. The two nematode genera differ in their life cycle. Nematodes and bacteria have an obligate mutualism, bacteria provide nutrients to the nematodes, produce antibiotics that inhibit competing microbes, and kill the host through septicemia. Nematodes may also contribute to host death through suppression of immune system or toxin production; importantly they serve as vectors for bacteria. Without the nematode the bacteria cannot survive in the natural environment. These nematodes possess a number of attractive qualities as biocontrol agents like a durable infective stage, host seeking ability, suitability to mass production, and safety to mammals and other non-target organisms, which allows exemption from registration.

12.1 Natural occurrence of EPN in India

Poinar *et al.* (1992) detected *Heterorhabditis indica* in sugarcane fields at Coimbatore. Extensive surveys have been taken up by PDBC since 1996 in different parts of the country and documented *S. carpocapsae*, *S. abbasi*, *S.bicornutum*, *S. tami*, Steinernema sp. SSL2 and H. indica in the country. Surveys have revealed natural occurrence of EPN from Kerala (Banu *et al.,* 1998), Rajasthan (Rajkumar *et al.*, 2001; Parihar *et al.*, 2002), IARI, New Delhi (Ganguly and Singh, 2000), Tamil Nadu (Joseph, 1993; Ambika, 1995) and Central India (Gokte Narkhedkar *et al.*, 2001). Three new species of *Steinernema* are described from India *Steinernema thermophilum* was described from Delhi by Ganguly and Singh (2000) and *S.seemae* and *S.masoodi* from Kanpur (Ali, 2005). Many other species are likely to add. India being a mega diversity country, all efforts is to be made to exploit the full potential of EPN in the country. The diverse agro-climatic zones in the country ensure the wide genetic variability necessary in the population. Exploratory surveys are to be conducted in many parts of the country like northeastern region and islands because these areas are by and large remained unexplored. And the indigenous isolates will be well adapted to the prevailing local conditions. Conservation and augmentation of naturally occurring entomopathogenic nematode populations through agricultural practices show promise for insect control.

12.1.1 Strains and Host Range

They are ubiquitous in distribution having been recovered from soils throughout the world. Most commonly recovered from soil by baiting with susceptible insects some have recovered from infected insects. Nematode strains can be maintained by repeated subcultures through *G. mellonella* larvae. Fifty species of *Steinernema* and a dozen of *Heterorhabditis* have been described from different parts world and also a number of them have been recorded from India (Annexure I).

The nematode-bacterium complex kills insects so rapidly that the nematodes do not form the intimate, highly adapted, host-parasite relationship characteristic of other insect-nematode associations, e.g., mermithids. This rapid mortality permits the nematodes to exploit a range of hosts that spans nearly all insect orders a spectrum of activity well beyond that of any other microbial control agent. In laboratory tests, *S. carpocapsae* alone infected more than 250 species of insects from over 75 families in 11 orders (Poinar 1979)

The nematodes attack a far wider spectrum of insects in the lab where host contact is assured, environmental conditions are optimal, and no ecological or behavioural barriers to infection exist (Kaya & Gaugler 1993, Gaugler *et al.*, 1997) For example, foliage feeding lepidopteran larvae are highly susceptible to infection in petridishes, but are seldom impacted in the field, where nematodes tend to be quickly inactivated by the environmental extremes (i.e., desiccation, radiation, temperature) characteristic of exposed foliage. Although EPNs possess an extremely broad laboratory host range, ecological and behavioural barriers restrict their natural host range. Temporal and spatial factors, such as synchronous life cycles and differing habitat preferences, also appear to determine the host range. Behavioural barriers also restrict nematode efficacy to a few selected hosts or host groups. Some species are sit-and –wait strategists or ambushers that tend to stay near the soil surface where their specialized foraging behaviours (nictation, jumping) allow them to efficiently infect mobile host species (e.g., *S. carpocapsae* and *S. scapterisci*). Other species are widely searching foragers or cruisers that distribute themselves actively throughout the soil profile and are well adapted to infecting sessile or less mobile hosts (e.g., *H. bacteriophora* and *S. glaseri*) (Grewal *et al.*, 1994) Cruisers, which are highly mobile, respond strongly to long-range host chemical cues, and are therefore best adapted to find sedentary hosts (Grewal *et al.*, 1994). Ambushers and cruisers differ in their responses to host associated cues. Contact cues (host faeces or cuticle) are short range and provide specific information about the host. Volatile host cues are long range and usually provide general, directional information for host location. Cruisers have strong attraction to volatile cues containing CO_2 whereas ambushers rely on host movement for successful parasitization.

12.1.2 Infection

Location and entry of the non - feeding IJs into haemocoel is the first step in infection. This is followed by the second stage where the recovery of the IJ and the release of the symbiotic bacteria into the haemolymph take place. The nematode exits from the dauer stage and reinitiates development called recovery. *Heterorhabditis* IJs develop into self-fertilizing hermaphrodites while *Steinernema* into amphimictic male or female adults. Food signals may control different stages of nematode development. It is likely that dominant recovery signals originate in the haemolymph of the insect. Signals that control nematode development are also produced. *In vitro* juvenile recovery depends on the signals secreted by the symbiotic bacteria. This food signal was not until the bacteria reached the late exponential phase of growth. Second stage is characterized by the growth of the bacteria to high cell density and the bioconversion of the insect cadaver into a nutrient source suitable for nematode reproduction and development. This takes place within 24-48 h after entry of IJ in the terrestrial organisms, adopted unique survival mechanisms to resist environmental extremes.

Nematodes can survive unfavorable haemolymph. The last stage in infection is distinguished by the formation of IJ stage and the colonization of the IJ by the partner bacteria. In EPN the formation of IJs is critical for symbiosis. Colonization requires specific interaction between the nematode gut and the bacterial cell surface.

12.2.3 Habitat preferences

All organisms have specific niche requirements that will be satisfied only in particular habitats. Also, the soil habitat has three dimensions, so preferences may extend to occurrence at particular depths in the soil. For example, *S. carpocapsae* was recovered primarily near the soil surface in turf grass and that recovery varied during the day., while *H. bacteriophora* was recovered uniformly throughout the upper 8cm of soil. Habitat preferences are likely to reflect the distribution of suitable hosts. There is now general support for the opinion that EPN are more restricted in their host range than lab infections suggest. Each species will be physiologically and behaviourally adapted for survival only in particular habitats. Thus *H.indica* occurred mainly in calcareous soils while *H.bacteriophora* was found in more acidic conditions. *H.indica* is not restricted by vegetation type .The association with sandy coastal soils remains the most robust correlation between the habitat and habitat and presence of heterorhabditids. However *H.bacteriophora* depends less on coastal region than either *H.indica* or *H.megidis* complex and is widely distributed in turf and weedy habitats . In another place *H.bacteriophora* was found at low altitudes and displayed no habitat preference as it was recorded from cropland, woodland, pasture and orchard and native vegetation.

12.2.4 Survival and Persistence

The natural habitat -the soil- is a difficult environment for persistence of EPN considering the complexity of components. However EPN have been isolated from soil from sub arctic to arid, and temperate to tropical climates. Unfavourable environmental conditions include absence of water, extreme temperatures, lack of oxygen, and osmotic stress affect their survival. Besides, the persistence of these nematode populations depends on the recycling of nematodes and dauer survival. The infective or dauer stage is the only free-living stage and is responsible for new infections. Dauer persistence in the soil varies among nematode species and strains, and those differences are related to physiological and behavioral adaptations. Nematode species that reside closer to the soil surface have evolved mechanisms that increase persistence despite desiccating conditions (high temperature and low moisture) and exposure to UV light. Nematode species that move actively beneath the soil surface have evolved mechanisms to decrease their susceptibility to abundant pathogens beneath the soil surface and to use energy efficiently. Extreme conditions (principally temperature and moisture) will affect dauer persistence negatively, but some species and strains are better adapted to extreme environmental conditions. Dauer juveniles of all nematode species are known to enter a state of physiological inactivity because of extreme environmental conditions, and, generally, this inactivity increases persistence. Natural enemies (pathogens, predators) affect dauers negatively, and competition among nematode species affects persistence by limiting access to microsites suitable for persistence. Of the various edaphic factors that affect dauer persistence, soil moisture and temperature appear to be the most important (Kaya 1990, Curran 1993). Desiccation does not affect all nematode species equally and the ambusher, *S. carpocapsae*, is less susceptible than the cruiser, *Heterorhabditis* species (Smits *et al.*, 1996). Temperatures above 30°C or below 9 to 10°C affect persistence negatively, but the severity of the impact depends on species and strain (Kaya 1990). The low persistence of nematodes after applications (either inundative or inoculative) generally results in low efficacy (Baur and Kaya 199). Dauer juvenile stage is adapted morphologically and physiologically to remain in the environment without feeding searching for a new food source and is also the infective juvenile stage in EPN, provides protection to symbiotic bacteria carried in intestine. Survival and the rate of water loss during desiccation are related to cuticle ultra structure. IJ external openings –mouth and anus- are closed preventing penetration by microbial antagonists and toxic chemicals. Ijs are dependent on the stored energy reserves accumulated in the body for energy as they do not feed. The quality and quantity of energy reserve in the form of lipid is another important factor which determines the survival and persistence of nematode IJs in the soil. Variation in nutritional status and energy stores of dauers within and between cohorts influences their ability to persist. The variation in persistence between species and strains of steinernematid and heterorhabditid nematodes results from differences in physiological and behavioral adaptations. Inactivity caused by low temperature or desiccation increases persistence of most entomopathogenic nematode species and strains.

12.2.5 Symbiotic bacteria

Association of EPN with bacteria of genera *Xenorhabdus* and *Photorhabdus* (Enterobacteriaceae) is a a pre requisite in pathogenicity .Bacteria produce virulence factors responsible for the death of the insect larva. They are gram-negative non-spore forming pathogens, long peritrichous rods and facultatively anaerobic. Nematode IJs carry the bacterial symbionts monoxenically within intestine or in a separate pouch. Invasion of insect releases the bacterium into the haemocoel. Bacteria proliferate, kill the insect, and establish conditions suitable for nematode reproduction. They may exist as two or more forms with differences in biochemical and physiological properties. Phase I bacteria produce antibiotics inhibiting other microorganisms proliferating within the cadaver, and thus limiting competition for nutrients. Phase I and II are distinguished as pest morphology, physiology, dye uptake, and antibiotic production. Phase variation is suggested for survival of bacteria, as phase I is more susceptible to phage virus. Nematodes do not reproduce in axenic cultures. Specificity of bacteria in the association has been known. Metabolites produced are anti fungal and anti bacterial and have been identified in most cases.

12.2.6 Mass production

Amenability of EPN to mass production *in vivo* as well as in *vitro* make them promising as commercially viable biopesticides. These nematodes were cultured more than 70 years ago and currently they are successfully being produced using three culture methods: *in vivo* and *in vitro* solid and liquid culture methods.

In vivo production is based on Dutky's adaptation of the White trap of 1927, with some modifications, in which nematode-killed hosts are placed above a water reservoir. The method exploits the tendency of infective nematodes to migrate from depleted host cadavers into the reservoir, which is decanted to collect infective juveniles. This system is appropriate for laboratory bench scale production of inoculums for experiments, but its labour-intensive nature makes it inefficient for large-scale production

The most common host insect used for lab and commercial EPN culture is the last instar of greater wax moth, *G. mellonella*, because of its high susceptibility to most nematodes, wide availability, ease in rearing and its ability to produce high yields. There are only two EPN that are not amenable to culture in *G.mellonella* due to extremes in host specificity. *S.kushidai* is most amenable to culture in scarab beetle larvae and *S. scapterisci* is most amenable to mole crickets (Mamiya,1980; Grewal *et al.*, 1999). Other hosts in which in which *in vivo* production has been studied include the navel orangeworm(*Amyelois transitella*), tobacco pink bollworm (*Heliothes virescens*), cabbage looper (*T.ni*), pink bollworm (*Pectinophora gossypiella*), beet armyworm(*Spodoptera exigua*), corn earworm *Helicoverpa zea*, gypsy moth, *Lymantria dispar*, house cricket *Acheta domesticus*, and various beetles including the yellow meal worm, *Tenebrio molitor* (Shapiro & Gaugler, 2002) elsewhere in the world. In India several workers have studied the suitability of different insect hosts for *in vivo* production of EPN. Hussaini *et al.*, (1998, 2000) studied the efficacy of final instar larvae of *G. mellonella, Corcyra cephalanica*, *Agrotis ipsilon*, *Spodoptera litura, H.armigera*, and *Plutella xylostella* for EPN multiplication. Karunakar *et al.* (1992) reared *S. feltiae, S. glaseri* and *H. bacteriophora* on *C. sacchariphagus indica* and *S. excerptali. In vivo* culturing of *H. bacteriophora* and *S. carpocapsae* was studied in larvae of silk worm, *Bombyx mori* (Zaki *et al.*, 2000). Rajkumar *et al.* al., (2002) and Singh and Yadav studied multiplication of local isolates of Rajasthan (*H. bacteriophora* and *Steinernema* spp.) in *G. mellonella.*

Glaser published first results of experiments on the in vitro culture of S. glaseri that time scientists were still unaware of the existence of the symbiotic bacteria. A significant breakthrough in the nematode research was the discovery of the symbiotic bacterium by Poinar and Thomas and the development of in vitro production techniques applicable to nematodes of both the genera. The first successful commercial scale monoxenic culture was developed by Bedding (1981, 1984) In this method, nematodes are cultured on a crumbed polyether polyurethane sponge impregnated with artificial medium along with symbiotic bacteria. In vitro mass production based on Bedding (1984) method was standardized for different indigenous isolates of steinernema and heterorhabditid nematodes (Hussaini *et al.*, 2002; Vyas *et al.* 1999).

Friedmann reported the development of a liquid fermentation technique for large scale production of nematodes (Friedmann, 1990). In this method, cost of production decrease rapidly. The current yields in liquid fermentation process average at about 2.5×10$_5$ infective juveniles/g in case of *S. carpocapsae*. Presently *S. carpocapsae*, *S. feltiae*, *S. glaseri* and *S. riobrave* can be consistently and efficiently produced in 7500-80,000 liter fermentors with a yield capacity as high as 1, 50,000 IJs/ml (Georgis and Manweiler, 1994). Though liquid fermentation process is highly efficient and economical involves high capital cost and technical expertise.

12.2.7 Product development - economics

Biological pesticide must be able to compete in relative terms with chemical pesticide in field efficacy, cost and application methods (Georgis, 1990b).

The development of cost effective mass production technology has led to the availability of nematode products comparable with standard insecticides in market. Quality control strategy has to ensure preservation of high nematode viability and virulence during large-scale production and formulation. Factors affecting quality included provision of essential nutrients and appropriate bacteria, delivery of oxygen, production and storage temperature and contamination control (Grewal, 1996).

12.2.8 Formulation and storage

Formulation of nematode into a stable product has played a significant role in commercialization. Active nematodes are immobilized to prevent depletion of their lipid and glycogen reserves. Immobilization has been accomplished by maintenance in aqueous suspension at low temperature (5-15 °C). Steinernematids can be stored at 4-10 °C for 6-12 months without much loss of activity whereas heterorhabditids do not store well and 2-4 months of storage at 4-10 °C is considered good. A major constraint of commercialization of these nematodes is their poor storage capacity. They can be stored under refrigeration in oxygenated water for up to 5 years (Friedman, 1990). But this methodology is not commercially feasible. Nematode metabolism is temperature driven and warm temperature (20-30 °C) increase metabolic activities, reducing nematode viability (Georgis, 1990a). Subsequently, nematodes formulated in various carriers such as alginate, clay, vermiculite and gel forming polyacrylamides can be stored for at least 6 months under refrigeration and at least 3 months at room temperature. Further advancement has been made with the development of a flowable concentrate. This technological advancement, if commercially successful, will revolutionize nematode applications and eliminate many of the constraints associated with current formulation, namely the dissolution of the carrier or separation of the nematodes from the carriers.

Formulation of nematodes is important and can affect handling, application, persistence and stability in storage. Most common formulation is made in alginate, polyacrylamide gels, clay, vermiculite, activated charcoal (Georgis, 1990b). Partially desiccating and immobilizing infective juveniles in specific carriers also achieve formulation stability.

Compared to many biocontrol agents nematodes as biopesticides have a good shelf life. Most can be stored under refrigeration in aqueous suspension under adequate supply of oxygen. Temperature requirement depends on species. General range for steinernematids is 5-10°C and 10-15°C for heterorhabditids (Georgis, 1990b). The introduction of commercial nematode formulations has accelerated in recent years, but new formulation technology that produces stable, effective and consistent products would expand markets even further (Georgis 1990a; Georgis and Hague, 1991). Spray application using a nematode suspension in water is straightforward, inexpensive and often effective. Novel alginate capsules containing nematodes (Kaya and Nelson, 1985; Poinar *et al* .1985; Kaya *et al.* 1987), wheat bran bait pellets (Capinera *et al.* 1988) and alfalfa-wheat bait pellets (Capinera and Hibbard, 1987) have been developed.

Kaya and Nelson (1985) suggested that encapsulation of entomophilic nematodes with seeds has potential to protect roots from insect attack. Nematodes escaped from the capsule within 7 days, as the seed germinated, it has an opening for the nematodes to escape. Pesta, granular product with entrapped nematodes based on cohesive dough made wheat flour, fillers and additives was developed by Connick *et al.* (1993).

12.2.9 Formulation

Several formulations have been developed for the storage and application of entomopathogenic nematodes. The shelf-life of different nematode-based products varies depending on the formulation, nematode species and temperature. Different carriers include polyether polyurethane sponge, peat, vermiculite, activated charcoal, clay and cedar shavings etc. Nematodes held on the sponge need to hand-squeezed into water before application, whereas from other carriers they may be applied directly to the soil. During the past decade, significant progress was made in developing nematode formulations with improved shelf-life and ease of use. A significant advancement was made with the advent of a water dispersible granule formulation consisting of different mixtures of various types of silica, clays, cellulose, lignin and starches (Silver *et al.*, 1995). Studies on nematode formulation are limited in India. Talc and sponge based formulation were developed at Project Directorate of Biological Control, Bangalore with a shelf-life of 3-6 months depending nematode species, dosage and storage (Hussaini *et al.*, 2002b). Storage and application of EPN in infected cadavers is a promising approach for insect pest suppression. Significant control of the banded cucumber beetle, *Diabrotica balteata* Le Conte (Creighton and Fassuliotis, 1985), and the sweet potato weevil, *Cylas formicarius* (F.) (Jansson *et al.*, 1993), were obtained in field with *Heterorhabditis* spp. Applied as cadaver formulation. Laboratory studies have indicated greater dispersal ability in *S. carpocapsae* and *H. bacteriophora* (Shapiro and Glazer, 1996) and greater infectivity (Shapiro and Lewis, 1999) in *H. bacteriophora* when applied in cadavers compared with application in aqueous suspension. Commercialization of nematode-infected cadavers has been hindered by limitations in storage and application (Koppenhofer, 2000) but formulation of cadavers can increase ease of handling and storage. Formulation of cadavers provides complete protection from rupture and sticking together. Application in infected cadavers would reduce production costs for *in vivo* nematode producers because labor-intensive steps such as harvesting and concentrating IJs would be eliminated. Our studies conducted with *H.indica* infected *Galleria mellonella* cadavers indicated the suitability of nematode infected cadaver formulation for storage and application (Husaini *et al.*, 2004).

A better understanding of the biochemistry, physiology and behaviour of the infective juveniles is a prerequisite in the selection of formulation type, ingredient, packaging size and storage conditions in development of formulations that are more stable and easier to use.

12.2.10 Persistence

Persistence is defined as the ability to remain viable and infectious in the environment despite inhospitable conditions, and can be divided into two categories: individual and population persistence. Population persistence includes persistence of individuals and a recycling step and is applicable only to the control of soil pests since nematodes applied against foliage pests will not recycle above the soil surface. Persistence of applied IJs is influenced by various intrinsic (e.g. behavioural, physiological (Womersley, 1993; Wright *et al.*, 1998) and genetic characteristics (Gaugler, 1993) and extrinsic factors including abiotic factors (extreme temperature, soil moisture, osmotic stress, soil texture, RH, UV radiation (Kaya, 1990; Glazer 1996; Smits, 1996) and biotic factors (antibiosis, competition, and natural enemies (Kaya and Koppenhofer, 1996)

For successful inoculative releases of EPN, long-term, multigenerational survival and recycling of the nematode populations are essential. To achieve this goal, several conditions are important including, presence of moderately susceptible insect hosts throughout the year,

High economic threshold level of the target insect pests and soil conditions favourable for nematode survival. The optimal release method for inoculative releases of EPN may depend on the system into which they are released. Release methods could involve high or low application rates that may be uniformly distributed or applied in points from which colonization of other areas can occur.

Periodic augmentative releases into established nematodes populations, and/or management of the susceptibility of the host/pest populations are two other approaches that may be used to boost or manage established nematode populations and warrant more attention. Persistence studies carried out at PDBC, Bangalore indicated the persistence of *H. indica* and *S. carpocapsae* for more than 6 months in Chickmagalur, Karnataka and Idukki, Kerala.

Quality assessment of commercially produced EPN: Gaugler *et al.*, (2000) had a detailed analysis of commercial nematode formulations marketed in USA. The most widely commercialized species, *S.carpocapsae* and *H. bacteriophora* was studied in this analysis. 30 shipments from six US suppliers were mailed to three locations. These suppliers comprise a cottage industry aimed in large part at a mail-order market. Most companies were accessible and reliably shipped pure populations of the correct species on time, in sturdy containers, and often with superb accompanying instructions. Nematodes were received in satisfactory condition with acceptable levels of viability. Consistency however was a problem, with each supplier having one or more weak spots to bolster. Most shipments did not contain the expected nematode quantity and one shipment had no nematodes. Pathogenicity of several products against *G.mellonella* larvae was not equivalent to control. *H.bacteriophora* was not always available when ordered. A few products contained mixed populations of both nematode species. Application rates provided by some suppliers appeared unsound. Improved reliability by the nematode industry will most likely be achieved via industry-generated agreement on standards for quality.

12.2.11 Application technology

Application of nematodes is mostly targeted to the soil and against cryptic habitats. Inundative and inoculative approaches have been tried (Parkman and Frank,1992). When applied in field certain biotic factors like host susceptibility, pest population levels, host behaviour, different nematode species and host cues affect the nematode behaviour. Other than this susceptibility to environmental extremes reduces their field efficacy (Kaya and Gaugler, 1993). In soil application one or two pre and post application irrigation have been tried to provide enough moisture (Shetlar *et al.* 1988).

Timing soil applications of *Steinernema riobrave* with the life cycle of target insect is a key factor. Cabanillas and Raulston (1996) found that the efficacy of *S. riobrave* was higher on *Helicoverpa zea* in corn when applied at 200,000 IJ/mt2 through furrow irrigation or post irrigation than when nematode were sprayed on to the soil before irrigation. Thus *S. riobrave* was effective for suppressing corn earworm under field conditions of high temperature with irrigation. Presently they are being applied through conventional equipment such as high volume sprayers, drip irrigation, food baits or through sound trap (Georgis *et al* .1989; Parkman and Frank, 1992). For foliar application of EPN, antidesiccants have been used successfully to retard evaporation of the nematode suspension on foliage and thus to reduce desiccation of nematodes (Welch and Briand, 1961; Webster and Bronskill, 1968; Shapiro *et al.* 1985; Glazer and Navon, 1990). Glycerin 10% has been the most effective adjuvant for increasing survival and activity of nematodes on foliage (Nash and Fox 1969). However with the high cost of glycerin and the risk of phytotoxicity at high temperatures, search for more suitable adjuvant is continued. Excellent control of Japanese beetle, *Popillia japonica* and the northern masked chaffer, *Cyclocephala borealis* was obtained with 1.25-5.0 billion IJs/ha when plots were irrigated before application and again within 24 hrs after treatment (Downing 1994). Bauer *et al* (1997) studied the effect of a number of adjuvant on entomopathogenic nematode persistence and efficacy against *Plutella xylostella.* In laboratory TX 7719, Rodspray oil, and Nufilm P provided the best antidesiccant activity and in greenhouse, TX 7719 + Blankophor BBH increased nematode persistence on watercress leaves and efficacy against *P. xylostella* significantly. Addition of phagostimulants enhanced the effectiveness of *S. carpocapsae* against the 4th instar larvae of *Spodoptera litura* feeding on sunflower head in laboratory and field tests (Sezhian *et al.* 1996).

12.3 Success stories

Commercially EPN are available throughout the World. These nematodes are most efficacious against insects in cryptic and soil habitats besides dipterans in mushroom houses. Not all soil insects are susceptible and failures might result from the use of unsuitable nematode species or strains. Poor field efficacy despite excellent lab results may often be due to use of a nematode poorly adapted to the target insect. Striking differences in field efficacy for the same host reflect differences in adaptation. *S. carpocapsae* is effective against mole crickets but provides poor results against white grubs. *H. bacteriophora* can be as effective against white grubs as chemical insecticides, but is ineffective against mole crickets. Control efficacy with EPN is influenced by nematode species, strain, production and storage conditions, persistence in the introduced habitat and susceptibility of target insects. Although they have successfully suppressed foliage-feeding insects under high humidity and sheltered conditions in which the nematodes are protected from desiccation and UV radiation. The addition of anti desiccants may increase the efficiency of the nematodes against foliage insects in greenhouses by prolonging their activity. Some examples of biological control using EPNs on small and large scale to control insects belonging to Coleoptera, Lepidoptera, Diptera and other groups are listed. In field tests, steinernematids or heterorhabditids had significantly suppressed pest population of Japanese beetle, *Popillia japonica,* tawny mole cricket, *Scapetriscus vicinus*, black vine weevil, *Otiorhynchus sulcatus*; cabbage maggot, *Delia radicum* and western corn rootworm, *Diabrotica virgifera.*

Factors that have contributed to the commercialization of EPN in citrus for control of *Diaprepes abbreviatus* and other root weevils are biological, ecological and economical in nature. From a biological standpoint, successful commercialization has been achieved because a proper match between the nematode and target pest was made. From an ecological standpoint, Florida citrus groves contain many benefits for applying EPNs. The soils in these groves have a high sand content and oxygen availability for nematode survival. Because *D. abbreviatus* occurs primarily under the tree canopy, nematodes only need to be applied within the drip line where shade protects them from harmful UV radiation. The importance of economic factors leading to successful commercialization of nematodes for *D. abbreviatus* control cannot be underestimated. In management of *D. abbreviatus* larvae, nematodes face little or no competition from other control agents. Inexpensive chemical insecticides have been eliminated due to regulatory pressures and, thus far, the newer chemicals that have been tested for *D. abbreviatus* larval control have not been able to compete with nematodes in price or efficacy. The most important success-contributing factor is nematode cost. The low cost of nematode application combined with the high value of citrus, adds greatly to the success of EPNs in Florida citrus.

Two basic elements are necessary for nematodes to be successful: a suitable nematode for the target pest and favourable economics. For example nematodes associated with control of *D. abbreviatus*, *O. sulcatus*, and sciarid pests are all highly suited to their hosts, are applied in relatively high value commodities, and face little or no competition from other control measures. Where EPNs have not succeeded, the causes are generally due to a poor match of nematode and host, or poor economic conditions. For example, *S. riobrave* proved to be highly efficacious towards pink bollworm, but in cotton, nematodes could not compete on an economic level with other control strategies. Other insects in row crops fall into the same category, e.g., *H. zea* and *Diabrotica* spp. in maize. In other instances a suitable match of nematode to target insect could not be found, e.g., wireworms and imported fire ants, *Solenopsis invicta*, in which case strategies to employ EPN are futile. Other factors that may affect the success of commercial ventures with EPN include efficacy of pest suppression relative to other available control tactics, and reliable provision of a nematode product of high quality.

Proper match of the nematode to the host entails virulence, host finding, and ecological factors. If a nematode does not possess a high level of virulence toward the target pest there is little hope of success. In rare cases persistence may compensate for moderate virulence. Ecological factors such as relative ability to withstand desiccation or temperature tolerance are also important in choosing the best-adapted nematode for a particular pest. The list of pests that commercial nematode-producing companies advocate as targets has frequently been inflated. Mortality caused under laboratory conditions has often been inappropriately extrapolated to field efficacy. Although EPN are highly efficacious in controlling pink bollworm, they are not a viable management strategy because they cannot compete against the current management tactics. Chemical pesticides have been relied upon heavily for pink bollworm management, but current trends make transgenic cotton the most popular option. BT-cotton is far more economical than nematodes, and also will provide control of several major cotton pests simultaneously, including pink bollworm, and tobacco budworm. EPN use and BT-cotton appear to have minimal impact on non-targets and the environment. Hence pink bollworm appears to be an important example of an insect that can be managed using EPN but for practical reasons nematodes will remain unused in conventional cotton. The only niche where EPN could play a role in cotton pest management is in the organic, coloured cotton market.

12.4 Control of other organisms

Other than insect pests, plant parasitic nematodes and mites can be controlled by EPN. They also hold promise in control of sucking pests, household pests and those of public health and veterinary importance.

EPN appear to be a promising safe alternative to nematicides due to their ability to control plant parasitic nematodes like root-knot nematode *Meloidogyne* sp., *Belanolaimus* sp. *Criconemella* etc., in different crops. Several greenhouse and field trials demonstrated PPN suppression by application of EPN. *M. javanica* was suppressed on tomatoes when one application of 5×10^6 or 10 daily application of 5×10^5 *S. glaseri* Ijs/ potted plant were applied. In turf, a mix of 2.47×10^9 *H. bacteriophora* and 2.08 × 10^9 *S. carpocapsae* IJs / ha, and a single application of 2.47 × 10^9 *S. riobrave* Ijs/ha had suppressed plant-parasitic nematodes in field experiments. These findings indicate that the level of PPN suppression by EPN is impacted by host-plant, application rate, species of PPN, and the species of EPN.

12.5 Genetic manipulation

EPN are extremely promising biological control agents but certain limitations like susceptibility to environment extremes, host finding behaviour hamper the complete utilization of EPN potential. One solution to this problem is to develop genetically improved strains of EPN with enhanced traits for biological control. Strategies for genetic manipulation include artificial selection, hybridization and genetic transformation. Molecular biology, in EPN research has received little emphasis till recently. Most effort has been directed towards the application of molecular methods for nematode identification (Curran 1990; 1993; Joyce *et al.* 1994; Hashmi *et al.* 1996). Genetic variability within species provides insight into their phylogenetic relationships and population dynamics. Although conventional genetic improvement techniques have provided some nematode enhancement (Gaugler *et al* 1989a; 1989b; Zioni *et al.* 1992; Rahimi *et al.* 1993), the introduction of specific characters require biotechnological approaches. The close relationship of these nematodes with *Coenorhabditis elegans*, the model organism in molecular genetics, provides an extraordinary opportunity to exploit this new technology. Progress has been made in the genetic transformation, and the functionality of *C.elegans* genes into *H.bacteriophora* have been demonstrated (Hashmi *et al.* 1995, Gaugler & Hashmi 1996; Hashmi *et al* 1996; Hashmi & Gaugler 1997). These approaches may fail if the available nematode population do not possess sufficient genetic diversity for the beneficial traits (Glazer *et al.*, 1991) and this may be overcome by surveying natural population for the desired traits (Gaugler, 1987). Selection is employed to find genetically superior nematode species. A heat tolerant strain of *H. bacteriophora* designated IS 5 was discovered in Negev desert, Israel in 1996 and was found superior to HP 88, a commercially available strain of *H. bacteriophora*. Selection was also used to enhance the resistance of EPN to nematicides (Glazer *et al.* 1997).

12.6 Integration with IPM

Compatibility with Pesticides EPN may contribute more if integrated with other methods of control than their use solely as biocontrol agents. Studies on compatibility of EPN with other methods of control will be appropriate for any IPM strategy against insect pests.

Significant differences existed in the levels of tolerance between the heterorhabditids and steinernematids in lab tests. *H.bacteriophora* and *H. heliothidis* were less tolerant to some pesticides than *Steinernema* spp. which showed a higher survival, greater ability to recover and maintain infectivity after pesticide exposure. The normal development of all species was adversely affected after chemical treatment with most pesticides except for *S. glaseri.* This strain was the most tolerant to the toxic effects of pesticides -organophosphates and carbamates and appeared to be the most resistant and thus suitable for integration with pesticides

On sweet potato, *S. carpocapsae* was very effective against *Diabrotica* sp. The EPN provided significant protection from damage by *Conoderus* spp., *Diabrotica balteata, D. undecimpunctata howardi, Systena blanda, S. elongata* and *S. frontalis* in field. Fonofos did reduce damage caused by these insect pests, more so when applied in combination with the nematode. The response of *S. carpocapsae* and *S. feltiae* to 75 commercial pesticides has been assessed. IJs of both species tolerated most of the tested chemicals. The response of the two species to the different pesticides appears to be very similar. Only dodine for the fungicides and alachlor and paraquat for the herbicides seriously affected the juveniles. By contrast a larger proportion of formulations in the insecticide-acaricide-nematicide group were harmful. Parathion, aldicarb, methomyl, flubenzimine, metam sodium and fenamiphos were the most toxic. Overall, results indicate the feasibility of an integrated use of these nematode species and chemical pesticides in crop protection. Compatibility of pesticides with indigenous isolates of EPN has been monitored. Isolates of both *Steinernema* sp. and *H. indica* were compatible with the pesticides (malathion, endosulfan, mancozeb, carbofuran, quinalphos, fenvalerate and neem) could be combined with those for field application or could be used safely in pesticide treated field to enhance the mortality of the pests (Hussaini *et al.*, 2001 a&b).

12.7 Natural enemies

Pathogens: *S.carpocapsae* and NPV (SeM *NPV*) in combination produced additive effect for control of *S. exigua* in soybean than either of the organisms alone (Gothama *et al.* 1995). *S.carpocapsae* may infect Se*NPV* infected larva, which is then unsuitable for the virus development since the nematode development will kill the larva in 1-2 days and one nematode is capable of killing the larva.

Treatment with *H. bacteriophora* and *B. bassiana* resulted in higher total mortality of *S. exigua* in soil than treatments with either nematode alone or fungus alone. For *S. carpocapsae* plus *B. bassiana*, insect mortality was not significantly different compared with *S. carpocapsae* alone. These results suggest that inundative release of EPN where *B. bassiana* occurs may result in greater control of soilborne insect pests than application of nematodes alone. The biological control agents, *Bacillus* and the nematodes, provided intermediate levels of control.

12.8 Defense mechanism

White grubs, as soil insects are predisposed to nematode attack. However as a result of co evolution with soil pathogens white grubs have developed various defensive mechanisms that have conferred to them varying degrees of nematode resistance among white grub species. To optimize nematode efficacy, nematode species to be identified that have adapted to white grub biology and ecology. Generally, highly mobile cruiser-type nematodes are better adapted to sessile insects such as white grubs. The differences in search strategy explain the superiority of *H. bacteriophora* and *S. glaseri* over *S. carpocapsae* in Japanese beetle control. Differences have also been found in the interactions between the nematode-bacterium complex and the white grubs' immune system. Nematode efficacy against white grubs varies with scarab and nematode species. For example, the Oriental beetle, *Exomala orientalis* is less susceptible to *H. bacteriophora* than *P. japonica*, whereas the European chafer, *Rhizotrogus majalis*, or the Asiatic garden beetle, *Maladera castanea* are resistant to this nematode.

Commercial EPN Products: There are no commercial firms marketing EPN products in India at present. There is a need for commercialization. However these products are available commercially in other countries. Progress in nematode commercialization during the 1990s has been phenomenal. Development of large-scale mass-production technology and innovative easy-to-use formulations led to the expanded use of nematodes. Improvements in the quality and mass production of several species, and the development of alginate, clay, and granular formulations were among the major milestones accomplished that enhanced consumer acceptance of nematode products. Despite this progress, the reality is that the insecticidal nematodes are not yet reducing reliance on chemical insecticides to any significant degree. This is at least partly because of the lack of information on the proper use of nematodes and other biologicals. In the past few years, a noticeable change has occurred in the commercialization of nematodes. As opposed to the typical chemical pesticide development model followed by erstwhile Biosys, a distinct small-scale cottage industry has emerged in the US and Europe. These small businesses cater mainly to the needs of the high-value home lawn and garden markets, and rely heavily on *in vivo* nematode production.

12.8.1 Strengths

Eco-friendly-Hazards associated with chemical pesticides are well known. EPN can be used as an effective alternative to chemical pesticides. They can be used under diverse agro-climatic conditions. Efficacy could be conditioned by local isolates. EPN are already known to be existing under diverse conditions and have been recorded from all the continents except Antarctica. The records in India are expanding and since India is a mega-diversity country it is an added advantage. EPN can be exploited as model organisms for basic research studies. The cost of treatment with EPN has been worked out to be on par with chemical control in developed countries.

Exempted from cumbersome registration requirements following prolonged protocol involving expenditure of money. Apart from their use in insect pest control in agriculture they could very well be used in control of pests of public health and veterinary importance.

12.8.2 Weaknesses

Presently taxonomy and identification of EPN is cumbersome and conservative. Mass production, formulation, storage and transport need special attention in respect of different isolates. EPN research needs to be strengthened in the country and many workers are to be identified collaborated. Presently EPN is earmarked for niche market throughout the world. Till broad based usage will be evolved this will be considered as a distinct disadvantage

Fumigants applied for plant parasitic nematode control in the soil is a threat. New chemical molecules are being found out which are eco-friendly (eg. Imidochlorpid) for use as pesticides hence competition is expected.

12.8.3 Future Thrusts

- Protocols for molecular characterization of indigenous isolates of EPN and associated bacteria.
- Protocols for economically viable mass production technology.
- Protocols for suitable formulation, storage and transport.
- Evolving technologies for utilization of EPN against foliar feeding insect pests.
- Better understanding of the biochemistry, physiology and behaviour of IJs.
- Nematode metabolism is temperature-dependent, IJs to remain viable and pathogenic temperature tolerance and adaptation to extreme environmental conditions to be induced.

- Genetic manipulation of EPN for desired traits.
- Energy utilization strategies to maximize their survival.
- Conservation of Genetic diversity by frequent surveys.
- Characterization and utilization of bacteria.
- Explore the possibility of utilizing EPN in public health/ veterinary/ against sucking pests.

12.9 Conclusion

Potential of EPN as a safer alternative to insect control is proved and utilized for control of several pests in many parts of the world. In the future, discovery of new species and strains, mass production of new species, and further improvements in formulations will be continued. The use of genetic engineering techniques will be explored to enhance nematode biological control potential. Research will also be focused on the mechanisms of suppression of plant-parasitic nematodes by entomopathogenic nematodes. Emphasis needs to be placed on the integration of nematodes in IPM systems and on the training of extension agents and public at large in the proper use of nematodes

Table 12.1. Differences between Steinernematids and Heterorhabditids

Character	Steinernematid	Heterorhabditid
Second stage cuticle	Easily lost	Good retention
Second stage cuticle visiblity	Loosely fitted Easily seen	Tightly fitted, Difficult to see
Develops to	Male or female	Hermaphrodite
Bacterial symbionts	Modified ventricular intestine	Throughout intestine
Luminescence	No	Yes
Colour of dead larvae	Ochre, black	Red, pink
Reproduction	Amphimictic	Hermaphroditic- 1st Amphimictic- 2nd
Bursa in male	No	Yes

Table 12.2. Use of EPN against insect pests in India

Pest/disease	Crop	Bioagent /Formulation	Reference
Tryporyza incertulus	Rice, Sugar cane	*S. carpocapsae*	Rao and Manjunath (1966)
Chilo partellus	Maize	*Steinernema* sp.	Mathur et al. (1966)
Tryporyza incertulus C. suppressalis *Cirphis compta Pseudaletia separata*	Paddy	*S. carpocapsae*	Israel et al. (1969)

Pest/disease	**Crop**	**Bioagent /Formulation**	**Reference**
Rice pests	Paddy	*S. carpocapsae*	Yadava and Rao (1970)
Arctiidae, Lymantridae, Noctuidae, Pyralidae		*S. carpocapsae*	Mathur *et al.* (1971)
T. incertulus	Rice	*S. carpocapsae*	Rao etal. (1971)
Athalia proxima, Aulacophora foveicollis, Diacrisia obliqua, Dysdercus cingulatus, Heliothis armigera, Leucinodes orbonalis, S. litura		*S. carpocapsae*	Singh and Bardhan (1974)
A. ipsilon, A. segetum, Amathes nigrum and *white grubs*		*Steinernema* sp.	Singh (1977)
Papilio demoleus		*S. carpocapsae*	Srivastava (1978)
Anomala sp.	Potato	*S. carpocapsae*	Rajeswari Sundarababu *et al.* (1984)
S. *litura, H. armigera, C. partellus*		*S. carpocapsae*	Gupta *et al.* (1987)
S. litura	Bioefficiency test	*S. carpocapsae, S. feltiae*	Narayanan and Gopalakrishnan (1987)
H. armigera	Mortality tests	*S. carpocapsae, S. feltiae*	Ghode *et al.* (1988)
Orthacris simulans, Drasterius sp. Papilio aristotelia and Ergolis merione C. cephalonica	Susceptibility tests	*H. bacteriophora*	Sivakumar *et al.* (1989)
Cnaphalocrosis medinalis	Rice	*S. carpocapsae*	Srinivas and Prasad (1991)
Papilio sp.	Citrus	*Stenernema* sp.	Singh (1993)
Amsacta albistriga		*S. carpocapsawe, H. indica, H. bacteriophora*	Bhaskaran *et al.* (1994)
Pieris brassicae, Alphitobius diaperinus, Oryzaephilus mercator		*S. feltiae, S. carpocapsae*	Mathur *et al.* (1994)
Basilepta fulvicorne Conognethes punctiferalis		*S. feltiae S. glaseri*	Balu and Varatharasan (1991)
Odoiporus longicollis		*H. indica*	Padmanabhan *et al.* (2002)

Pest/disease	Crop	Bioagent /Formulation	Reference
Leucinodes orbonalis	Brinjal	*Steinernema carpocapsae Heterorhabditis indica/ spray formulation*	Hussaini *et al.* (2002a)
A. ipsilon	Tomato	*S. bicornutum/ bait & alginate capsule*	Hussaini *et al.*, 2001
A. ipsilon		*S. carpocapsae S. abbasi and H. indica/ talc based*	Hussaini *et al.*, 2002
H. armigera	Pigeon pea	*H. indica*	PDBC2001
H. armigera	Cotton	*H. indica*	PDBC2001
S. litura	Tobacco Nurseries	*S. carpocapsae*	Sitaramaiah *et al.*, 2002

Table 12.3. Some Available Commercial Products Containing *Steinernema* and *Heterorhabditis*

Formulation	Nema spp.	Product	Company
Alginate gel	*S. carpocapsae*	Mioplant	Novartis, Vienna, Austria
	S. carpocapsae	Boden-Niitzlinge	Rhone-Poulenc, Celaflor, Germany
Flowable gel	*S. carpocapsae*	BioSafe	SDS Biotech, Minato-Ku, Tokyo, Japan
	S. feltiae	Exhibit	Novartis, Basel, Switzerland
	S. feltiae	Stealth	Novartis, Macclesfield, Chester, UK
Clay	*H. megidis*	Nemasys-H	MicroBio, Cambridge, UK
		LarvaNem	Koppert B. V., Berkel en Rodenrigs, Netherland
	S. feltiae	Nemasys	MicroBio
		Entonem	Koppert BV
	S. scapterisci	Proactant Ss	BioControl, Gainesville, FL
Water- dispersible granules	*S. carpocapsae*	Biosafe	ThermoTrilogy, Columbia, MD
		Biosafe N	Thermotrilogy
		Biovector	Thermotrilogy
		Vector TL	Lesco, Lansing MI
	S. carpocapsae	Helix	Novartis, Mississauga, Canada
	S. carpocapsae	X-GNAT	E. C. Geiger, Harleysville, PA
	S. feltiae	Magnet	Amycel-Spawn Mate, Watsonville, CA
	S. riobravis	BioVector	ThermoTrilogy
		Vector MC	Lesco

Table 12.4. Host suitability of EPN

Pest	Nematode species	Host suitability
Black vine weevil *(Otiorhynchus sulcatus)*	*H. bacteriophora* *S. feltiae* *S. carpocapsae*	Good Good Fair
Colorado potato weevil *(Leptinotarsa decemlineata)*	*S. carpocapsae*	Fair
Corn root worms *(Diabrotica* spp.)	*S. carpocapsae*	Good
Japanese beetle *(Popillio japonica)*	*H. bacteriophora* *S. carpocapsae* *S. glaseri*	Excellent Fair Good
White grubs *(Phyllophaga* spp.)	*H. bacteriophora*	Good
Chafers *(Cyclocephala* spp.)	*H. bacteriophora*	Fair
Sciaridae *(Lycoriella* spp. and *Bradysia* spp.)	*S. feltiae*	Excellent
Leaf miner *(Liriomyza trifolii)*	*S. carpocapsae*	Good
Black cutworm *(Agrotis ipsilon)*	*S. carpocapsae*	Excellent
Diamondback moth *(Plutella xylostella)*	*S. carpocapsae*	Fair
Corn earworm *(Helicoverpa zea)*	*S. riobrave*	Excellent
Borers *(Synanthedon* spp.)	*S. feltiae*	Excellent
Spodoptera spp.	*S. carpocapsae*	Poor
Imported fire ant *(Solenopsis invicta)*	*S. carpocapsae*	Poor

Table 12.5. Field applications of EPN in soil ecosystems

Crop	Pest	Nematode	Dosage (Ijs)	Efficacy
Apple	*Cydia pomonella* *Conotrachelus nenuphar*	*S. carpocapsae* *S. carpocapsae*	5 b/ha $2x10^9$/ha	83% 85%
Citrus	*Diaprepes abbreviatus* *Pachneus litus*	*S. riobrave* *S. riobrave*	$2x10^6$/tree	79-90%
Groundnut	*Maladera matrida*	*H. bacteriophora*	0.25-$1x10^6$/sq.mt	50-90%
Strawberry	*Otiorhynchus sulcatus*	*S. carpocapsae*	$7.6x10^9$/ha	49.5-65%
Sweet potato	*Cylas formicarius*	*S. carpocapsae* *H. bacteriophora*	$1-3x10^9$/ha $1.1-4.9x10^9$/ha	Significant reduction in population
Turf field	*Popilia japonica*	*S. glaseri*	$5x10^9$/ha	44-66%
Cabbage	*Delia radicum*	*H. bacteriophora*	$1x10^2$-$2x10^5$/plant	Reduced fly pupae and root damage

Crop	Pest	Nematode	Dosage (Ijs)	Efficacy
Cauliflower	*Delia radicum*	*S. feltiae*	$3.5x10^4$/plant	2-3 fold increase in yield
Lettuce	*Liriomyza huidobrensis*	*S. feltiae*	$2x10^{10}$–$1x10^{11}$/ha	82%
Bean	*L. trifolii*	*S. carpocapsae*	$9x10^9$/ha	>65%
Mushroom	*Lycoriella auripila*	*S. feltiae*	$3x10^6$/tray	8 and 11 % increased weight and number
	L. solani	*S. feltiae*	$1\text{-}3x10^6/m^2$	91-93% fly emergence reduction
	L. mali	*S. feltiae*	$1x10^6/m^2$	82.8%
Household pest	*Blattella germanica*	*S. carpocapsae*	$2x10^6$/bait	Significant reduction

References

Ali, S.S., Shaheen, A., Parvez, R., and Abid Hussain, M. 2005. *Steinernema masoodi* sp.n . and *S. seemae* sp.n. (Nematoda : Rhabditidae : Steinernematidae) from India *International Journal of Nematology* 15: 89-99.

Ambika, V. 1995. Taxonomy, distribution and biology of native entomopathogenic nematodes. Dissertation submitted to the Tamil Nadu Agricultural University, Coimbatore, India, for the award of M. Sc. (Ag.) degree in Plant Nematology, 68 pp.

Balu, A. and Varatharasan, S. 1991. Evaluation of entomogenous nematode *Steinernema* spp. on insect pests of cardomom (abs.) *National Seminar on biological control in plantation crops*, Kottayam, Kerala, India.

Banu, J., Sosamma, V.K. and Koshy, P.K. 1998. Natural occurrence of an entomopathogenic nematodes, *Heterorhabditis indicus* (Poinar, Karunakar and David) from Kerala, India. *Tissaan proceeding of the third International Symposium of Afro-Asian Society of Nematologists,* April 16-19, Coimbatore. 274-279.

Bauer M E, Kaya H K, Gaugler R and Tabashnik B. 1997. Effects of adjuvants on entomopathogenic nematode persistence and efficacy against *Plutella xylostella. Biocontrol Science and Technology* 7 : 513-525.

Bedding R A. 1981. Low cost *in vitro* mass production of *Steinernema* (=*Neoaplectana*) and *Heterorhabditis* species (Nematoda) for field control of insect pests. *Nematologica* 27:109-114.

Bedding R A. 1984. Large scale production, storage and transport of the insect-parasitic nematode *Steinernema* (=*Neoaplectana*) spp. and *Heterorhabditis* spp. *Annals of Applied Biology* 104 : 117-120.

Bhaskaran, R. K. M., Sivakumar, C. V. and Venugopal, M. S. 1994. Biocontrol potential of entomopathogenic nematode in control of red hairy caterpillar, *Amsacta albistriga* of groundnut. *Indian Journal of Agricultural Sciences* 64 : 655-657

Cabanillas H E and Raulston J R. 1996. Evaluation of *Steinernema riobravis, S. carpocapsae* and irrigation timing for the control of corn earworm, *Helicoverpa zea. Journal of Nematology* 28 : 75-82.

Capinera J L and Hibberd B E. 1987. Bait formulations of chemical and microbial insecticides for suppression of crop feeding grasshoppers. *Journal of Agricultural Entomology* 4 : 337-344.

Capinera J L, Pelissier D, Menout G S and Epsky N D. 1988. Control of black cutworm, *Agrotis ipsilon* with entomogenous nematodes. *Journal of Invertebrate Pathology* 52 : 427-435.

Connick W J Jr., Nickle W R and Vinyard B T. 1993. Pesta : New granular formulations for *Steinernema carpocapsae. Journal of Nematology* 25 : 198-203.

Creighton, C. S., and Fassuliotis, G. 1985. *Heterorhabditis* sp. (Nematoda: Heterorhabdilidae: A nematode parasite isolated from banded cucumber beetle *Diabrotica balteata. Journal of Nematology* 17: 150-153.

Curran , J. 1993 In : Entomopathogenic nematodes in Biological Control , Gaugler, R., and Kaya, H.K., (eds.) C.R.C. Press, Boca Raton, Fl .,pp.365

Downing, A S. 1994. Effect of irrigation and spray volume on efficacy of entomopathogenic nematodes against white grubs. *Journal of Economic Entomology* 87 : 643-646.

Dutky, S.R.; Thompson, J.V. and Cantwell, G.E. 1964. A technique for mass propagation of the DD-136 nematode. *Journal of Insect Pathology* 6: 417

Friedman, M.J. 1990. Commercial production and Development. In: *Entomophilic nematodes in Biological Control.* Boca Raton, FL: CRC Press, pp. 153-72

Ganguly, S., and Singh, L.K. 2000. *Steinernema thermophilum* sp.n (Rhabditida: Steinernematidae) from India .*International Journal of Nematology* 10 : 183-191.

Gaugler R and Georgis R. 1991. Culture method and efficacy of entomopathogenic nematodes (Rhabditida: Steinernematidae and Heterorhabditidae). *Biological Control* 1 : 269-74.

Gaugler, R. 1987. Entomogenous nematodes and their prospects for genetic improvement. In: '*Biotechnology in Invertebrate Pathology and Cell Culture*'. Maramorosh K (eds.), Academic Press, San Diego, CA, pp. 457-484

Gaugler, R., Grewal, P., Kaya, H.K. and Fiola, D.B. 2000. Quality assessment of commercially produced Entomopathogenic nematodes. *Biological control Theory and Application in Pest Management* 17: 100-109.

Georgis R Wojcik W F and Shetlar D J. 1989. Use of *Steinernema feltiae* in a bait for the control of blackcutworm *Agrotis ipsilon* and tawny mole cricket *Scaptericus visinus. Florida Entomologist* 72 : 203-204.

Georgis R and Hague N G M. 1991. Nematodes as biological insecticides. *Pesticide Outlook* 2 : 29-32.

Georgis R and Manweiler S A. 1994. Entomopathogenic nematodes: Infectivity a developing biological control technology. In: *Agricultural Zoology Reviews.* Evans K (ed.), Andover, Intersept, pp. 63-94.

Georgis R. 1990a. Commercialization of steinernematid and heterorhabditid entomophilic nematodes. *Brighton Crop Protection Conference Insecticides Fungicides.* 1: 275-80.

Georgis R. 1990b. Formulation and application technology. In : *Entomopathogenic nematodes in Biological Control.* Gaugler R and Kaya H K (eds.), CRC Press, Boca Raton, Fl.

Ghode, M. K., Mishra, R. P., Swain, P. K., Nayak, U. K. and Pawar, A. D. 1988. Pathogenicity of *Steinernema feltiae* Weiser to *Heliothis armigera* Hubner. *Plant Protection Bulletin* 40: 7-8.

Glazer I and Navon A. 1990. Activity and persistence of entomoparasitic nematodes tested against *Heliothis armigera. Journal of Economic Entomology* 83 : 1795-1800.

Glazer I, Gaugler R. and Segal D. 1991 Genetics of the nematode *Heterorhabditis bacteriophora* strain HP88 : The diversity of beneficial traits. *Journal of Nematology* 23 : 324-333.

Glazer I, Salme L and Segal D. 1997. Genetic enhancement of nematode resistance in Entomopathogenic nematodes. *Biocontrol Science and Technology* 7 :199-512.

Glazer-I 1996 Survival mechanisms of entomopathogenic nematodes. Biocontrol-Science-and-Technology 6: 3, 373-378

Grewal P S. 1996. Quality of Entomopathogenic Nematode products. Third International Nematology Congress, Guadelope (abs), p. 54.

Grewal, P.S., Converse,V., and Georgis, R. 1999. Influence of production and bioassay methods on infectivity of two ambush foragers (Nematoda : Steinerenmatidae) *Journal of Invertebrate Pathology* 73: 40-44

Grewal, P.S., Selvan S., and Gaugler, R. 1994. Thermal adaptation of entomopathogenic nematodes: niche breadth for infection, establishment, and reproduction. *J.Thermal Biol.* 19: 4, 245-253.

Gupta, M., Desai, P. and Pawar, A. D. 1987. The feasibility of using DD-136 strain of *Steinernemacarpocapsae* for the control of insect pests. *Plant Protection Bulletin*, 39: 6-19.

Hashmi S, Ghazala Hashmi and Gaugler R. 1995. Genetic transformation of an entomopathogenic Nematode by microinjection. *Journal of Invertebrate Pathology* 66 : 293-296.

Hashmi,G., and Gaugler, R. 1998. Genetic diversity in insect-parasitic nematodes (Rhabditida: Heterorhabditidae). *Journal of Invertebrate Pathology*. 72: 185-189. Hashmi,G., Glazer, I., and Gaugler, R. 1996. Molecular comparisons of entomopathogenic nematodes using randomly amplified polymorphic DNA (RAPD) markers. *Fundamenatal-and Applied Nematology*. 19: 399-406

Hussaini, S. S., Kavita J. Satya and Abid Hussain, M. 2001a. Tolerance of Entomopathogenic nematode isolates to pesticides and their effect on multiplication *Current Nematology* 12 : 29-34

Hussaini, S. S., S. P. Singh, R. Parthasarathy and V. Shakeela . 2002. *In vitro* production of entomopathogenic nematodes in different artificial media. *Indian Journal of Nematology*. 32: 44-46. (ISSN 0303- 6960)

Hussaini, S. S., Singh, S. P. and Shakeela, V. 2002b. Variable efficacy of different formulations of entomopathogenic nematodes against black cutworm *Agrotis ipsilon* (Hufnagel) larvae. In: *Biological Control of Lepidopteran Pests* , P.L. Tandon, C.R. Ballal, S.K. Jalali ,R.J. Rabindra (Eds.,) Precision Phototype Services, Bangalore, pp. 103-107.

Hussaini, S. S., Singh, S. P. and Nagesh, M. 2002a. *In vitro* and field evaluation of some indigenous isolates of *Steinernema* and *Heterorhabditis indica* against shoot and fruit borer, *Leucinodes orbonalis. Indian Journal of Nematology*, 32: 63-65.

Hussaini, S. S., Singh, S.P., and Shakeela, V. 2001b. Compatibility of Entomopathogenic nematodes (Steinernematidae, Heterorhabditidae: Rhabiditoidea) with selected pesticides and their influence on some biological traits . *Entomon* 26: 37-44.

Israel, P., Rao, Y. R. V. J., Prakash Rao, P. S. and Verma, A. 1969. Control of paddy cutworms by DD-136, a parasitic nematode. *Current Science* 16: 390-391.

Jansson, R. K, Lecrone, S. H, and Gaugler, R. 1993. Field efficacy and persistence of entomopathogenic nematodes (Rhabditida: Steinernematidae, Heterorhabditidae) for control of sweet potato weevil (Coleoptera: Apionidae) in southern Florida. *Journal of Economic Entomology* 86, 1055-1063

Joyce S A, Reid A, Driver F and Curran J. 1994. Application of polymerase chain reaction (PCR) methods to the identification of entomopathogenic nematodes. *In* : *Proceedings of Symposium and Workshop*. Burnell, A M, Ehlers, R U and Masson J P (eds.), St. Patrick's College, Maynooth, Co. Kildare, Ireland, Luxemberg, European Commission, DG XII:178 - 187.

Karunakar, G., David, H. and Easwaramoorthy, S. 1992. Influence of dosage of *Steinernema carpocapsae, S. glaseri* and *Heterorhabditis indicus* on mortality of the host and multiplication of infective juveniles in sugarcane internode borer, *Chilo sacchariphagus indicus. Journal of Biological Control,* 6: 26-28.

Kaya H K and Nelson C E. 1985. Encapsulation of steinernematid and heterorhabditid nematodes with calcium alginate. A new approach for insect control and other applications. *Environmental Entomology* 14 : 572-574.

Kaya H K, Mannion C M, Burlando T M and Nelson C E. 1987. Escape of *Steinernema feltiae* from alginate capsules containing tomato seeds. *Journal of Nematology* 19 : 287-291.

Kaya, H. K; Koppenhofer, A. M 1996. Effects of microbial and other antagonistic rganism and competition on entomopathogenic nematodes. Biocontrol-Science-and-Technology 6: 3, 357-371

Kaya, H.K. and Gaugler, R. 1993. Entomopathogenic nematodes. *Annual Review of Entomology* 38: 181-206.

Kaya, H.K., 1990. Soil Ecology In: *Entomopathogenic nematodes in Biological Control,* Boca Raton, FL: CRC Press pp. 93-115

Koppenhofer, A. M. 2000. Nematodes. *In: Field Manual of Techniques in Invertebrate Pathology* (L. A. Lacey and H. K, Kaya, Eds.), pp. 283-301. Kluwer Academic, Dordrecht,

Mamiya, Y. 1989. Comparison of infectivity of *Steinernema kushidai* (Nematoda: Steinernematidae) and other steinerenmatid and heterorhabditid nematodes for three different insects. *App Entomol Zool* 24: 302-308

Mathur, S. B., Srivastava, P. and Srivastava, K. L. 1971. Pathogenicity of DD-136 an insect parasitic nematode to five pests of agricultural and horticultural importance. *Proceedings of National Academy of Sciences* (India), 41: 325-327.

Mathur, S. B., Srivastava, R. P. and Subba Rao, B. R. S. 1976. Record of nematode parasitising *Chilo zonellus* Swinhoe, stem borer of maize. *Indian Journal of Entomology* 28: 414-415.

Narayanan, K. and Gopalakrishnan, C. 1987. Effect of entomogenous nematode *Steinernema feltiae* (Rhabditida: Steinernematidae) to the pre-pupa, pupa and adult of *Spodoptera litura* (Noctuidae; Lepidoptera). *Indian Journal of Nematology* 17: 273-276.

Nash R F and Fox R C. 1969. Field control of the pinetip moth by the nematode DD-136. *Journal of Economic Entomology* 62 : 660-663.

Padmanabhan, B., Sundararaju, P., Cannayane, I. and Hussaini, S. S. 2002. Effect of Entomopathogenic nematode, *Heterorhabditis indica* (PDBC EN13.3) on banana stem weevil, *Odoiporus longicollis in vitro. Indian Journal of Nematology* 32: 209.

Parkman J P and Frank J H. 1992. Use of sound trap to inoculate *Steinernema scapterisci* into pest mole cricket population. *Florida Entomologist* 76 : 75-82.

Poinar G O Jr, Thomas G M, Lin K C and Mookerjee P. 1985. Feasibility of embedding parasitic nematodes in hydrogels for insect control. *IRCS Medical Science* 13 : 754-755

Poinar G O Jr. 1979. *Nematodes for biological control of insects*, CRC Press, Boca Raton, Fla., pp. 277.

Rajeswari Sunderababu, Vasudeva Menon, P. P. and Sivagami Vadivelu. 1984. Preliminary tests with DD -136 for the control of potato chafer grub, *Anomala* sp. *Indian Journal of Nematology*, 14: 187-189.

Rajkumar, M., Parihar, A. and Siddiqui, A.U. 2002. Mass multiplication of indigenous entomopathogenic nematodes from Udaipur. *National Symposium on Biodiversity and Management of Nematodes in Cropping Systems for Sustainable Agriculture*, held at ARS, Durgapura (Jaipur), Rajasthan from 11-13 Nov., 2002: p. 77

Rao, V. P. and Manjunath, T. M. 1966. DD 136 nematode that can kill many insect pests. *Indian Farming*, 16: 143-44.

Rao, Y. R. V. J., Prakash Rao, P. S., Verma, A. and Israel, P. 1971. Tests with an insect parasitic nematode DD-136 against the rice stem borer, *Tryporyza incertulas. Indian Journal* of *Entomology*, 33: 215-217.

Sezhian N, Sivakumar C V and Venugopal M S. 1996. Alteration of effectiveness of *Steinernema carpocapsae* against *Spodoptera litura* larvae on sunflower by addition of an insect phagostimulant. *Indian Journal of Nematology* 26 : 77-81.

Shapiro M, Mc Lane W and Bell R. 1985. Laboratory evaluation of selected chemicals as antidesiccants for the protection of the entomogenous nematode, *Steinernema feltiae* against *Lymantria dispar. Journal of Economic Entomology* 78 : 1437-1441.

Shapiro, D. I., and Glazer, I . 1996. Comparison of entomopathogenic nematode dispersal from infected hosts versus aqueous suspension. *Environmental Entomology* 25 : 1455-1461.

Shapiro, D. I., and Lewis. E.E. 1999. Comparison of entomopathogenic nematode infectivity from infected host versus aqueous suspension. *Environmental Entomology* 28: 907-911.

Shetlar-DJ; Suleman-PE; Georgis-R. 1988 Irrigation and use of entomogenous nematodes, *Neoaplectana* spp. and *Heterorhabditis heliothidis* (Rhabditida: Steinernematidae and Heterorhabditidae) for control of Japanese beetle (Coleoptera: Scarabaeidae) grubs in turfgrass. *Journal-of-Economic-Entomology* 81: 5, 1318-1322

Singh, J. and Bardhan V. 1974. Effectiveness of DD-136, an entomophilic nematode against insect pests of agricultural importance. *Current Science*, 43: 662.

Singh, S. P. 1993. Effectiveness of an indigenous nematode against citrus butterfly. *Journal of Insect Science* 6: 107-108.

Singh, S.P. 1977. Isolation of an entomophilic nematode from potato cutworms. *Current Science*, 46: 454-455.

Singh, V. and Yadava, C.P.S. 2002. Mass production technique of entomogenous nematode, *Heterorhabditis bacteriophora* on Greater Wax moth, *Galleria mellonella* for insect control. *National Symposium on Biodiversity and Management of Nematodes in Cropping Systems for Sustainable Agriculture*, held at ARS, Durgapura (Jaipur), Rajasthan from 11-13 Nov., 2002: p. 80

Sitaramaiah, S., Gunneswara Rao, S., Hussaini, S. S., Venkateswarlu, P. and Nageswara Rao, S. 2002. Use of entomopathogenic nematode *Steinernema carpocapsae* against *Spodoptera litura* Fab. intobacco nursery. In: *Biological Control of Lepidopteran Pests*, P.L. Tandon, C.R. Ballal, S.K. Jalali, R.J. Rabindra (Eds.) Precision Phototype Services, Bangalore, pp. 211-213.

Sivakumar, C. V., Jayaraj, S. and Subramanian, S. 1989. Observations on Indian population of the entomopathogenic nematode *Heterorhabditis bacteriophora*. *Journal of Biological Control* 2: 112-113.

Smith, D.C., Gill, S.A and Way, R.G., 1996 Evaluation of entomopathogenic nematodes as biological control against the banded ash clearwing borer. *Journal of Environmental Horticulture* 14 : 67-71

Srinivas, P. R. and Prasad, J.S. 1991. Record of DD-136 nematode infection on rice leaf folder, *Cnaphalocrosis medinalis*. *Indian Journal of Agricultural Science*, 61: 348-349.

Vyas, R.V., Patel, N.S. and Patel, D.J. 1999. Mass production technology for entomopathogenic nematode, *Steinernema* sp. *Indian Journal of Nematology*, 29:178-181.

Webster ,J .M and Bronskill , J .F. 1968. Use of Gelgard M and an evaporetardant to facilitate control of larch sawfly by nematode-bacterium complex. *Journal of Economic Entomology* 61 : 1370-1373.

Welch H E and Briand L J. 1961. Tests of the nematode DD-136 and an associated bacterium for the control of Colorado potato beetle. *Canadian Entomologist* 43 : 75.

Wright, R.J; Villani, M.G.; Agudelo Silva, F., Silva, F.A. 1988. Steinernematid and heterorhabditid nematodes for control of larval European chafers and Japanese beetles (Coleoptera: Scarabaeidae) in potted yew. *Journal-of-Economic-Entomology*. 81:152-157

Yadava, C. P. and Rao, Y. S. 1970. On the effectiveness of the entomophilic nematode DD-136 in the biological control of insect pests of rice. *Oryza*, 7: 131-136.

Zaki, F.A., Mantoo, M. A. and Gul, S. 2000. *In vivo* culturing of entomopathogenic nematodes *Heterorhabditis bacteriophora* and *Steinernema carpocapsae* on silkworm (*Bombyx mori*) and their effect on some lepidopterous insects. *Indian Journal of Nematology*, 30: 41-45.

SPIDERS AS NATURAL ENEMIES TO CONTROL INSECT PESTS

13

A.K. Bhowmick and D. P. Abrol

13.0 INTRODUCTION

Integrated pest management (IPM) uses multiple approaches to keep pest levels below economic injury levels (Kogan 1998). Accomplishing this requires an understanding of ecological processes in agroecosystems and how they are modified by disturbances such as the application of pesticides or harvesting (Kogan 1998; Brown 1999). Although agroecosystems are more disturbed than natural habitats, the interactions between organisms, the characteristics of the habitat, and immigration and emigration of species at local and regional scales are common factors structuring any community. Perennial crops such as orchards are more stable temporally and spatially than annual crops. (Brown 1999).

Integrated pest management strategies must consist of both preventative and curative methods (Kogan 1998). Preventative methods include conservation biological control: manipulation of the agroecosystem environment to favour the immigration, establishment, survival and reproduction of biocontrol agents such as parasitoids and generalist predators (insects and spiders) (Landis *et al.* 2000). For preventative biological control to work, we need to understand the ecology of natural enemies in the agroecosystem, including their interactions with pest species and how to manage the agroecosystem to increase and sustain natural enemy populations. Curative (i.e. active, immediate) pest control methods should minimally disrupt these natural enemy assemblages.

Although spider assemblages have been well characterized, only a few studies have explored their biocontrol potential in orchards

(Mansour *et al.* 1980; Haddad *et al.* 2004; Rodriguez *et al.* 2005; Sarvary *et al.* 2007). However, considerable theoretical and empirical advances on the role of spiders in biocontrol have been made in other crop systems.

Spiders (Araneae) are abundant and species-rich generalist predators in agroecosystems, especially in perennial crops such as apple orchards (Dondale *et al.* 1979; McCaffrey and Horsburgh 1980; Olszak *et al.* 1992a). Spiders are among the first arthropods to colonize disturbed agricultural areas, primarily arriving by ballooning rather than cursorial dispersal (Bishop and Riechert 1990; Ehmann 1994; Marc *et al.* 1999). Higher proportions of non-crop habitat and greater heterogeneity of the landscape are correlated with the persistence of populations and higher ground-dwelling spider abundances and species richness in agricultural fields (Halley *et al.* 1996; Clough *et al.* 2005; Thorbek and Topping 2005; Schmidt and Tscharntke 2005b; Isaia *et al.* 2006). Spiders are negatively affected by insecticide applications (Bostanian *et al.* 1984; Wisniewska and Prokopy 1997; Pekár 1999a; Pekár 1999b; Bajwa and Aliniazee 2001), although populations are more stable in perennial crops (e.g. orchards) than annual crops because of the absence of physical disturbance through harvesting (Riechert and Lockley 1984). Enhancing spider recolonization after populations are reduced by disturbances such as harvesting or insecticide application requires a knowledge of which natural habitats are potential sources (Marc *et al.* 1999; Sunderland and Samu 2000).

Notwithstanding the positive correlation between regional habitat diversity and spider abundance and species richness, the source of spiders colonizing crops is not clear; many studies have found that habitats adjacent to crops have different spider assemblages than the crop, suggesting that most colonization is through long-distance dispersal (Bishop and Riechert 1990; Samu *et al.* 1997; Topping and Lovei 1997; Samu and Szinetár 2002). Some spider species are specialized for habitats that have high frequencies of structural disturbance (Topping and Lovei 1997; Samu and Szinetár 2002; Schmidt and Tscharntke 2005a), and are sensitive to habitat structure (Greenstone 1984; Rypstra *et al.* 1999; Langellotto and Denno 2004).

Although no comparisons of spider community composition have been done between orchards and adjacent habitats, spider species found in the foliage of apple trees have been collected from adjacent deciduous shrubs and trees (Olszak *et al.* 1992a; Olszak *et al.* 1992b; Miliczky and Horton 2005). The relative abundance of spiders in orchard trees decreases with increased distance of the orchard from the adjacent habitat (Miliczky and Horton 2005) and within the orchard on the distance of the tree to the edge of the orchard (Bogya *et al.* 2000).

13.1 Spiders in biocontrol

There is theoretical and empirical support that generalist predators, including spiders, can contribute to pest control, in contradiction to the long-held assumption that only natural enemies that exhibit density dependent tracking and prey specificity are relevant (Riechert and Lockley 1984; Symondson *et al.* 2002; Snyder and Ives 2003; Stiling and Cornelissen 2005). Spiders can prevent the early increase in pest populations by acting as a buffer on pest populations rather than reacting to pest outbreaks in a density dependent fashion (Riechert and Lockley 1984; Riechert *et al.* 1999). This potential to buffer pest populations at low levels requires particular characteristics, many of which spiders exhibit: polyphagy, self-limiting populations (Wise 1993), the ability to exploit changes in temporal availability of prey (Riechert *et al.* 1999), and being part of a diverse assemblage of species; multi-species generalist predator assemblages may consume prey at a higher level and with more temporal stability than one species alone (Riechert and Lawrence 1997; Riechert *et al.* 1999; Stiling and Cornelissen 2005; Tscharntke *et al.* 2005), but see Rosenheim (1998). Additionally, spiders can respond quickly to changes in pest densities as they are among the first predators to colonize a crop (Gut *et al.* 1988; Rathman and Brunner 1988; Bishop and Riechert 1990; Ehmann 1994) and are not dependent upon the presence of particular prey to maintain their populations within crops (Symondson *et al.* 2002). Spiders also kill more prey than they consume, through both wasteful killing (Wise 1993) and indirect effects on prey (e.g. inducing prey to leave the plant) (Mansour *et al.* 1981; Gastreich 1999; Cronin *et al.* 2004).

Spiders exhibit a variety of hunting behaviours and, as such, occupy a diversity of microhabitats. Uetz *et al.* (1999) classify agricultural spiders into guilds that reflect similarities in hunting strategy. The primary division is between web spinners and hunting spiders (the latter hunt without webs). Within the hunting spider guild, there are the running spiders (actively wander searching for prey), stalkers (wandering followed by stalking prey), and ambushers (remain still and ambush prey). Web-spinners are classified into guilds depending on the type of web they build. These include space-web builders (three-dimensional web occupying a defined space), tangle-web weavers (a looser, less defined web) and the orb-web weavers (the archetypal, two-dimensional orb web). Due to their different foraging modes the diet of the spiders in different guilds varies: web-spinners eat almost entirely insects, while hunting spiders have a wider diet breadth, including up to 25% other spiders (Nyffeler 1999).

Spider predation on pests has been studied extensively in field crops, where they can reduce pest populations (Nyffeler and Sunderland 2003), cascading down to a reduction in plant damage (Riechert and Bishop 1990; Carter and Rypstra 1995; Halaj *et al.* 2000; Lang 2003). Few studies have manipulated spider populations in orchards to investigate their biocontrol potential. However, Mansour *et al.* (1980; 1981) found spiders, primarily the hunting spider *Cheiracanthium mildei* L. Koch (Miturgidae), controlled populations of *Spodoptera littoralis* (Boisduval) (Lepidoptera: Noctuidae) in apple orchards. More recently, Sarvary *et al.* (2007) found that the predation rate (by all predators) on *C. rosaceana* larvae in orchard plots ranged from 20-35% and did not differ between conventionally treated plots and those treated with reduced-risk insecticides.

The impact of spiders and other generalist predators on pest populations is often unpredictable as they also feed on economically unimportant arthropods and other natural enemies within the crop (i.e. intra-guild predation, IGP), which can dampen the top-down influence on herbivores or even cause increased herbivore populations (Rosenheim 1998; Hodge 1999; Symondson *et al.* 2002). In some cases, even with documented IGP, the activity of generalist predators still cascades down to decreases in pest numbers (Snyder and Ives 2001; Denno *et al.* 2004) and increases in plant yield or quality (Lang 2003). Whether or not IGP dampens biocontrol of pests is difficult to predict as it depends on a variety of factors, such as natural enemy characteristics (e.g. foraging mode) (Rosenheim and Corbett 2003), habitat complexity (Finke and Denno 2002), and the community composition and phenology of the arthropods within the crop over the season (Snyder and Wise 2001). Despite the complexities inherent in predicting the success of biocontrol with IGP as a factor, a review of manipulative field experiments showed that natural enemy assemblages, including generalist predators, caused a significant reduction in the pest population in 77% (n=181) of the reviewed cases (Symondson *et al.* 2002).

13.1.1 Habitat manipulations

Generalist predators work most efficiently as assemblages of multiple species (Riechert and Lawrence 1997; Stiling and Cornelissen 2005), but because the inundative release of generalist predators is largely impractical, other methods to increase and stabilize generalist predator populations (and other natural enemies) are important. Diversification of the agroecosystem is an effective way of increasing natural enemy populations, as these alternative habitats are a source of natural enemies for periodic recolonization and provide extra microhabitats, protected microclimates, overwintering sites, and alternative resources (Landis *et al.* 2000; Sunderland and Samu 2000). Habitat manipulations include changes within the agroecosystem, surrounding the agroecosystem, or at a landscape scale (Symondson *et al.* 2002). Habitat manipulations within the agroecosystem such as reduced tillage, mulching, and the addition of straw bales increase the densities and species richness of predators like spiders and carabids (Halaj *et al.* 2000; Landis *et al.* 2000). Noncrop plants planted in rows through the crop or surrounding the crop provide alternative resources (i.e. nectar, pollen, shelter) for parasitoids (Bostanian *et al.* 2004; Norris and Kogan 2005) and predators (Lemke and Poehling 2002; Prasifka *et al.* 2006). Natural habitats surrounding the crop, such as woodland and fields, may be sources of natural enemies to recolonize the crop after disturbance (i.e. harvesting or pesticide application) (Landis *et al.* 2000). The regional landscape can affect species richness and density of natural enemies in agroecosystems, as higher landscape diversity has been correlated with higher abundance and species richness of generalist predators and parasitoids (Landis *et al.* 2000; Sunderland and Samu 2000; Bianchi *et al.* 2006). There are problems associated with increased diversification of the agroecosystem, including providing refuges and source populations of pests (MacHardy 2000; Vakenti *et al.* 2001; Prokopy 2003; Norris and Kogan 2005) and keeping land out of production (Landis *et al.* 2000).

For epigeic spiders in arable crops, increased density and species richness of spider populations, through mulching or the addition of crates or straw bales between the rows, reduces pest populations and plant damage (Riechert and Bishop 1990; Carter and Rypstra 1995; Halaj *et al.* 2000). In perennial crops such as orchards, the planting of non-crop strips, usually flowering annuals, may augment the diversity of spiders inhabiting the foliage of the trees. In some cases, spider abundance increases in orchard plots containing flowering strips (Wyss 1995), but in other cases there is no significant effect, and the species overlap between spiders on the orchard foliage and herbaceous plants is low (Samu *et al.* 1997; Jenser *et al.* 1999). A potential disadvantage of intercropping is that spiders may not consistently move from the non-crop strips into the crops if the resources are superior in the non-crop strips (Riechert and Bishop 1990; Samu 2003).

A viable alternative to intercropping to increase the immigration of spiders into orchard foliage is to design agroecosystems to include natural habitats that harbour similar spider assemblages to the orchard (i.e. landscape approach). The amount of movement and establishment of arthropods between patches in a landscape will vary depending on relative population densities, species dispersal abilities and propensities, and habitat affinities (Hillebrand and Blenckner 2002). For many arthropods there is some overlap of species between the crop and adjacent fields and woodlands (Bedford and Usher 1994; Rieux *et al.* 1999; Tixier *et al.* 2000; Thomas *et al.* 2001; Varchola and Dunn 2001).

However, for spiders, it is not clear if these adjacent habitats are a source of colonists to the crop, because the epigeic spider assemblages as a whole are dissimilar in crops and their adjacent habitats (Bishop and Riechert 1990; Samu *et al.* 1997; Topping and Lovei 1997; Samu and Szinetár 2002), suggesting that long-distance dispersal, rather than local, plays a major role in the colonization of crops. Arable crops have a distinct "agrobiont" spider fauna; species that make up a large proportion of the spider community in crops, are less dominant in natural habitats, and are adapted for frequent dispersal between ephemeral habitats (Wissinger 1997; Samu and Szinetár 2002). However, the density of agrobiont species per unit area may actually be equivalent or higher in adjacent perennial grasslands, indicating that the dominance of these species in crops may be due to low densities of other species rather than numerical dominance of the agrobionts *per se* (Schmidt and Tscharntke 2005a). Arable crops and their adjacent habitats often differ in several key factors that influence the composition of spider assemblages: vegetation structure (Greenstone 1984; Rypstra *et al.* 1999; Langellotto and Denno 2004) and frequency of structural disturbance (i.e. annual tilling of fields) (Topping and Lovei 1997; Wissinger 1997; Samu and Szinetár 2002; Schmidt and Tscharntke 2005a). To have similar spider assemblages in a crop and its adjacent habitat, both vegetation structure and the frequency of structural disturbance (through harvesting) should be similar (Bishop and Riechert 1990; Samu and Szinetár 2002).

Perennial crops are structurally more stable than annual crops, and, as such, the spider assemblages may be similar to the assemblages in perennial habitats such as deciduous forest, which may also have a similar vegetation structure. Studies have shown that there is species overlap between orchards and adjacent trees and shrubs (Olszak *et al.* 1992a; Olszak *et al.* 1992b; Miliczky and Horton 2005), and that there is a negative correlation between the relative abundance of spiders in orchard foliage and distance of the orchard to the adjacent habitat (Miliczky and Horton 2005). These studies suggest that habitats that are similar to the orchard in both vegetation structure and temporal stability may be sources of colonists to the orchard.

13.1.2 Spider dispersal and colonization

Spiders can potentially move from patch to patch in a landscape via two modes of dispersal: cursorial (walking) and ballooning. The main mode of colonization for both epigeic and foliage-dwelling spiders is ballooning (Bishop and Riechert 1990; Weyman 1993; Ehmann 1994; Weyman *et al.* 2002). For ballooning, strands of silk from the spinnerets create drag with which spiders are pulled aloft, and they drift passively with the wind. Spiders can theoretically balloon up to tens of kilometers per day (Thomas *et al.* 2003). The total distance travelled is probably less and varies with the propensity of the spider to re-balloon after alighting, related to species, environmental conditions (e.g. air flow), population density, and prey availability (Morse 1993; Weyman and Jepson 1994; Weyman *et al.* 1995; Weyman *et al.* 2002). The ability of spiders to balloon is limited by size; ballooning spiders are typically less than 2 mg in mass (Greenstone *et al.* 1987). Spiders in some families are small as adults and can balloon during all life stages (e.g. Linyphiids), which may be an adaptation to living in unpredictable, ephemeral habitats (Weyman *et al.* 2002). Linyphiids are often the most common spiders caught in aerial traps (Greenstone *et al.* 1987; Pearce *et al.* 2005). In most other families the spiders tend to be larger as adults and therefore balloon as juveniles (Duffey 1956), often during the spring and summer (Marc *et al.* 1999).

The ability of spiders to disperse long distances by ballooning indicates that the regional landscape, in addition to the habitats surrounding the agroecosystem, may influence spider populations within a crop. Early models demonstrated the importance of non-crop habitats as sources and refuges for spiders to recolonize agricultural habitat (Halley *et al.* 1996; Samu *et al.* 1999). Subsequent studies have confirmed the importance of landscape: the species richness of epigeic spiders in cereal is positively correlated with the percentage of non-crop area within radii of 500 m (Clough *et al.* 2005), and the abundance of Linyphiids in winter wheat also increases with higher percentages of non-crop habitats in a 1-to-3 km radius (Schmidt and Tscharntke 2005a). In winter wheat the abundance of epigeic spiders is higher in organic as compared to conventionally managed fields (Schmidt *et al.* 2005), but species richness of spider assemblages is not always related to the management regime (Clough *et al.* 2005; Schmidt *et al.* 2005). According to models, the landscape diversity (i.e. number of different types of non-crop habitats) is more important to epigeic spider diversity than the spatial arrangement of these habitats (Thorbek and Topping 2005).

Habitat diversification at all spatial scales clearly affects the species richness and abundance of spiders, but how dispersal at these various spatial scales affects species composition of assemblages is poorly understood. Additionally, most research into the dispersal and colonization of spiders into crops has focused on epigeic species, primarily the Linyphiidae. The colonization dynamics of foliage-dwelling spiders into perennial crops such as orchards are largely unexplored.

13.2 Effects of pest management on spiders

The species composition of a community results from the regional influences from dispersal and colonization and from factors within the habitat such as biotic interactions, environmental and habitat conditions, and disturbance to the assemblages through changes in habitat structure (such as seasonal growth or senescence of plants) (Ricklefs 1987). Agricultural crops differ from natural habitats in that there are also periodic disturbances to the system from the application of pesticides, which have varying effects on the arthropod community (Hu *et al.* 1996; Prokopy *et al.* 1996; Desneux *et al.* 2007). The abundance and diversity of spider populations are reduced by conventional (broad-spectrum) insecticide application to orchard foliage (Croft and Whalon 1982; Hull and Starner 1983; Bostanian *et al.* 1984; Olszak *et al.* 1992a; Wisniewska and Prokopy 1997; Pekár 1999a; Pekár 1999b; Bogya *et al.* 2000). Insecticides used in conventional management also reduce the number of overwintering spiders in orchards (Pekár 1999c; Bogya *et al.* 2000; Horton *et al.* 2001). Spiders are sensitive to insecticides from multiple chemical families; broad-spectrum organophosphates, pyrethroids and carbamates decrease spider populations (Bajwa and Aliniazee 2001), although when applied at reduced frequencies the effects are not as pronounced (Wisniewska and Prokopy 1997; Bajwa and Aliniazee 2001). Some insecticides such as insect growth regulators or *Bacillus thuringiensis* formulations do not affect spider populations (Gurr *et al.* 1999; Bajwa and Aliniazee 2001). Spider guilds differ in susceptibility to pesticides partially because their different foraging modes lead to different degrees of contact with the insecticide. In laboratory assays hunting spiders are more sensitive than web-spinners, as the latter are protected to a certain degree by their webs and are not in constant contact with the sprayed foliage (Pekár 1999b). In the field, the response of particular guilds and families to insecticide applications is not consistent, indicating that the vulnerability to the insecticide is not the only important factor, but that different rates of resistance development and competitive release may also contribute to observed patterns (Wisniewska and Prokopy 1997; Pekár and Kocourek 2004; Cardenas *et al.* 2006). Spiders can recognize and avoid insecticide residues within one day of application, but not after, leading to potentially harmful contact (Pekár and Haddad 2005).

Because of the negative impact of chemical insecticides on natural enemies, as well as on human health and the environment (Wilson and Tisdell 2001), alternatives to broad-spectrum insecticides are in increasing demand. Alternatives to chemical methods include physical methods, which use a variety of techniques to alter the physical environment of the pest and thereby reduce its threat to the crop (Vincent *et al.* 2003). Recently, a physical method, consisting of particle spray formulations of kaolin clay, has been developed for pest management (Glenn *et al.* 1999). This technique is unusual in that it is applied as a spray (much like a pesticide), but its mode of action is physical rather than chemical.

Spiders play an important role in regulating insect pests in the agricultural ecosystem. There are a large number of species, many of them with high population densities. They limit the availability of habitats open to insect pests of rice by occupying various microhabitats. They have a wide range of prey species, catch significant numbers of prey and use various foraging strategies. Most of the spiders in rice fields seem to evacuate the field after the application of insecticides and move back into the field later. Their predatory capacity can have a synergistic effect in suppressing densities of insect pests when they are used to complement the effect of insecticides.

For many decades, insecticides have been widely used to control rice pests. However, the continuous use of a wide range of pesticides has caused many side effects, including loss of biodiversity, the problem of secondary pests, insecticide resistance, residual toxicity, the resurgence of insect pests, and environmental pollution. Recently, many efforts have been made to combine various non-chemical control methods with insecticides in systems of Integrated Pest Management (IPM). One such effort is the combined use of natural enemies with insecticides. Many uses of parasitic and predatory natural enemies to control agricultural pests have been reported (Van den Bosch *et al.* 1992).

Spiders are carnivorous arthropods. They are classified into 40,000 species, and are found all over the world in almost every kind of habitat. They mainly prey on insects, although they may also feed on various other kinds of prey. The population densities and species abundance of spider communities in agricultural fields can be as high as in natural ecosystems (Turnbull 1973, Tanaka 1989, Riechert 1981).

Spiders consume a large number of prey, and do not damage plants. They can achieve an equilibrium in pest control, after which their own numbers are suppressed by their territoriality and cannibalism. For some time, spiders have been considered important predators which help regulate the population densities of insect pests (Pickett *et al.* 1946, Dondale 1956, Duffey 1962, Kajak *et al.* 1968, Fox and Dondale 1972, Tanaka 1989). In particular, spider communities in areas with a temperate climate achieve equilibrium in the control of agricultural pests (Riechert 1982). In spite of this, they have not usually been treated as an important biological control agent, because there is so little information on the ecological role of spiders in pest control (Turnbull 1973, Riechert and Lockley 1984).

However, research has shown that spiders in rice fields can play an important role as predators in reducing the densities of planthoppers and leafhoppers (Hamamura 1969, Sasaba *et al.* 1973, Gavarra and Raros 1973, Samal and Misra 1975, Kobayashi 1977, Chiu 1979, Holt *et al.* 1987, Tanaka 1989). Much work has been done on the use of spiders for control of rice pests in Korea. The details of findings are reproduced as given below: Since E. Strand first described Ganphosa koreae in 1907, there have been many publications about spiders in Korea. Currently, a total of 46 families, 222 genera and 626 species of spiders are recorded from Korea (Namkung *et al.* 2000).

Up until 1970, most of the research on spiders concentrated on identification. From the early 1970s, researchers began to study the basic ecological and biological characteristics of spiders as biological control agents. Spiders in rice fields have been studied more than spiders on other crops (Park *et al.* 1972, Paik and Kim 1973, Paik *et al.* 1974, Choi and Namkung 1976, Okuma *et al.* 1978, Paik *et al.* 1979, Yun and Namkung 1979, Paik and Namkung 1979, Paik and Namkung 1979, Kim *et al.* 1990, Kim 1992, Lee *et al.* 1993a, 1993b, Song and Lee 1994, Kim and Kim 1995, Kim 1995a, 1995b, Im and Kim 1996, Lee *et al.* 1997, Yun 1997, Yang *et al.* 1998, Kim 1998. Im and Kim 1999). However, most of these studies were limited to the identification of spiders, and to investigating the dominant spider species, their regional distribution, seasonal fluctuations and the effect of insecticides. There were few studies on the spatial distribution of spiders, how this is related to their ecological role, and how many insect pests they consume in rice fields.

13.3 Spiders in Korean Rice Fields

The spiders living in rice fields in Korea comprise 27.96% of all Korean spiders, and can be classified into 22 families, 99 genera and 175 species (Table 13.1). It has been reported that *P. subpiraticus*, *P. clercki* and *G. dentatum* are dominant spider species in rice fields (Paik *et al.* 1979). However, more recent studies have shown that there is no regional pattern in the distribution of rice field spiders, and that *P. subpiraticus* is the overall dominant species in Korea.

13.4 Seasonal occurrence of spiders in Korean rice fields

Seasonal fluctuations of spiders in rice fields in Korea have been reported to follow an "M" shaped curve (Im and Kim 1996, Choi and Namkung 1976, Hokyo *et al.* 1976, Kim *et al.* 1990, Paik *et al.* 1979, Yang *et al.* 1998). Other surveys which began at a different time and covered a different period have reported that fluctuations in the density of spider communities followed a curve shaped like a "W" (Choi and Namkung

1976, Yun and Namkung 1979). During the rice growing period in Korea, rice field spiders generally show two peaks. One is when spiders which have overwintered nearby migrate into the paddy, and the other is when the second generation of the year appears. However, annual fluctuations in numbers of rice field spiders showed several peaks when overwintering periods are included.

Spiders in rice fields in Korea start to build up their populations in the middle of July, 40-45 days after transplanting, (Kim 1992, Yun 1997). The same pattern is seen in Japan, which has a similar agricultural environment (Kobayashi and Shibata 1973). Hamamura (1969) has reported that the density of spiders in rice fields in Japan is always low when transplanting is taking place. Lee *et al.* (1997) suggested that the different time taken by different spider species to immigrate from levees into paddy fields is the cause of the low initial population densities, in spite of the abundance of prey. However, Kim (1998) showed that the immigration rate of overwintering spiders from levees into paddy fields was low. He suggested that most of the dominant spider species immigrated from outside areas into the rice fields by ballooning (Note: "Ballooning" is the term used to describe the habit of young spiders of sailing through the air borne by silk strands on wind currents). He also suggested that the immigration of the spiders was delayed after transplanting because the young rice plants are too small to provide a suitable habitat for either hunters or web builders.

Even though web builders were more numerous than hunters in some areas (Choi and Namkung 1976, Yun and Namkung 1979, Lee *et al.* 1993a, 1993b, Kim and Kim 1995), seasonal fluctuations in rice field spiders in Korea are thought to depend mainly on the density and species composition of hunting species of spider (Table 2, Park *et al.* 1972, Paik and Kim 1973, Paik *et al.* 1974, Okuma *et al.* 1978, Paik *et al.* 1979, Paik and Namkung 1979, Kim 1992, Song and Lee 1994, Kim and Kim 1995, Im and Kim 1996, Yang *et al.* 1998, Im and Kim 1999). Kim (1998) pointed out that if spiders are being used as biological control agents, it is very important to understand their life styles. Both web builders and hunters follow a foraging strategy. Song and Lee (1994) suggested that web builders such as *P. clercki* were better able to suppress insect pests than hunters such as *P. subpiraticus*. However, hunters are usually active predators which follow a "pursue and kill" foraging strategy, while web builders follow a passive "sit and wait" strategy (Enders 1974). Furthermore, the webbing sites of web builders are easily affected by environmental factors. In addition, when the web spaces overlap, there is competition with and between species of web builders. Therefore, hunters probably are more effective predators than web builders.

13.5 Suppression of rice brown planthoppers by spiders

All the organisms in a rice field, as in other kinds of ecosystem, are involved in interactions within and between species, including prey-predator and host-parasite relationships (Hijii 1984). In particular, the interaction of prey and predator show a constant numerical interaction. Information about these relationships is fundamental to biological control. Yamano (1977) suggested that spiders are the most important biological control agents regulating insect populations in rice fields, including insect pests. Predators tend to cluster in stable sites where many species of prey are maintained at a high density (Schmitt 1987). Seasonal fluctuations in the numbers of spiders in rice fields showed a constant numerical interaction with those of other arthropods, including insect pests.

The brown planthopper (*Nilaparvata lugens* Stål) is a pest which feeds only on rice plants (Otake and Hokyo 1976). Insecticides have been widely used to control brown planthoppers in Asian countries, including Korea (Nagata 1985). Integrated pest management programs to control this pest have been proposed, using natural enemies of planthoppers. Studies have reported numerous parasitic and predatory natural enemies of planthoppers. For predators alone, there are 14 species belonging to 13 genera in Korea (Okuma 1958, Suenaga 1963, Gavarra and Raros 1973, Samal and Misra 1975, Hamamura 1971, Chiu 1979, Kuno and Dick 1984, Shepard *et al.* 1987, Chua and Mikil 1989, Tanaka 1989, Kim *et al.* 1991, Choi *et al.* 1992, Choi and Lee 1990, Shepard and Ooi 1991, Ooi and Shepard 1994, Choi *et al.* 1996)(Table 13.2). For effective control of brown planthoppers, predators should inhabit the same part of the rice plant, to reduce the time spent searching for prey. Brown planthoppers prefer to live on rice plants about 10 cm above the surface of the irrigation water (Lin 1970). This part of the plant is usually shaded, and has a high relative humidity and a relatively low temperature, both of which are favorable for reproduction of the pest. Leasar and Umzicker (1978) showed that humidity and temperature are the most important factors affecting the distribution of spiders. Most of the dominant species of spiders in Korean rice fields live on the lower 20 cm of rice plants (Table 13.3).

Spiders represent more than 90% of the natural enemies of brown planthoppers living in paddy fields in Korea (Lee *et al.* 1997). Because of this fact, most of the studies related to biological control of planthoppers have focused on spiders. The studies revealed that of planthoppers each day Subadults of *P. subpiraticus* consumed the greatest number, although the total numbers consumed by C. kurilensis and G. dentatum were also high. Spiders belonging to the species *P. subpiraticus* showed the greatest variation in daily consumption, while the number of planthoppers consumed each day by *C. kurilensis* and *G. dentatum* varied by only 0-4 planthoppers (Lee *et al.* 1997).

It appears that the difference in the numbers of planthoppers consumed by hunters and by web builders is the result of differences in the efficiency of their foraging strategies. Generally, female spiders consumed more prey than males, except in the case of *G. dentatum*. It is thought that females usually need more energy for oviposition and brood care, while males need only the essential energy for survival. Subadults had a higher consumption rate than adults, which implies that the energy needs for growth are higher than those needed for reproduction.

13.6 Compatibility of spiders and insecticides

Most studies of the effect of insecticides on natural enemies in the paddy field have focused on spiders. As we might expect, the results have shown that insecticides have a negative effect on the population densities of rice field spiders (Kuno 1968, Kuno and Hokyo 1970, Kawahara *et al.* 1971, Choi *et al.* 1978, Paik *et al.* 1979, Jang 1981, Ryu *et al.* 1984, Kim *et al.* 1984, Paik and Hwang 1990, Lee *et al.* 1993b, Bae *et al.* 1994).

The relative toxicity to spiders of various insecticides used against the brown planthopper was tested in the laboratory. Carbofuran, which is widely used in rice fields, is very toxic to spiders (Bae *et al.* 1994). Choi *et al.* (1978) have suggested that the root zone placement of Carbofuran reduces the density of spiders in rice fields by direct lethal action, and by poisoning their food chain.

All the insecticides which were most lethal to spiders were dust or granule types. Many of those tested, such as sprays and wettable powders, were selective, in that *P. subpiraticus* had a higher survival rate after a longer period of exposure. This was probably because of the active behavior of this species. It is recommended that if there is no difference in the effect of different types of insecticides on brown planthoppers, the use of dust or granule type insecticides should be avoided. Several studies have shown that the application of insecticides reduces the population density of spiders in rice fields (Park *et al.* 1972, Kawahara *et al.* 1971, Paik *et al.* 1979, Kim 1992, Yun 1997). They have a negative effect, not only on numbers, but also on species diversity (Lee *et al.* 1993a, 1993b). Even insecticides which generally protect natural enemies often have a negative effect on spiders (Clausen 1990). Hunting species tend to suffer more damage than web builders (Specht and Dondale 1960, Legner and Oatman 1964, Bostanian *et al.* 1984), while active hunters suffer more than less active ones (Bostanian *et al.* 1984).

It has been shown that reduction of spider populations in rice fields after insecticide applications is not because the spiders were killed by the insecticides, but also because their prey were killed. Lee *et al.* (1993a) showed that the density of hunters was higher than that of web builders in the period after insecticides were applied. They suggested that during the recovery period, the active hunters played an important role in rebuilding spider densities at this time.

Factors which produce mortality in brown planthoppers include high temperatures, predation and parasitism by natural enemies, nutritional deficiencies, and the maturity of crops (Cheng and Holt 1990).

If the density of rice field spiders is high enough before the immigration of brown planthoppers, the spiders can control the planthoppers effectively and the number and quantity of insecticide applications can be reduced. Other ecological and biological characteristics of spiders also need to be understood. Chiu (1979) suggested that it takes longer for spiders to rebuild their population densities after the application of insecticides than planthoppers and leafhoppers, because spiders have a longer generation interval. Also, the development of selective insecticides, the effect of insecticides on spiders, and appropriate timing and quantities of insecticide applications, should all be considered.

13.7 Rearing spiders as biological pest-control agents

Spiders are the largest group of arachnids, comprising more than 30,000 species distributed over 60 families worldwide. Its members are among the best known and most widely distributed of all the arachnids. Spiders are also unique in that they have inhabited almost all environments, including one at 23,000 ft on Mount Everest. They have also invaded the world of water despite the fact that none can breathe in water although one species from Europe, *Argyroneta aquatica,* is totally aquatic.

Spiders are an important group of predators in various ecosystems. However, their role in pest control and crop protection has not been utilized properly in India. This report describes the mass rearing technology that has been developed to rear one species of household spider (giant crab spider) which is an efficient predator of cockroaches. This technology can also be extended to other spiders with certain modifications. Besides this, a few practical tips are provided which could be combined with mass rearing and release for conservation and augmentation of these spiders. The innovation is part of a general scheme designed to utilise biological control agents for control of pests without recourse to toxic pesticides the impact of which on health is well known.

The problem or situation being addressed by the practice/ innovative experience Various living organisms lived in perfect harmony and balance with each other in different ecosystems before chemical agriculture and chemical pest control came into the picture. Today, we see a kind of agricultural development which is based on monoculture and extensive use of chemicals. This has led to the emergence of several pests which has in turn led to the failure of crops. The number of pests developing resistance to pesticides over the years has been increasing at a very alarming rate. The pesticides have proven to be extremely toxic and have led to a number of side effects: impact on public health, toxic residues in food and disturbance of local ecosystems. Use of such chemicals has also led to the eradication of natural predators such as spiders which normally keep pest populations in balance.

Use of chemical pesticides has killed off natural predators not only in the agroecosystems but also in households. Several toxic chemicals are often recommended to control pests in households. Often, these chemicals are mistakenly ingested by children, domestic pets or even birds. The work with spiders detailed in this report was designed to address the problems faced by farmers and homemakers. Mass rearing technologies have been worked out for certain species which can be extended to other groups. This can form a basis for biological control. One of the main reasons many people kill spiders on sight is the ugly looking web, which they construct for many purposes like catching prey, attracting the opposite sex, etc. But the fact is that not all spiders construct webs. There are certain spiders called "Wanderers" which do not construct webs to catch their prey but actively chase them instead. Spiders are beneficial to human beings in the sense that they feed on the pests of man such as cockroaches, flies, etc. Predatory arachnids such as spiders are an important group of biological control agents. They are able to survive periods of starvation when prey is difficult to find. As the use of chemical insecticides gained currency, however, the role of spiders as natural pest controllers was greatly neglected and even forgotten. Today, however, people are becoming increasingly aware that in the long run, the harmful effects of the insecticides far outweigh the advantage of temporary pest control. What people are not so aware of is that even creatures commonly found in their own homes can play an important role in controlling domestic pests. The giant crab spider is one such example of a beneficial "house guest".

Spiders are voracious predators of insects. They are well adapted to certain habitats because of their ability to withstand periods of low food availability and also to take advantage of periods of prey abundance. Spiders are important predators of pests of cotton, rice, apple, banana and various other crops and plantations. In households, a particular spider known as the giant crab spider has been identified as an effective "tool" in controlling cockroaches and other insect pests found in the domestic environment. The mass multiplication of such spiders can be effected without any sophisticated equipment.

13.8 Giant crab spider - the cockroach controller

In India and other tropical countries, the giant crab spider, *Heteropoda venatoria,* is a commonly found predator. It gets its name from its size, which ranges between 20 mm and 25 mm, the curvature of its legs and the way it scuttles sideways like a crab. Giant crab spiders are brown in colour and have a flat body. While they feed mainly on cockroaches, in human dwellings they also prey on house crickets and other domestic pests. These hunting spiders do not use webs to capture prey. They seek out their prey in a special way. Their flat body enables them to get into small cracks and crevices inhabited by cockroaches. Their great speed and strong fangs enhance their hunting efficiency. Giant crab spiders attack cockroaches from the rear, a clever way to prevent being bitten by the powerful mouth parts of the prey. After having captured the cockroach, they bite into its neck and paralyse it with their poison. They then begin to feed on it, sucking its body fluids and discarding the remains.

These spiders use a multiple-capture technique. While feeding, if another prey passes by, they capture that too and hold it under their abdomen. The number of victims a giant crab spider can hold at any given moment is proportional to the size of the individual spider and of the prey. The prey-handling organs of these spiders are so designed that they can handle only a particular size of prey. Larger spiders generally catch large cockroaches. As a result, competition with the young ones for prey is avoided.

13.9 Mass rearing of giant crab spiders

Male and female giant crab spiders can be collected from inside houses, where they are found in plenty. Normally, one can come across many of them in tiled houses and houses with thatched roofs. One male and one female can be put in a 2-litre container and allowed to mate. After mating, the male is removed. The females construct an egg sac within two weeks which hatches out in about 20 days. Females which have constructed egg sacs can also be caught from the wild. The spiderlings that hatch resemble the adults except for being smaller in size. They undergo 13 to 14 moults to attain maturity, with males sometimes requiring only 13 moults and females, 15 moults.

13.9.1 Feed

Spiderlings feed on different prey, hunting them actively. All the instar stages and adults feed only on live and mobile prey. They feed on *Tribolium* larvae, housefly larvae and adults, and *Drosophilu* larvae and adults. Mass rearing of these spiders does not require any modern sophisticated technology. What is needed is just a few bottles, prey and a lot of patience. The newly hatched spiderlings are separated in small containers, the lid of which has a few small holes for aeration and one moderately big hole for feeding. A small strip of paper should be inserted into the container to give additional support to the spiderling. The bigger hole should be plugged with cotton.

The first instar spiderlings should be fed with small larvae, like the second instar larvae of houseflies, *Drosophilu* and *Tribolium.* Once these moult to second instar, they can be fed with slightly bigger larvae of *Tribolium* and very small-sized cockroaches. These larvae should be given with the help of a brush. Once the spider reaches the seventh or eighth instar stage, it can be fed adult houseflies and medium-sized cockroaches apart from the other prey given earlier. From the 10th and 11th instar onwards, adult cockroaches can be given. The hunger rate increases with each instar stage and reaches its maximum in the pre-adult stage. In the adult stage, predation is maximum after mating and after the egg sac is dropped once the spiderlings have hatched. During these periods, the spiders should be provided with enough prey.

The spiderling can be grown up to the fifth instar stage in this container after which it should be transferred into a 1-litre container where it is kept until the 10th instar. Tenth to fourteenth instar stages and adults should be grown in 2-litre containers. The containers should be numbered properly and a regular register should be maintained to note the date of moulting and other observations. The containers should be cleaned regularly. The spider should first be transferred to another container and the old one should be cleaned with soap water and sun-dried. The spiders should be protected from ants by either putting the containers in water trays or putting a cup of oil or water under the legs of the table on which the containers are kept.

13.9.2 Rearing of prey

As has already been mentioned, spiders feed only on live insects. *So* when an attempt is made to rear them, we need to also learn how to rear the insects on which these spiders could feed. The technology of rearing prey insects like houseflies, *Drosophilu* and cockroaches is hereby described.

13.9.3 Laboratory rearing of houseflies

Materials: A cage covered on all sides with a wire mesh, petri dishes, milk powder, cotton, rice bran, groundnut oilcake and buckets.

Method: Houseflies that are collected in the field are introduced into the cage. Both males and females feed on the same food such as milk, sugar solution, etc. Hence milk made from milk powder with sugar is kept soaked in cotton for the flies to feed on. The females should be provided with ovitraps inside the cage which should simulate the breeding conditions of the houseflies. Rice bran and groundnut oil paste, in a ratio of 40:60, kept in the cage attracts the females to oviposit. The milk food and ovitrap must be changed every day. The ovitrap kept in a small petri dish should be transferred into a bucket with more of the breeding medium. After four days of larval development, the larvae pupate inside the bucket. After three days of development inside the pupa, the adult emerges. But before this takes place, the bucket should be filled with water, so that the pupae float in it. They are collected and kept in the cage for emergence. The cage should be emptied once a week and cleaned with spirits.

Life cycle of houseflies:

a. **Egg:** The eggs are banana-shaped, white in colour and of 1 mm size. Their morphology can be studied under a dissection microscope. The incubation period is 12-14 hours.
b. **Larva:** There are three larval instars. The larvae feed on the decaying material present in the breeding medium and need moisture for their survival. Larval duration lasts for 3-5 days and when the development is over, the larva moves out of the moist medium and pupates. To avoid this, buckets are used so that the larvae will not be able to crawl up the bucket and will thus pupate inside, making the collection of pupae easier.
c. **Pupa:** These are initially white in colour and slowly darken to coffee brown after a day or two. Imagines emerge out of the pupae, breaking the anterior part.
d. **Adult:** Two days after emergence, the adults mate and then start feeding.

13.9.4 Rearing of cockroaches

Materials: Buckets with lids, muslin cloth, wire mesh, chart papers, biscuits and cotton.

Method: Holes are made on three sides of the bucket (2-inch diameter) and are covered with wire mesh and muslin cloth. Chart papers are glued to the bottom of the bucket in a circular fashion. A hole is made on the lid of the bucket and plugged with cotton to provide enough light and oxygen to the cockroaches. Cockroaches and oothecae are introduced into the bucket and biscuits are provided for the roaches to feed upon. Any edible material can be given to the cockroaches, as they are scavengers.

Life cycle: The oothecae are purse-shaped, consisting of around eight eggs. Small nymphs hatch out of the oothecae in about a week which resemble the adults in all aspects except wings and reproductive organs. It takes three months for the nymphs to attain maturity. The life cycle continues inside the bucket.

13.9.5 *Drosophila* rearing

Materials: 50-ml. containers with tight lids, agar, jaggery, wheat powder, distilled water.

Preparation of the medium: 420 ml of distilled water is boiled and 25 g of jaggery is dissolved in it. To this solution is added 25 g of agar and wheat powder. Stirring should be done while heating. After cooling to 40°C, yeast powder is smeared on the top to avoid fungal infection.

Rearing: The prepared medium is poured into 50-ml containers and after 24 hours, adult Drosophila sp. are introduced into the containers. Females deposit their eggs in the medium and the larvae that come out feed on the medium. Male and female Drosophila adults also feed on the same medium.

Life cycle: The whole life cycle is completed in a week. The eggs laid by the female hatch within a day. The larval instars take three days to enter into the pupal stage where they remain for 3-4 days. The adults which emerge mate and feed in two days, thus continuing the life cycle.

13.10 Points to remember while rearing spiders

a. Spiders have cannibalistic tendencies, i.e., when there are many of them in a single container, they may tend to attack or prey upon others. Hence, they should be separated and maintained in individual containers.

b. Small air-holes should be provided in the containers. A bigger hole, plugged with cotton, should be made for introducing prey.

c. For the smaller instars, food like *Tribolium* larvae (the small worms found in rava, maida flour), fruitfly larvae and housefly larvae can be given. After they have grown a little bigger, cockroaches can be given.

d. Ensuring the cleanliness of the containers is a must. The prey remains and the moulted skin should be removed periodically.

e. Two sets of containers should be kept. After transferring the spiders to fresh containers, the old containers can be cleaned with soap and water and dried.

f. Regular observations like on the date of moulting, pattern of feeding, etc. should be made.
g. It is very important to keep ants away. The containers in which the spiders are reared can be kept in a plate on a tray filled with water. This plate should not be allowed to float on the water. If the containers are to be placed on shelves, then the legs of the shelves should be placed in small cups containing water or oil. The mass rearing technology described above is for the giant crab spider, which is a cockroach controller. This technology has currently been extended to other spiders found in other ecosystems. While extending this technology to other spiders, certain modifications have to be made to suit the needs of a particular species. However, the basic technology can be the same for various species.

Description of the institution responsible and its organizational aspects

The Centre for Indian Knowledge Systems is an institution devoted to exploring and developing the contemporary relevance and application of traditional Indian knowledge systems. One of the major activities of the centre has been action research and training programmes on the use of natural products for pest control and crop protection. A number of plant products, including neem, have been tested out in farmers' fields. The institution currently conducts a fairly large number of training programmes for farmers and nongovernmental organisations (NGOs) on the use of non-chemical methods for pest control and crop protection. Besides this, it produces a number of publications on pest control and sustainable agriculture. The centre also works on setting up rural gene banks, development of the use of biological control agents such as spiders, research on the applications of Vrkshayurveda (Indian plant science), and work on indigenous breeds of cattle. The centre is also involved in research and publications on medicinal plants. From time to time, training programmes on the use of traditional medicine for various groups are conducted. The centre also currently publishes newsletters, manuals and monographs on traditional health care and traditional agriculture.

13.11 Problems or obstacles encountered and how they were overcome

People either get repulsed by or ignore the sight of a spider. But these spiders do not deserve such a response. Most spiders are harmless, beneficial, colourful predatory creatures. The initial problem faced in using spiders as biological control agents was thus how to clear the misconceptions regarding spiders. Detailed awareness programmes had to be conducted amongst schoolchildren, rural folk and the general public. They were taught that most spiders in India are not fatal to human beings and are in fact very efficient predators in a variety of agroecosystems. After the awareness programmes were conducted, the fear amongst most people disappeared and they were willing to accept spiders as an organism with which they can live in harmony.

13.12 Effects of the practice/innovative experience

The use of spiders in the domestic setting for control of cockroaches would bring down the use of the extremely poisonous substances conventionally used to control such pests. Surveys conducted in Madras revealed that nearly 60% of people interviewed were willing to introduce natural pest controllers like spiders as an alternative to insecticides. The public-health benefits of a more widespread use of such natural agents cannot be doubted. In agricultural areas, a programme based on the use of spiders would naturally lead to a decline in the use of poisonous chemicals. In any case, both (spiders and toxins) cannot go together as the pesticides would eliminate the spiders. Studies on the effect of pesticides on spiders have shown that most of the pesticides currently used in agroecosystems kill off many spiders. Studies have also been made on the effect of various plant-based pesticides on spiders.

These are not fatal to spiders. Spiders can also be conserved in agroecosystems by providing hedges which can serve as a habitat when there are no standing crops. To increase the number of spiders in an agroecosystem, spider egg sacs can also be collected and covered with hay and placed around the standing crop. When the young ones hatch, they go to the crop in search of pests. An agricultural programme that is based on the use of predators like spiders would also create suitable environments for the return of several other beneficial insects which are unable to tolerate the presence of pesticides.

Currently there are a number of people who use these spiders in the household setting to control cockroaches. There is another group of people who do not put spiders to use specifically but leave them alone whenever they encounter one. They have learned to live in harmony with the spiders. In the agricultural fields also, farmers have started accepting and acknowledging the role of spiders as biocontrol agents. The misconception that all spiders are deadly has been cleared by the awareness programmes and farmers have started recognising their importance. Farmers also realise that once they stop using pesticides, the population of spiders in their fields is found to increase.

13.13 Suitability and possibility for upscaling

The multiplication of biological control agents is location-specific. Once local species are identified, they can be multiplied as per the available space and number of people. The technology is simple and the practice can be implemented in a small garage or one's own backyard. This technology can be important to rural women and schoolchildren. The former can undertake this as a private entrepreneurial activity which can serve as a source of additional income for them. This technology can also be taken up by schoolchildren as a hobby which earns them pocket money.

13.14 Significance for (and impact on) policy-making

The activities of man play a decisive role in the survival of spiders. Indiscriminate use of pesticides has wiped off whole populations of several species. The destruction of whole habitats such as the species-rich tropical rainforests threatens the survival of numerous spider species as well as all other animal inhabitants of those areas. Although this is not very apparent since man is at the apex of most food chains today, his dangerous activities will eventually boomerang on him. If these disastrous activities are not curtailed, stopped and reversed, man is likely to head the list of early departures of life forms from this planet. Hence it is very essential to take up efforts to conserve biocontrol agents such as spiders immediately. The use of biocontrol agents for control of pests is already recognised public policy in several countries, including India. Use of spiders as biological control agents has already proved to be a great success in countries like China and Japan. Once this technology gains visibility and acceptance, it can be incorporated in the national biological control programme.

Table 13.1. Overview of spiders in Korean rice fields

Family	No.of genera	No. of species	Life style
Titanoecides	1	1	Web builder
Dictnifae	1	1	Web builder
Uloboridae	3	3	Web builder
Pholcidae	9	16	Web builder
Thericidae	1	1	Web builder
Nesticidae	13	23	Web builder
Linyphidae	10	22	Web builder
Araneidae	6	20	Web builder
Tetragnathidae	1	4	Web builder
Agelenidae	3	3	Web builder

Family	No.of genera	No. of species	Life style
Hahniidae	1	1	Web builder
Pisauridae	2	3	Hunter
Lycosidae	6	20	Hunter
Oxyopidae	1	4	Hunter
Liocranidae	3	3	Hunter
Clubionidae	2	6	Hunter
Corinnidae	1	1	Hunter
Ctenidae	1	1	Hunter
Gnaphosidae	8	11	Hunter
Thomisidae	8	13	Hunter
Philodromidae	2	5	Hunter
Salticidae	16	22	Hunter
Total	**99**	**175**	

Table 13.2. Important spider predators of rice pests

Family	Species	Habitat
Theridae	Enoplognatha transversifoveata (Bosenberg et Strand, 1906)	Paddy field
	Theridion subpallens Bosenberg et Strand, 1906	Paddy field and levee
Linyphiidae	Gnathonarium dentatum (Wider, 1834)	Paddy field
	Umbellate insecticeps (Bosenberg et Strand, 1906)	Paddy field
Araneidae	Argiope bruennichi (Scopoli. 1772)	Paddy field
	Lariniodes cornutus (Clerck, 1757)	Paddy field
Tetragnathaidae	Pachygnatha clercki Sundevall, 1823	Paddy field
	Tetragnatha maxillosa Thorell, 1895	Paddy field
Pisauridae	Dolomedes sulfurous L. Koch, 1877	Paddy field and levee
Lycosidae	Arctose ebicha Yaginums,1960	
	Arctose subamylacea Bosenberg et Strand, 1906	
	Pardosa astrigera L Koch, 1878	Paddy field and levee
	Pardosa laura Karsch, 1879	
	Pirata subpiraticus (Bosenberg et Strand, 1906)	
Clubionidae	Clubiona kurilensis Bosenberg et Strand, 1906	Paddy field
Saticidae	Marpissa magister (Krsch, 1879)	Paddy field and levee

13.15 Details of some of the important spiders is as given below :

13.15.1 The ridiidae, cobweb, cobweb weaver, or combfooted spiders

Feed on insects that walk or fly into their webs. Almost always found hanging upside down by their claws in irregularly spun, sticky webs, waiting for prey. The spider is usually concealed in a corner of the web, in a silken tent, or behind debris. This group includes the black widow spider, which produces relatively thick silk that feels rough and sticky. They generally have a soft, round, bulbous abdomen and slender legs without spines. Over 200 species in North America.

13.15.2 Linyphiidae (=Microphantidae), dwarf spiders

Prey on insects that fall, walk, or land in their web. Diurnal (day active) spiders occurring in the plant canopy and among litter on the ground. They produce sheetlike webs on the surface of plants or soil and are common in some field and vegetable crops. Most are relatively small. Several hundred species in North America.

13.15.3 Araneidae, orb weavers or garden spiders

Feed on insects that fly, fall, or are blown into web. Elaborate silken webs are spun in concentric circles. Spiderlings often make symmetrical webs; mature spiders may spin a more specialized design that is helpful in identifying certain species. The spider rests at the center of its web or hides in a shelter near the edge, waiting for prey to become entangled. Orb weavers generally have poor vision and rely on web vibrations to locate and identify prey. About 200 species in North America.

13.15.4 Lycosidae, wolf spiders

Prey on insects that are walking or resting on the ground. Actively hunt in the open during the day and night, often observed on the ground in litter and on low vegetation. Can occur in burrows and under debris on soil. Instead of spinning webs to catch prey, make a small, thick web where they rest. Have a distinctive pattern of eyes: four small eyes in front in a straight row, one middle pair of larger eyes, and one rear pair of widely spaced eyes on top of the head. They have long hairy legs. They are usually black and white or strongly contrasting light and dark, which can make them difficult to discern unless they are moving. About 200 species in North America.

13.15.5 Clubionidae (including Corinnidae), sac spiders or twoclawed hunting spiders

Stalk and capture prey that is walking or resting on surfaces. They spin silken tubes or sacs under bark, among leaves, and in low plants or on the ground, where they hide during the day or retreat after hunting. Commonly are nocturnal, medium-sized, pale spiders with few markings. About 200 species in North America.

13.15.6 Salticidae, jumping spiders

Day-active hunters in plants or on the ground. They make no web; instead they stalk and pounce on prey by jumping distances many times their body length. Jumping spiders have a distinctive pattern of eyes in three rows: the first row of four eyes, with large and distinctive middle eyes; a second row of two very small to minute eyes; and a third row of two medium sized eyes. They usually have an iridescent, metallic-colored abdomen and black carapace. About 300 species in North America.

13.15.6 Oxyopidae, lynx spiders

Stalk and capture resting or walking insects. Active hunters with good vision. Most have spiny legs and a brightly colored body that tapers sharply toward the rear. They have four pairs of eyes grouped in a hexagon. About 2 dozen known species in North America

13.15.7 Thomisidae, crab spiders or flower spiders

Stalk and capture insects walking or resting on surfaces. Diurnal hunters that do not spin webs. Front two pairs of legs are enlarged and extend beyond the side of their flattened body, making them look like tiny crabs. Their small eyes occur in two slightly curved rows, with the top row often much wider than the lower row. Over 200 species in North America.

13.15.8 Agelenidae, funnel weavers or grass spiders

Sit-and-wait predators feed during the day and night on the ground in most types of vegetation, including low-growing plants and trees. Spin funnel-shaped webs, often with several-inch-wide, flat extension covering plants or soil. The spider waits in the hole of its web. When it detects vibrations from an insect that flew or walked into the web, the spider runs out, captures and bites the prey, then carries it back into the funnel to be eaten. Webs on low vegetation become conspicuous in morning light after collecting dew. Funnel weavers have six or eight eyes, all about the same size, arranged in two rows. About 300 species in North America.

Table 13.3. Important arthropod predators of brown planthopper (*Nilaparvata lugens*) in paddy fields in Korea

Predatory species	Stage of BPH attacked
Paederus fusiceps (coleoptra: staphylinidae)	Nymph, adult
Microvelia douglasi (Hemiptera: Vellidae)	Nymph, adult
Mesovelia vittigera (Hemiptera: Mesovellidae)	Adult
Gerris lacustris (Hemiptera: Gerridae)	Adult
Aquaris paludum (Hemiptera: Gerridae)	Adult
Cytorhinus lividipennis (Hemiptera: Miridae)	Egg, nymph
Tytthus chinensis (Hemiptera: Miridae)	Egg
Staccia *diluta* (Hemiptera: Reduviidae)	Nymph
Oncocephalus sp. (Hemiptera: Reduviidae)	Nymph
Hydrometra sp. (Hemiptera: Hydrometridae)	Adult
Sepdon sp. (Diptera: sciomizidae)	Adult
Pardosa pseudoannulata (Araneae: Lycosidae)	Nymph and adult
Pirata subpiraticus (Araneae: Lycosidae)	Nymph and adult
Ummeliata inscticeps (Araneae: Linphylidae)	Nymph and adult

13.16 Conclusion

Large numbers of a wide range of spider species inhabit agricultural fields. Their presence limits the habitats open to insect pests. Spiders threaten insect pests with various foraging strategies. As well as consuming large numbers of insect pests as prey, they have the trait of killing all insects living in their territory. For this reason, spiders are a favorable biological control agent in the agricultural ecosystem. Fagan *et al.* (1998) indicated that treatments which combine the augmentation of natural enemies with insecticide applications may be counterproductive. However, spiders still play an important role in reducing the numbers of insect pests in agricultural fields, even when insecticides are used. In fact, spiders may be responsible for a significant proportion of insect deaths which were thought to be from insecticide applications (Carter and Brown 1973). A quantitative analysis of the capacity of spiders to suppress insect pests, including the spatial distribution of major species of spiders and pests, should be carried out in the field on a large scale, so that spiders can be successfully used as biological control agents.

References

Bae, Y.H., J.H. Lee, and J.S. Hyun. 1994. A Systematic Application of Insecticides to Manage Early Season Insect Pests and Migratory Planthoppers on Rice. Kor. Jour. Appl. Entomol. 33, 4: 207-215.

Bajwa, W. I., and Aliniazee, M. T. 2001. Spider fauna in apple ecosystem of western Oregon and its field susceptibility to chemical and microbial insecticides. J. Econ. Entomol. 94: 68-75

Bedford, S. E., and Usher, M. B. 1994. Distribution of arthropod species across the margins of farm woodlands. Agric., Ecosyst. Environ. 48: 295-305.

Bianchi, F., Booij, C. J. H., and Tscharntke, T. 2006. Sustainable pest regulation in agricultural landscapes: a review on landscape composition, biodiversity and natural pest control. Proc. R. Soc. Lond., Ser. B: Biol. Sci. 273: 1715-1727.

Bishop, L., and Riechert, S. E. 1990. Spider colonization of agroecosystems - mode and source. Environ. Entomol. 19: 1738-1745.

Bogya, S., Markó, V., and Szinetár, C. 2000. Effect of pest management systems on foliage- and grass-dwelling spider communities in an apple orchard in Hungary. Int. J. Pest Manage. 46: 241-250.

Bostanian N.J., Dondale, C.D., Binns, M.R. and D. Pitre. 1984. Effects of Pesticide on Spiders (Araneae) in Quebec Apple Orchards. Canadian Entomologist 116: 663-675.

Bostanian, N. J., Goulet, H., O'Hara, J., Masner, L., and Racette, G. 2004. Towards insecticide free apple orchards: flowering plants to attract beneficial arthropods. Biocontrol Sci. Technol. 14: 25-37.

Brown, M. W. 1999. Applying principles of community ecology to pest management in orchards. Agric., Ecosyst. Environ. 73: 103-106.

Cardenas, M., Ruano, F., Garcia, P., Pascual, F., and Campos, M. 2006. Impact of agricultural management on spider populations in the canopy of olive trees. Biol. Cont. 38: 188-195.

Carter, P. E., and Rypstra, A. L. 1995. Top-down effects in soybean agroecosystems - spider density affects herbivore damage. Oikos 72: 433-439.

Cheng, J.A. and J. Holt. 1990. A system analysis approach to brown planthopper control on rice in Zhejiang Province, China. I. Simulation of Outbreaks. Jour. Appl. Ecol. 27: 85-99.

Chiu, S.C. 1979. Biological control of the brown planthopper. In: Brown Plant-hopper, Threat to Rice Production in Asia. International Rice Research Institute, Los Baños, Laguna, Philippines, pp. 335-355.

Choi, B.R., K.L. Heong, J.O. Lee, J.K. Yoo and C.G. Park. 1996. Effects of Sublethal Doses of Insecticides on the Brown Planthopper, Nilaparvata lugens Stål (Homoptera: Delphacidae) and Mirid predator, Cyrtorhinus lividipennis Reuter (Hemiptera: Miridae). Kor. Jour. Appl. Entomol. 35, 1:52-57.

Choi, J.S., H.G. Goh, K.B. Uhm, K.M. Choi and C.Y. Hwang. 1992. Life cycle of the Mirid predator, Cyrtorhinus lividipennis Reuter (Hemiptera: Miridae). Kor. Jour. Appl. Entomol. 31, 4: 329-337.

Choi, K.M. and M.H. Lee. 1990. Use of Natural Enemies to Control Agricultural Pests in Korea. In: Use of Natural Enemies to Control Agricultural Pests. Food and Fertilizer Technology Center for the Asian and Pacific Region, Taipei, Taiwan, ROC pp. 40-50.

Choi, S.S. and J. Namkung. 1976. Survey on the Spiders of the Rice Paddy Field (I). Kor. Jour. Plant Protection 15, 2: 89-93.

Choi, S.Y., H.R. Lee and J.K. Ryu. 1978. Effects of Carbofuran Root-zone Placement on the Spider Populations in the Paddy Fields. Kor. Jour. Plant Protection 17, 2: 99-103.

Chua, T.H. and E. Mikil. 1989. Effects of prey number and stage on the biology of Cyrtorhinus lividipennis (Hemiptera: Miridae): A predator of Nilaparvata lugens (Homoptera: Delphacidae). Environ. Entomol. 18: 251-255.

Clausen, I.H.S. 1990. Design of research work based on a pilot study dealing with the effect of pesticides on spiders in a sugar-beet field. Acta Zool. Fennica 190: 69-74.

Clough, Y., Kruess, A., Kleijn, D., and Tscharntke, T. 2005. Spider diversity in cereal fields: comparing factors at local, landscape and regional scales. J. Biogeogr. 32: 2007-2014.

Croft, B. A., and Whalon, M. E. 1982. Selective toxicity of pyrethroid insecticides to arthropod natural enemies and pests of agricultural crops. Entomophaga 27: 3-21.

Cronin, J. T., Haynes, K. J., and Dillemuth, F. 2004. Spider effects on planthopper mortality, dispersal, and spatial population dynamics. Ecology 85: 2134-2143.

Denno, R. F., Mitter, M. S., Langellotto, G. A., Gratton, C., and Finke, D. L. 2004. Interactions between a hunting spider and a web-builder: consequences of intraguild predation and cannibalism for prey suppression. Ecol. Entomol. 29: 566-577.

Desneux, N., Decourtye, A., and Delpuech, J. M. 2007. The sublethal effects of pesticides on beneficial arthropods. Annu. Rev. Entomol. 52: 81-106.

Dondale, C. D., Parent, B., and Pitre, D. 1979. 6-Year study of spiders (Araneae) in a Quebec apple orchard. Can. Entomol. 111: 377-380.

Dondale, C.D. 1956. Annotated list of spiders (Araneae) in Nova Scotia apple orchards. Canadian Entomologist 88: 697-700.

Duffey, E. 1956. Aerial dispersal in a known spider population. J. Anim. Ecol. 25: 85-111. Ehmann, W. J. 1994. Organization of spider assemblages on shrubs – an assessment of the role of dispersal mode in colonization. Am. Midl. Nat. 131: 301-310.

Enders, F.A. 1974. Vertical stratification in orb web spiders (Araneae: Araneidae) and a consideration of other methods of coexistence. Ecology 55: 317-328.

Fagan, W.F., A.L. Hakim, H. Ariawan, and S. Yuliyantiningsih. 1998. Interactions between Biological Control Efforts and Insecticide Applications in Tropical Rice Agroecosystems: The Potential Role of Intraguild Predation. Biological Control 13:121-126.

Finke, D. L., and Denno, R. F. 2002. Intraguild predation diminished in complex-structured vegetation: implications for prey suppression. Ecology 83: 643-652.

Fox, C.J.S. and C.D. Dondale. 1972. Annotated list of spiders (Araneae) from hayfields and their margins in Nova Scotia. Canadian Entomologist 104: 1911-1915.

Gastreich, K. R. 1999. Trait-mediated indirect effects of a theridiid spider on an ant-plant mutualism. Ecology 80: 1066-1070.

Gavarra, M. and R.S. Raros. 1973. Studies on the biology of the predator wolf spider, Lycosa pseudoannulata Bös. st Str. (Araneae: Lycosidae). Philippine Entomologist 2, 6: 427.

Glenn, D. M., Puterka, G. J., Vanderzwet, T., Byers, R. E., and Feldhake, C. 1999. Hydrophobic particle films: A new paradigm for suppression of arthropod pests and plant diseases. J. Econ. Entomol. 92: 759-771.

Greenstone, M. H. 1984. Determinants of web spider species-diversity - vegetation structural diversity vs prey availability. Oecologia 62: 299-304.

Greenstone, M. H., Morgan, C. E., Hultsch, A. L., Farrow, R. A., and Dowse, J. E. 1987. Ballooning spiders in Missouri, USA, and New South Wales, Australia - family and mass distributions. J. Arachnol. 15: 163-170.

Gurr, G. M., Thwaite, W. G., and Nicol, H. I. 1999. Field evaluation of the effects of the insect growth regulator tebufenozide on entomophagous arthropods and pests of apples. Aust. J. Entomol. 38: 135-140.

Gut, L. J., Westigard, P. H., and Liss, W. J. 1988. Arthropod colonization and community development on young pear trees in southern Oregon. Melanderia 46: 1-13.

Haddad, C. R., Louw, S. V., and Dippenaar-Schoeman, A. S. 2004. An assessment of the biological control potential of *Heliophanus pistaciae* (Araneae: Salticidae) on *Nysius natalensis* (Hemiptera: Lygaeidae), a pest of pistachio nuts. Biol. Cont. 31: 83-90.

Halaj, J., Cady, A. B., and Uetz, G. W. 2000. Modular habitat refugia enhance generalist predators and lower plant damage in soybeans. Environ. Entomol. 29: 383-393.

Halley, J. M., Thomas, C. F. G., and Jepson, P. C. 1996. A model for the spatial dynamics of linyphiid spiders in farmland. J. Appl. Ecol. 33: 471-492.

Hamamura, T. 1969. Seasonal fluctuation of spider populations in paddy fields. Acta Arachnol. 22, 2: 40-50.

Hamamura, T. 1971. Ecological studies of Pirata subpiraticus (Bös. st Str. 1906) (Araneae: Lycosidae) I. Acta Arachnol. 23: 29-36.

Hamamura, T. 1977. Ecological studies of Pirata subpiraticus (Bös. st Str. 1096) (Araneae: Lycosidae) II. Acta Arachnol. 27 (Special number): 229-238.

Herbivore populations. Annu. Rev. Entomol. 43: 421-447.

Hijii, N. 1984. Arboreal Arthropod Fauna in a Forest. II. Presumed Community Structures based on Biomass and Number of Arthropods in a Chamaecyparis Obtusa Plantation. Jap. Jour. Ecol. 34: 187-193.

Hillebrand, H., and Blenckner, T. 2002. Regional and local impact on species diversity - from pattern to processes. Oecologia 132: 479-491.

Hodge, M. A. 1999. The implications of intraguild predation for the role of spiders in biological control. J. Arachnol. 27: 351-362.

Holt, J., A.J. Cook, T.J. Perfect and G.A. Norton. 1987. Simulation analysis of brown planthopper (Nilaparvata Lugens) population dynamics on rice in the Philippines. Jour. Applied Ecology 24: 87-102.

Horton, D. R., Miliczky, E. R., Broers, D. A., Lewis, R. R., and Calkins, C. O. 2001. Numbers, diversity, and phenology of spiders (Araneae) overwintering in cardboard bands placed in pear and apple orchards of central Washington. Ann. Entomol. Soc. Am. 94: 405-414.

Hu, X. P., Prokopy, R. J., and Mason, J. 1996. Populations of predatory and pest mites in first-level and second-level commercial apple orchard blocks in Massachusetts. J. Appl. Entomol. 120: 47-51.

Hull, L. A., and Starner, V. R. 1983. Impact of 4 synthetic pyrethroids on major natural enemies and pests of apple in Pennsylvania. J. Econ. Entomol. 76: 122-130.

Im, M.S. and S.T. Kim. 1996. Study on the ecology of the spiders as natural enemy on insect pest of main crops I, the fauna and population structure of the spiders at rice paddy field and levee. Jour. Life Sciences, Konkuk Univ. 3: 37-72.

Im, M.S. and S.T. Kim. 1999. Studies on the ecological characteristics of the spiders as natural enemy for rice insect pest at rice paddy field in Koesan area, Chungcheongbuk-do. Journal of Development of Agricultural Resources, Konkuk Univ. 21: 57-66.

Isaia, M., Bona, F., and Badino, G. 2006. Influence of landscape diversity and agricultural practices on spider assemblage in Italian vineyards of Langa Astgiana (Northwest Italy). Environ. Entomol. 35: 297-307.

Jang, Y.D. 1981. Studies on the protection and utilization of the natural enemy against rice insect pests. Res. Reports, Agri. Tech., Chungnam Nat. Univ. 8,1: 19-29.

Jenser, G., Balazs, K., Erdelyi, C., Haltrich, A., Kadar, F., Kozar, F., Markó, V., Racz, V., and Samu, F. 1999. Changes in arthropod population composition in IPM apple orchards under continental climatic conditions in Hungary. Agric., Ecosyst. Environ. 73: 141-154.

Kajak, A., L. Andrzejeuska and Z. Wojcik. 1968. The role of spiders in the decrease of damages caused by Acridoidea on meadows — Experimental investigations. Ekol. pol. (A)16: 755-764.

Kawahara, S., K. Kiritani and T. Sasaba. 1971. The selective activity of rice-pest insecticides against the green leafhopper and spiders. Botyu-Kagaku 36: 121-128.

Kim, H.S. 1992. Suppressive Effects of Wolf Spider, Pirata subpiraticus (Araneae: Lycosidae) on the Population Density of Brown Planthopper (Nilaparvata lugens Stål). Unpublished Ph.D. Thesis, Dongkuk Univ, Seoul, Korea. 69pp.

Kim, H.S., D.J. Im, D.R. Choi and H.K. Ko. 1991. In: Encyclopedia of Identification of Natural Enemy. Rural Development Administration, Suweon, Korea, pp. 25-52.

Kim, H.S., Y.I. Lee and H.P. Lee. 1990. Influence on the levee-burning on the fauna of insect pests and their natural enemies. Kor. Jour. Appl. Entomol. 29,3: 209-215.

Kim, J.B., D. J. Cho, C.H. Kim, K.B. Uhm and S.D. Chang. 1984. Studies on the seasonal fluctuation and the timing of control for the brown planthopper (Nilaparvata lugens Stål) and white-backed planthopper (Sogatella furcifera Horvath) in southern rice cultural area. Res. Reports, ORD (Office of Rural Development, Korea) 26-2 (S.P.M.U): 16-20.

Kim, J.P. 1995a. Studies of spiders in paddy fields and collecting methods. Kor. Arachnol. 11,2: 115-122.

Kim, J.P. 1995b. A list of Korean spiders in paddy fields. Kor. Arachnol. 11,2: 123-133.

Kim, J.P. and J.C. Kim. 1995. Influence of the levee and paddy burning on the fauna of overwintering spider. Kor. Arachnol. 11,2: 59-71.

Kim, S.T. 1998. Studies on the Ecological Characteristics of the Spider Community at Paddy Field and Utilization of the Pirata subpiraticus (Araneae: Lycosidae) for Control of Nilaparvata lugens (Homoptera: Delphacidae). Unpublished Ph.D. Thesis, Konkuk Univ., Seoul, Korea, 90 pp.

Kobayashi, S. 1977. Changes in population density of spiders in paddy field during winter. Acta Arachnol. 27 (special number): 247-251.

Kobayashi, S. and H. Shibata. 1973. Seasonal changes in population density of spiders in paddy fields, with reference to the ecological control of the rice insect pests. Jap. Jour. Applied Ent. Zool. 17,4: 193-202.

Kogan, M. 1998. Integrated pest management: historical perspectives and contemporary developments. Annu. Rev. Entomol. 43: 243-270.

Kuno, E. 1968. Studies on the population dynamics of rice leafhopper in a paddy field. Bull. Kyushu Agri. Exp. Stn. 14,2: 131-246.

Kuno, E. and N. Hokyo. 1970. Comparative analysis of the population dynamics of rice leafhopper, Nephotettix cincticeps Uhler and Nilaparvata lugens Stål, with special reference to natural regulation of their numbers. Res. Popul. Ecol. 12: 154-184.

Kuno, E. and V.A. Dyck. 1984. Dynamics of Philippines and Japanese populations of the brown planthopper: Comparison of basic characteristics. In: Proceedings, ROC-Japan Seminar on the Ecology and the Control of the Brown Planthopper. National Science Council, Rep. of China, pp. 1-9.

Landis, D. A., Wratten, S. D., and Gurr, G. M. 2000. Habitat management to conserve natural enemies of arthropod pests in agriculture. Annu. Rev. Entomol. 45: 175-201.

Lang, A. 2003. Intraguild interference and biocontrol effects of generalist predators in a winter wheat field. Oecologia 134: 144-153.

Langellotto, G. A., and Denno, R. F. 2004. Responses of invertebrate natural enemies to complex-structured habitats: a meta-analytical synthesis. Oecologia 139: 1-10.

Leasar, C.D. and J.D. Umzicker. 1978. Life history, habits and prey preferences of Tetragnatha laboriosa (Araneae: Tetragnathidae). Environ. Entomol. 7: 879-884.

Lee, H. P., J.P. Kim, and J.R. Jun. 1993a. Utilization of insect natural enemies and spiders for the biological control in rice paddy field, community structure of insect pest and spiders, suppress effect on insect pest by natural enemies, and their overwintering habitats in rice paddy field. RDA Jour. Agri. Sci. (Agri. Inst. Coop.) 35: 261-274.

Lee, H.P., J.P. Kim, and J.R. Jun. 1993b. Influences of the insecticidal application on the natural enemies and spider community. Jour. Indus. Tech. Grad. Sch. Dongguk Univ. 1: 295-307.

Lee, J.H., K.H. Kim and U.T. Lim. 1997. Arthropod community in small rice fields associated with different planting methods in Suwon and Icheon. Kor. Jour. Appl. Entomol. 36, 1: 55-66.

Legner, E. F. and E.R. Oatman. 1964. Spiders on apple in Wisconsin and their abundance in a natural and two artificial environments. Canadian Entomologist 96:1202-1207.

Lemke, A., and Poehling, H. M. 2002. Sown weed strips in cereal fields: overwintering site and "source" habitat for *Oedothorax apicatus* (Blackwall) and *Erigone atra* (Blackwall) (Araneae: Erigonidae). Agric., Ecosyst. Environ. 90: 67-80.

Lin, K.S. 1970. Studies on the microclimatic factors in relation to the occurrence of the rice planthopper. Plant Protection Bulletin (Taiwan) 12,4: 184-189.

MacHardy, W. E. 2000. Current status of IPM in apple orchards. Crop Prot. 19: 801-806

Mansour, F., Rosen, D., and Shulov, A. 1980. A survey of spider populations (Araneae) in sprayed and unsprayed apple orchards in Israel and their ability to feed on larvae of *Spodoptera littoralis* (Boisd). Acta Oecol. 1:189-197.

Mansour, F., Rosen, D., and Shulov, A. 1981. Disturbing effect of a spider on larval aggregations of *Spodoptera littoralis*. Entomol. Exp. Appl. 29: 234-237.

Marc, P., Canard, A., and Ysnel, F. 1999. Spiders (Araneae) useful for pest limitation and bioindication. Agric., Ecosyst. Environ. 74: 229-273.

McCaffrey, J. P., and Horsburgh, R. L. 1980. The spider (Arachnida, Araneae) fauna of apple trees in central Virginia. Environ. Entomol. 9: 247-252.

Miliczky, E. R., and Horton, D. R. 2005. Densities of beneficial arthropods within pear and apple orchards affected by distance from adjacent native habitat and association of natural enemies with extra-orchard host plants. Biol. Cont. 33: 249-259.

Morse, D. H. 1993. Some determinants of dispersal by crab spiderlings. Ecology 74: 427-432

Nagata, T. 1985. Chemical control of the brown planthopper in Japan. Jap. Agri. Res. Quart. 18: 176-181.

Namkung, J, S.T. Kim and J.H. Lee. 2000. Revision of the Fauna of Korean Spiders. Insecta Koreana 17,4: 303-343.

Norris, R. F., and Kogan, M. 2005. Ecology of interactions between weeds and arthropods. Annu. Rev. Entomol. 50: 479-503.

Nyffeler, M. 1999. Prey selection of spiders in the field. J. Arachnol. 27: 317- 324.

Nyffeler, M., and Sunderland, K. D. 2003. Composition, abundance and pest control potential of spider communities in agroecosystems: a comparison of European and US studies. Agric., Ecosyst. Environ. 95: 579-612.

Okuma, C. 1958. Some observations on the habits of Oedothorax insecticeps Bösengerg et Strand (Araneae). Acta arachnol. 15: 21-23.

Okuma, C., M.H. Lee and N. Hokyo. 1978. Fauna of spiders in a paddy field in Suweon, Korea. Esakia 11: 81-88.

Olszak, R. W., Luczak, J., and Zajac, R. Z. 1992b. Species composition and numbers of spider communities occurring on different species of shrubs. Ekol. Polska 40: 287-313.

Olszak, R. W., Luczak, J., Niemczyk, E., and Zajac, R. Z. 1992a. The spider community associated with apple trees under different pressure of pesticides. Ekol. Polska 40: 265-286.

Ooi, P.A.C. and B.M. Shepard. 1994. Predators and Parasitoids of Rice Insect Pests. In: Biology and Management of Rice Insects. E.A. Heinrichs, (ed.). Wiley, New Delhi, India, pp. 585-612.

Otake, A. and N. Hokyo. 1976. Rice Plant and Leafhopper Incidence in Malaysia and Indonesia. Report of a Research Tour January to March 1976. Shiryo No. 33, Tropical Agriculture Research Center, Tokyo. 64 pp.

Paik J.C., Y.B. Lee, H.R. Lee and K.M. Choi. 1979. Studies on the physiology, ecology and control of the rice insect pests. Res. Report R.D.A.: 341-367.

Paik, J.C. and J.Y. Hwang. 1990. Preliminary survey of Natural Enemies of Rice Pests for the Development of Integrated Control Programs. Res. Report RDA. 33: 153-171.

Paik, K.Y. and J.S. Kim. 1973. Survey on the spider-fauna and their seasonal fluctuation in paddy fields of Taegu, Korea. Kor. Jour. Plant Protection 12.3: 125-130.

Paik, W.H. and J. Namkung. 1979. Studies on the Rice Paddy Spiders from Korea. Seoul National University, Korea 101 pp.

Paik, W.H., J. Namkung and H.K. Kim. 1974. A list of spiders collected in the rice paddy at Milyang. Kor. Jour. Plant Protection 13,1: 24.

Park, J.S., S.C. Lee, B.H. Lee, Y.I. Kim, K.T. Park and K.J. Ahn. 1972. Influences of the insecticides on the rice insect pests. Res. Report R.D.A.: 146-169.

Pearce, S., Zalucki, M. P., and Hassan, E. 2005. Spider ballooning in soybean and non-crop areas of southeast Queensland. Agric., Ecosyst. Environ. 105: 273-281.

Pekár, S. 1999a. Effect of IPM practices and conventional spraying on spider population dynamics in an apple orchard. Agric., Ecosyst. Environ. 73: 155-166.

Pekár, S. 1999b. Foraging mode: a factor affecting the susceptibility of spiders (Araneae) to insecticide applications. Pestic. Sci. 55: 1077-1082.

Pekár, S. 1999c. Some observations on overwintering of spiders (Araneae) in two contrasting orchards in the Czech Republic. Agric., Ecosyst. Environ. 73: 205-210.

Pekár, S., and Haddad, C. R. 2005. Can agrobiont spiders (Araneae) avoid a surface with pesticide residues? Pest Manage. Sci. 61: 1179-1185.

Pekár, S., and Kocourek, F. 2004. Spiders (Araneae) in the biological and integrated pest management of apple in the Czech Republic. J. Appl. Entomol. 128: 561-566.

Pickett, A.D., N.A. Patterson, H.T. Stultz and F.T. Lord. 1946. The influence of spray programs on the fauna of apple orchards in Nova Scotia: I. An appraisal of the problem and a method of approach. Scient. Agric. 26: 590-600.

Prasifka, J. R., Schmidt, N. P., Kohler, K. A., O'Neal, M. E., Hellmich, R. L., and Singer, J. W. 2006. Effects of living mulches on predator abundance and sentinel prey in a corn-soybean-forage rotation. Environ. Entomol. 35: 1423-1431.

Prokopy, R. J. 2003. Two decades of bottom-up, ecologically based pest management in a small commercial apple orchard in Massachusetts. Agric., Ecosyst. Environ. 94: 299-309.

Prokopy, R. J., Mason, J. L., Christie, M., and Wright, S. E. 1996. Arthropod pest and natural enemy abundance under second-level versus first-level integrated pest management practices in apple orchards: A 4-year study. Agric., Ecosyst. Environ. 57: 35-47.

Rathman, R. J., and Brunner, J. F. 1988. Abundance and composition of predators on young apple, *M. domestica*, within sagebrush and riparian species pools in north central WA. Melanderia 46: 66-81.

Ricklefs, R. E. 1987. Community Diversity - Relative Roles Of Local And Regional Processes. Science 235: 167-171.

Riechert, S. E., and Bishop, L. 1990. Prey control by an assemblage of generalist predators - spiders in garden test systems. Ecology 71: 1441-1450.

Riechert, S. E., and Lawrence, K. 1997. Test for predation effects of single versus multiple species of generalist predators: spiders and their insect prey. Entomol. Exp. Appl. 84: 147-155.

Riechert, S. E., Provencher, L., and Lawrence, K. 1999. The potential of spiders to exhibit stable equilibrium point control of prey: tests of two criteria. Ecol. Appl. 9: 365-377.

Riechert, S.E. 1981. The consquences of being territorial: Spiders, a case study. Amer. Nat. 117: 871-892.

Riechert, S.E. and T. Lockley. 1984. Spiders as biological control agents. Annual Rev. Entomol. 29: 299-320.

Rieux, R., Simon, S., and Defrance, H. 1999. Role of hedgerows and ground cover management on arthropod populations in pear orchards. Agric., Ecosyst. Environ. 73: 119-127.

Rosenheim, J. A. 1998. Higher-order predators and the regulation of insect.

Rosenheim, J. A., and Corbett, A. 2003. Omnivory and the indeterminacy of predator function: can a knowledge of foraging behavior help? Ecology 84: 2538-2548.

Rypstra, A. L., Carter, P. E., Balfour, R. A., and Marshall, S. D. 1999. Architectural features of agricultural habitats and their impact on the spider inhabitants. J. Arachnol. 27: 371-377.

Ryu, J.K., S.Y. Choi, H.R. Lee and Y.H. Song. 1977. Root-zone placement of carbofuran for control of rice insect pests. Kor. Jour. Plant Protection 16,4: 217-220.

Samal, P. and B.c. Misra. 1975. Spiders: The most effective natural enemies of the brown planthoppers in rice. Rice Entomol. Newsletter 3: 31.

Samu, F. 2003. Can field-scale habitat diversification enhance the biocontrol potential of spiders? Pest Manage. Sci. 59: 437-442.

Samu, F., and Szinetár, C. 2002. On the nature of agrobiont spiders. J. Arachnol. 30: 389-402.

Samu, F., Racz, V., Erdelyi, C., and Balazs, K. 1997. Spiders of the foliage and herbaceous layer of an IPM apple orchard in Kecskemet-Szarkas, Hungary. Biol. Agric. Hortic. 15: 131-140.

Samu, F., Sunderland, K. D., and Szinetár, C. 1999. Scale-dependent dispersal and distribution patterns of spiders in agricultural systems: a review. J. Arachnol. 27: 325-332.

Sarvary, M. A., Nyrop, J., Reissig, H., and Gifford, K. M. 2007. Potential for conservation biological control of the obliquebanded leafroller (OBLR) *Choristoneura rosaceana* (Harris) in orchard systems managed with reduced-risk insecticides. Biol. Cont. 40: 37-47

Sasaba, T., K. Kiritani and S. Kawahara. 1973. Food preference of Lycosa in paddy fields. Bull. Kochi Inst. Agr. For. Sci. 5: 61-63.

Schmidt, M. H., and Tscharntke, T. 2005a. The role of perennial habitats for Central European farmland spiders. Agric., Ecosyst. Environ. 105: 235- 242.

Schmidt, M. H., and Tscharntke, T. 2005b. Landscape context of sheetweb spider (Araneae: Linyphiidae) abundance in cereal fields. J. Biogeogr. 32: 467-473.

Schmidt, M. H., Roschewitz, I., Thies, C., and Tscharntke, T. 2005. Differential effects of landscape and management on diversity and density of grounddwelling farmland spiders. J. Appl. Ecol. 42: 281-287.

Schmitt, R.J. 1987. Indirect interactions between prey: apparent competition, predator aggregation and habitat segregation. Ecology 68: 1887-1897.

Shepard B.M. and P.A.C. Ooi. 1991. Techniques for evaluating predators and parasitoids in rice. In: Rice Insects: Management Strategies, E.A. Heinrichs and T.A. Miller (Eds.). Springer-Verlag, Germany, pp. 197-214.

Shepard, B.M., A.T. Barrion and J.A. Litsinger. 1987. Helpful Insects, Spiders, and Pathogens. International Rice Research Institute, Los Baños, Philippines. 127 pp. Snyder, W. E., and Ives, A. R. 2001. Generalist predators disrupt biological control by a specialist parasitoid. Ecology 82: 705-716.

Snyder, W. E., and Ives, A. R. 2003. Interactions between specialist and generalist natural enemies: parasitoids, predators, and pea aphid biocontrol. Ecology 84: 91-107.

Song, Y.H. and Y.G. Lee. 1994. Studies on the spider fauna in the paddy fields of Chinju and Namhae areas. Korean Jour. Environ. Agric. 13,1: 98-110.

Specht, H.B. and C.D. Dondale. 1960. Spider populations in New Jersey apple orchards. Jour. Econ. Ent. 53: 810-814.

Stiling, P., and Cornelissen, T. 2005. What makes a successful biocontrol agent? A meta-analysis of biological control agent performance. Biol. Cont. 34: 236-246.

Strand, E. 1907. Süd und ostasiatische Spinnen. Abh. Naturf. Ges. Gölitz. 25: 107-215. Suenaga, H. 1963. Analytical studies on the ecology of two species of planthoppers, the white-backed planthopper, (*Sogata furcifera Horvath*) and the brown planthopper (*Nilaparvata lugens*), with special reference to their outbreaks. Bull. Kyushu Agric. Expt. Stn. 8, 1: 1-152.

Sunderland, K., and Samu, F. 2000. Effects of agricultural diversification on the abundance, distribution, and pest control potential of spiders: a review. Entomol. Exp. Appl. 95: 1-13.

Symondson, W. O. C., Sunderland, K. D., and Greenstone, M. H. 2002. Can generalist predators be effective biocontrol agents? Annu. Rev. Entomol. 47: 561-594.

Tanaka, K. 1989. Movement of the spiders in arable land. Plant Protection 43, 1: 34-39.

Thomas, C. F. G., Brain, P., and Jepson, P. C. 2003. Aerial activity of linyphiid spiders: modelling dispersal distances from meteorology and behaviour. J. Appl. Ecol. 40: 912-927.

Thomas, C. F. G., Parkinson, L., Griffiths, G. J. K., Garcia, A. F., and Marshall, E. J. P. 2001. Aggregation and temporal stability of carabid beetle distributions in field and hedgerow habitats. J. Appl. Ecol. 38: 100-116.

Thorbek, P., and Topping, C. J. 2005. The influence of landscape diversity and heterogeneity on spatial dynamics of agrobiont linyphiid spiders: an individual-based model. Biocontrol 50: 1-33.

Tixier, M. S., Kreiter, S., and Auger, P. 2000. Colonization of vineyards by phytoseiid mites: their dispersal patterns in the plot and their fate. Exp. Appl. Acarol. 24: 191-211.

Topping, C. J., and Lovei, G. L. 1997. Spider density and diversity in relation to disturbance in agroecosystems in New Zealand, with a comparison to England. N. Z. J. Ecol. 21: 121-128.

Tscharntke, T., Klein, A. M., Kruess, A., Steffan-Dewenter, I., and Thies, C. 2005. Landscape perspectives on agricultural intensification and biodiversity - ecosystem service management. Ecol. Lett. 8: 857-874.

Turnbull, A.L. 1973. Ecology of the true spiders (Araneomorphae). Ann. Rev. Entomol. 18: 305-348.

Uetz, G. W., Halaj, J., and Cady, A. B. 1999. Guild structure of spiders in major crops. J. Arachnol. 27: 270-280.

Vakenti, J. M., Cossentine, J. E., Cooper, B. E., Sharkey, M. J., Yoshimoto, C. M., and Jensen, L. B. M. 2001. Host-plant range and parasitoids of obliquebanded and three-lined leafrollers (Lepidoptera: Tortricidae). Can. Entomol. 133: 139-146

Van den Bosch, R., Messenger, P.S. and A. P. Gutierrez. 1982. An Introduction to Biological Control. Plenum Press, New York, USA. 247 pp.

Varchola, J. M., and Dunn, J. P. 2001. Influence of hedgerow and grassy field borders on ground beetle (Coleoptera: Carabidae) activity in fields of corn. Agriculture Ecosystems & Environment 83: 153-163.

Vincent, C., Hallman, G., Panneton, B., and Fleurat-Lessard, F. 2003. Management of agricultural insects with physical control methods. Annu. Rev. Entomol. 48: 261-281

Weyman, G. S. 1993. A review of the possible causative factors and significance of ballooning in spiders. Ethol. Ecol. Evol. 5: 279-291.

Weyman, G. S., and Jepson, P. C. 1994. The effect of food supply on the colonization of barley by aerially dispersing spiders (Araneae). Oecologia 100: 386-390.

Weyman, G. S., Jepson, P. C., and Sunderland, K. D. 1995. Do seasonal changes in numbers of aerially dispersing spiders reflect population density on the ground or variation in ballooning motivation? Oecologia 101: 487-493

Weyman, G. S., Sunderland, K. D., and Jepson, P. C. 2002. A review of the evolution and mechanisms of ballooning by spiders inhabiting arable farmland. Ethol. Ecol. Evol. 14: 307-326.

Wilson, C., and Tisdell, C. 2001. Why farmers continue to use pesticides despite environmental, health and sustainability costs. Ecol. Econ. 39: 449-462.

Wilson, M. J. 2003. Clay mineralogical and related characteristics of geophagic materials. J. Chem. Ecol. 29: 1525-1547.

Wise, D. H. 1993. Spiders in Ecological Webs. USA: Cambridge University Press.

Wisniewska, J., and Prokopy, R. J. 1997. Pesticide effect on faunal composition, abundance, and body length of spiders (Araneae) in apple orchards. Environ. Entomol. 26: 763-776.

Wissinger, S. A. 1997. Cyclic colonization in predictably ephemeral habitats: a template for biological control in annual crop systems. Biol. Cont. 10: 4-15.

Wyss, E. 1995. The effects of weed strips on aphids and aphidophagous predators in an apple orchard. Entomol. Exp. Appl. 75: 43-49.

Yamano, T. 1977. Seasonal fluctuation of population density of spiders in paddy field in Kyoto City. Acta Arachnol. 27 (Special Number): 253-260.

Yang, K.J., M.S. Im. and S.T. Kim. 1998. Studies on the ecology of spiders as natural enemy for rice insect pest at rice paddy field in Keumsan area, Chungcheongnam-do. The Research Bulletin of Agricultural Life Resource Science Institute Joongbu Univ. 6: 46-56.

Yoon, J. C. 1997. Arthropod Community Structure and Changing Patterns in Rice Ecosystems of Korea. Unpublished Ph.D. Thesis, Seoul National University, Seoul, Korea, 105 pp.

Yoon, J.K. and J. Namkung. 1979. Distribution of spiders on paddy fields in the suburbs of Kwangju City. Kor. Jour. Plant Protection 18, 3: 137-141.

INTEGRATED PEST MANAGEMENT AND REMOTE SENSING

14

D. P. Abrol

14.0 INTRODUCTION

The agro-ecosystems all over the world are rapidly changing due to advances in farm management and technology. New crop varieties are being introduced which are highly responsive to fertilizers, irrigation and pesticide treatment. Various other factors such as monoculture (cultivation of one type of crop over large areas), intensive cultivation of crops, changing weather conditions and indiscriminate use of pesticides have resulted in frequent outbreaks of crop pests and diseases causing huge crop losses. Weeds, rodents and birds also cause heavy crop losses. In our country, the annual crop loss due to pests is estimated to be much higher than the combined loss due to drought and flood. Minimizing these avoidable losses is one way of enhancing food grain production. Disease and pest problems are more severe in tropics than in the temperate zones. Unlimited food supply, lack of predators and favourable climate and weather conditions enable the insects, rodents etc., to acquire the status of pest. There have been many instances where pests and diseases, which were considered minor, have assumed major importance and have become highly destructive as a result of change in crop production programmes. In Andhra Pradesh, for example, *Helicoverpa armigera*, an insect pest of cotton, pigeon pea (tur) and several other crops, utilizes irrigated tomatoes grown in dry season, a practice that was unknown a few years ago, as a new host and builds up its population and attacks various other crops in the ensuing Kharif season. Another example is the brown planthopper, a major pest only in the cooler rice growing countries of Asia, namely, Japan, China and Korea till recently, has become a major threat to rice crop in India as well wherever cultivation of high yielding rice varieties became popular.

Monsoon and typhoons also help in the dispersal of pests and pathogens over large areas. Monitoring the weather systems and the change in agro-ecosystems, therefore, is necessary to provide the farmers with pest control advisories that help them in avoiding sole reliance on pesticides. Applications of remote sensing in entomology and plant protection have been reported earlier (Riley 1989, Rao, 1991) but much less in the context of Integrated Pest Management (IPM). The primary objective here is to give an introduction to the concept of IPM in agriculture and to update the users on the possible uses of remote sensing in IPM in the Indian scenario.

14.1 Integrated Pest Management (IPM)

IPM is a pest population management system that uses all suitable techniques (physical, chemical and biological) ensuring the least possible damage to the agricultural system. This involves the understanding of the life forms of the pest and their development in relation to environment. The IPM methods are aimed at lowering the population density of the pest to a threshold level that is no more considered economically important. Total eradication of the pest is rarely the objective of IPM. The IPM is advocated as the safest to reduce crop losses from pests with emphasis on following methods of pest control:

14.1.1 Physical methods

These include mechanical removal, use of high temperature treatments and electromagnetic energy for searing, injuring, etc. Mechanical control is more ideal in a smaller area. Collection and destroying of egg masses and caterpillars in case of cutworms and hairy caterpillars are effective in suppressing the pest population on many field crops. Uprooting and burning of stem borer infested plants especially in coffee and citrus plantation have been common practices in Karnataka. Use of hand nets for flying insects, sticky bands for mealy bugs, light traps for moths, digging the trenches against army worms and locusts, use of poison baits against rodents are some of the mechanical methods.

14.1.2 Cultural methods

These are regular farm operations designed to destroy or prevent pests from causing damage of economic proportions and often these are the best methods of control since they combine effectiveness along with minimum labour and cost. Following are some of the cultural methods of control:

- Change in time of sowing and harvesting in such a way to close the crop season at a critical time in the life cycle of a pest.

- Adoption of proper crop rotation practices (*e.g.,* cotton followed by leguminous crops)
- Implementation of special cultivation methods like fallow ploughing to control root grubs, cutworms, weevil grubs etc. Similarly ploughing immediately after the harvest and trimming of bunds in paddy fields destroy the hibernating pupae of stem borer and the egg pods of grasshoppers respectively.
- Removal and destroying of weeds that serve as alternate hosts.
- Use of trap crops and smoother crops. Growing of two rows of mustard, one row 15 days before and the other row 25 days after transplanting of cabbage in the ratio of 2:25 is known to reduce the pest population on cabbage to a large extent. Growing "bhendi" as a trap crop for cotton boll worms, and cowpea, soybean and castor in the ratio of 1:25 along the borders of groundnut crop reduces the infestation on the main crop.
- Use of healthy seeds of new varieties of crops that are resistant to pests.
- Avoiding application of high doses of nitrogen fertilizers, etc. An example of change in varietal composition, which avoided a major outbreak of rust, has been reported by Nagarajan *et al* (1983). These workers have observed that a change from local to improved varieties of wheat between the crop years 1973-75 throughout Indo-Gangetic plain delayed and lessened the development of brown and yellow rust. This change averted severe rust epidemic that would have otherwise occurred in 1975.

14.1.3 Chemical methods

Chemical methods are very quick in response and predictable results could be obtained. A growing trend in pesticide consumption indicates the continuous adoption of this method of control. Chemicals of different modes of action like repellents, stomach poisons, contact and systemic poisons, anti-feedants, growth retardant, fumigants, etc. are in use for pest control. Recently, insecticides derived from plant species have become more popular among the farming community. For example, the neem-based products are found to act on various insects, as repellent, feeding deterrent and inhibitor of larval growth. However, chemicals used in pest control rarely kill all the pests and the few which survive usually gain resistance and pose a fresh problem. Moreover chemical control is essentially repetitive in nature and has to be applied again and against each fresh outbreak of pests. The use of chemicals for pest control leads to environmental pollution and also affects the natural enemies of the pest species.

14.1.4 Biological methods

In the strict sense, this method involves the introduction of predators, parasites, and pathogens to reduce the pest population. Simple techniques like installing 4 or 5 bird perchers per hectare of cropland help the predatory birds to catch the larvae, pupae and adult stages of insect pests. Many predaceous beetles are available against mealy bugs and scale insects. Release of green lacewings has become a common practice for the control of aphids, whiteflies, etc. in vegetables and ornamental crops. Among pathogens, Nuclear Polyhedrosis Virus (NPV) is effective on caterpillars of insects that developed resistance to pesticides. Of recent development in the biological control of pests is the use of male sterilization in the control of insect pests and use of genetic manipulation and antibiotics to check the crop diseases.

14.1.5 Other methods

The other methods of pest control are the implementation of strict quarantine laws during export and imports. Effective IPM practices have been developed for major pests of rice, sugarcane, cotton, tobacco, pulses, oilseeds and a few horticultural crops like grapevine.

14.2 Role of remote sensing

Remote sensing is providing an important new dimension in the detection and quantification of damage to plants, assessment of the distribution of the principal or alternate host plants. Satellite remote sensing plays an important role in the detection of conditions favourable to the proliferation of pests. Extraction of information from remotely sensed data is carried out either through visual interpretation or computer aided image processing. Identification of crops and damage caused by pests are done through a photointerpretation key developed using a combination of tone, texture, pattern and knowledge of crop calendars or through spectral and spatial pattern recognition techniques using computers.

14.2.1 Crop stress detection and loss assessment

Regular assessment of crop loss is necessary to plan the crop protection strategies and to delineate the endemic zones. Crop loss estimation requires information on the type of crop, extent of damage caused by the pest, value of the crop etc. Damage caused by pests can result in either overall reduction in the yield or change in colour, offensive odour, etc, reducing the net value of the crop. Stress due to crop pests and diseases result in reduction in leaf area and light interception or loss of chlorophyll (chlorosis) and thus affecting the photosynthetic efficiency of the plants. As a result, the reflectivity and emissivity of the crop canopies are affected. The plant leaves have both diffuse and specular reflectance characteristics. The diffuse leaf reflectance emanates primarily from the interior of the leaf through multiple scattering. The specular character of leaf reflectance at the surface of the leaf is primarily affected by the topography of the cuticular waxes and leaf hairs. Decrease in infrared reflectance is one of the earliest symptoms of reduction in vigour in many plants. The spongy and palisade mesophyll cells become flaccid resulting in reduction in the infrared reflectance. In the case of fungal infection, the leaf air space may be invaded by fungal hyphae, further reducing the infrared reflectance from the leaves. In India, work on the application of remote sensing techniques in crop condition assessment began with the coconut wilt disease detection experiment in Kerala in 1969. Helicopter based multiband photography indicated the feasibility of identification of different plant species and detection of coconut plants affected by wilt. Airborne multiband and multitemporal photography around Mandya in Karnataka state during early 70s was found promising in the identification of sugarcane crop affected by leaf-blight. Multi-band airborne scanner data was used during late 1970s to study the growth stages and disease affected rice and sugarcane crops. During 1980-84, under the Joint Experiments Programme with Indian Council of Agricultural Research, ground-based sensors were used to evaluate many parameters related to crop growth and condition assessment using remote sensing systems. Work on application of remote sensing in plant protection and crop production forecasting has been intensified using the data received from Indian Remote Sensing Satellites (IRS), with the participation of many central and state agencies. In a project carried out jointly by Indian Agricultural Research Institute (IARI) and ISRO, satellite data collected during two years (1978- epidemic year and 1977- normal year) have been analysed using computer aided techniques to detect changes in the wheat growing region of West Punjab affected by yellow rust epidemic. In the early 1980s the rice crop in the Bhadra command area in Karnataka, covering about two lakh acres and valued at about Rs. 40 crores at that time, was damaged by the brown planthopper (BPH). Similar damages of rice crop in many parts of the country have occurred in the past (Krishna, Godavari delta in Andhra Pradesh in 1981; Gobichettipalyam in Tamil Nadu in 1975; Kuttanad and Kole area in Kerala in 1974). The main causes for BPH attacks were found to be use of higher dose of nitorgenous fertilizers, close spacing and micro climatic conditions. Landsat MSS data was used to assess the rice crop area affected by BPH (Rao and Rao, 1982). This early study found it possible to detect anomalies in the tonal level of rice crop canopies but was not able establish the causal factors. It was not possible to distinguish the anomalies due to BPH and that of crop management practices. High spatial resolution data now available from IRS-1C/1D could be used for assessing the conditions favourable for BPH attack on rice. False colour composites of Landsat TM data have been used for the assessment of cotton crop affected by white-fly. Areas of moderate (50 per cent) and severe (80%) crop loss have been identified. The damaged crop area estimated from the satellite data was found to be within 12% of the area reported by the Directorate of Agriculture, Government of Andhra Pradesh. The scientists at the Regional Remote Sensing Service Centre, Nagpur, have used remotely sensed data to delineate diseased and healthy stock of orange plantations. The area under diseased orchards accounted for 5 per cent of the total area of the orange plantation. Another interesting case study was reported by Ranganath *et al.* (2002) on detection of diseased rubber plantations using satellite remote sensing. Their study evaluated the potential of multi-date IRS-1C LISS-III data for detection, mapping and monitoring of rubber plantations affected by Corynospora and Gloeosporium fungi, that caused leaf spot and leaf fall. They had applied image enhancement and classification techniques to identify and extract information on the spatial extent and distribution of healthy and diseased rubber plants. The diseased rubber plantations have shown considerable reduction in the near-infrared reflectance followed by a rise in the reflectance in red and short wave infrared. Vegetation index images generated for different periods have shown the progress of disease incidence, severity and recovery of rubber plantations after fungicidal spraying. There have been a few studies showing a shift in the reflectance in the red edge (670-750 nm) as a result of stress induced by moisture and nutrient deficiency. The wavelength of the inflexion point in the red edge range from 683 nm for low chlorophyll concentrations (stressed leaves), to a maximum of 715 nm for healthy leaves. Similar behaviour was observed in the blue-green (452-548 nm) region as a function of chlorophyll concentration and in the middle infrared (1340-1446 nm and 1800-1922 nm) as a function of the leaf water content (Baret *et al.* 1992). In the case of rice crop the shift towards shorter wavelengths extending to 37 nm in the extreme of stressed conditions was reported (Jayaraman and Srivastava, 2002). Zhang *et al.* (2002) have used a spectral ratio wherein they used spectral response of infected tomatoes as numerator and that of healthy tomatoes as denominator to identify the wavelengths most sensitive to detect the late blight disease.

Evaluation of physiographic features for aerial spraying of pesticides

In the context of chemical methods of pest control, the pollution potentials of the pesticides and their source and sink relationships have to be studied. One such study has been carried out by Rao *et al.* (2002) in the Kasargod district of Kerala state. In this district aerial spraying of pesticides on cashew plantations is a common practice till recently to control Tea Mosquito Bug, an insect pest of the Cashew nut plantations. Over the years, many people in the cashew nut-cultivated region have been found to be suffering from abnormal health problems. At the initiative of the National Institute of Occupational Health, the RRSSC, Bangalore took up a remote sensing based survey of the region using IRS-LISS III data and evaluated the physiographic disposition of the area affected by the health problems. It was found that the affected villages are located in the micro watersheds that have steep slopes, nearly circular in shape, high runoff potential and the first order streams of drainage originating from cashew nut plantations that have been sprayed with pesticides for so many years. The micro watersheds of the affected region have terrain characteristics favourable for transporting the pesticide from the sources of spraying through the entire drainage system resulting in its accumulation in the food chain in the region. This first of its kind study clearly showed the potential of remote sensing in evaluating the suitability of a terrain for aerial spraying of pesticides. Such studies could help the decision-makers to appreciate the environmental implications of large-scale pesticide spraying. There are a few studies showing that the remote sensing techniques could be applied i) to implement the legislative enactment to stop continuous crop growing in a specified area, ii) to demarcate areas requiring crop rotation, iii) to monitor the time of sowing and harvest of crops so that decisions on breaking the life cycle of the pest could be taken, and iv) to estimate the acreage under primary (e.g.; cotton) and alternate hosts (tomato, chilli, tobacco, etc.) for assessing the carry over potential of the pest in the off season. All these are important components in the IPM context.

14.2.2 Remote sensing of weather variables favourable for pest occurrence

Weather conditions such as temperature, humidity (moisture), sunshine hours (light) and wind play major influence on the densities of pest population and their natural enemies. Among the weather parameters that can be remotely sensed, type of cloud, extent of cloud cover, cold cloud duration (a surrogate for rainfall) are the most easily retrievable. Such information was used by phytopathologists to study rust diseases of wheat crop. They found that epidemics of wheat rust over the Indian subcontinent are influenced by western disturbances and the depressions that pass over the southern hills (Nilgiri and Palani Hills). They have shown that the paths of transport of wheat stem rust uredospores are associated with the movement of visible stratus-type clouds from Nilgiri and Palani hills to Central and North India. The daily cumulative cloud cover monitored by weather satellite was used as an index to provide early warning on occurrence of diseases (Nagarajan and Singh 1973).

14.2.3 Desert locust forecasting

Desert locust plagues affect about 20 per cent of the earth's surface spreading across Africa, the middle east, and south-west Asia. Its recession area is 16 million sq.km from Mauritania to northwest India and its area of invasion could be about 30 million sq.km. They breed in areas that have sufficient soil moisture and vegetation to support the early stages of this insect (viz. egg laying and hopper development). They migrate from west to east along with the passage of troughs moving in the westerlies and northward and southward along with the Inter-tropical Convergence Zone (ITCZ). Locust control programmes largely depend on collection of data on weather, habitat condition and swarm size from different countries and transmission of such data to various centres for analysis and to plan control measures. Remotely sensed vegetation indices and rainfall estimates based on cold cloud duration and other cloud indexing techniques are the only cost-effective methods to survey the vast streches of desert locust habitat. Many researchers had studied the desert locustinfestation in relation to meteorology (for example Maske and Rathore, 1982 Dubey *et al.* 1987). Their analysis of the fortnightly data of egg laying, hoppers, scattered adults and swarm development showed chronological orderliness with the spatial progress of the monsoon rains. The egg laying starts with pre-monsoon rains, hopper emergence peaking up after a month of rainfall, the swarm intesity reaching a peak during first fortnight of August and September. The main weather systems bringing rainfall favourable for the development of desert locust are western disturbances, depressions over Arabian sea and a few depressions developed over land. A few studies have focussed on the collection of historical data on weather and habitat conditions with the dynamics of locust developmental stages and synthesized the data using Geographical Information System (GIS) and evolved decision support systems (Healey *et al.* 1996). Dutta *et al.* (2002) from the Regional Remote Sensing Service Centre, Jodhpur had developed a GIS-based information system called Geo-LIMIS (Geographically Encoded Locust Impact Minimization Information System) for Karnataka. This system integrates remotely sensed landform soil texture, soil moisture, vegetation density with the daily weather data to forecast the suitable breeding sites and time of onset of locust upsurge in and around the study area.

14.2.4 Use of rainfall nowcasting in IPM

Application of pesticides in the IPM context requires weather information ahead of a few hours. Rainfall after pesticide spray or dusting may wash away the pesticide or atleast dilute the recommended lethal dose of the pesticide and becomes inefficient in controlling the pest. Pesticide application could be preponed or postponed and if necessary hastened if weather information ahead of a 6-12 hrs is available to the farmer. Although such forecasts are incorporated into the farmer's weather bulletins issued by Regional Forecasting Offices of IMD during abnormal weather situations, there is still scope for improvement in their areal coverage, content and applicability to day to day agricultural operations. Many satellite meteorologists reported the effects of early morning cloud cover on afternoon thunderstorm development. Regions, which have no cloud in the morning generally, have strong convection during the afternoon. Hour-by-hour study of convective cloud development is possible by analysis of satellite data from geostationary orbit. We have many satellites of this orbital type including the most recently launched METSAT. Animation of METSAT images for tracking of cloud types, their movement and development using computerized superimposition of successive infrared images and differential colour coding would enable demarcation of the changes in the cloud contours and thus help in predicting rainfall events in advance of a few hours. The man-machine interaction is very useful to develop very short range forecasting techniques. The utility of microwave sensors in deriving cloud top temperature, atmospheric profiles and accurate measurement of wind flow, necessary for giving accurate short-range prediction, is very high. Techniques should be developed for combined use of data from different sensors.

14.3 Conclusions

Satellite based remote sensing and GIS technologies have been found promising in monitoring the growth and spread of plant diseases and pest infestation in terms of their overall effects on crop vigor and development has now been established. Use of satellite data for monitoring conditions favourable for pest development also has been found possible. Remote sensing can also play a "Watch-dog" role to check the violations of cropping patterns and crop rotation practices in any pest/disease endemic areas of the country. This encouraging start points to a promising future in effectively tackling the pests and diseases that are taking a heavy toll of our crops every year.

References

Baret, F., Jacquemond, S., Guyot, G and Leprieur, C. 1992, Modeled Analaysis of the Biophysical Nature of Spectral shifts and comparison with Information content of Broad Bands. Remote Sensing of Environ, 41: 133 _ 142.

Dubey, R.C., Maska, S.J., Ghosh, R., and Kulkarni, S.S. 1987, Desert Locust (Shistocera gregaria Forsk) infestation in relation to meteorology and sunspot cycles in India during 1945-1980, Mausam, 38(1): 53-58.

Dutta, D., Bhatavdekar, S., Chandrasekharan, B., Sharma, J. R., Adiga, S., Wood, D., and Mc Cardle, A 2002, Geo-LIMIS-A Decision Support System for Minimizing Locust Impact in Kazakhstan. J. Indian Soc. Of Remote Sensing (Communicated).

Healey, R.G., Robertson, S.G., Magor, J.I., Pender, J. and Cressman, K., 1996 A GIS for Desert Locust Forecasting and Monitoring, Int. J. Remote Sensing, 10(1): 117-136.

Jayaraman, V. and Srivastava, S.K. 2002, The Invariance of Red-edge Inflection Wavelengths arrived from Ground-based Spectro-radiometer and Space-borne IRS-P3 MOS-B Data, International Journal of Remote Sensing 23(14):2741-2765.

Maske, S.J. and Rathore, L.S. 1982, Desert locusts, a meteorological perspective, Vayu Mandal, 12 (3/4): 51-54.

Nagarajan, S 1983, Plant Diseas Epidemology, Oxford and IBH publishing Co., New Delhi.

Ranganath, B.K. Pradeep, N., Manjula, V.B, Balakrishna, G., Rajanna, M.D., Damodar, S., Rao Nageswara P.P. 2002, Detection of Diseased Rubber Plantations using Remote Sensing, J. Indian, Soc. of Remote Sensing (Communicated).

Rao Nageswara, P.P. and Rao, V.R. 1982. Identification of Bronn Planthopper and Bacterial Leaf Blight Affected Rice Crop on LANSAT false colour composites. Proc. Of 3 rd Asian Conference on Remote Sensing, December 4-7, 1982, Dhaka, Bangladesh, P.P A-4: 1-12.

Rao Nageswara, P.P., Jayaraman, V., Chandrasekhar, M.G. 1991, Application of Remote Sensing in Plant Protection, Technical Report, ISRO-NNRMS-TR-89-91.
Riley, J.R. 1989, Remote Sensing in Entomology, Ann.Rev. Entomol, 34:247-271.

Zhang, M., Liu, X., and O' Neil, M. 2002, Spectral Discrimination of Phytopthora Infestation Infection on Tomato based on Principal Component and Cluster Analysis. Internation Journal of Remote Sensing, 23(6):1095-1107.

EMERGING PEST PROBLEMS IN RELATION TO CLIMATIC CHANGE

15

Uma Shankar and D.P. Abrol

15.0 INTRODUCTION

Climate change is the complex global phenomenon primarily induced by the interventions of intensified human activities. Forest land was converted into human habitat and agricultural lands. Concentration of greenhouse gases, namely carbon dioxide, methane, Chlorofluorocarbons and nitrous oxide in the atmosphere started increasing. Increased concentration of these gases started absorbing long wave radiation emitted by the earth surface and the atmosphere got progressively warmer, which is called global warming.

Losses attributable to insect damage are likely to increase as a result of decreased crop diversity and increased incidence of insect pests resulting from climate change. Current estimates of changes in climate indicate an increase in global mean annual temperatures of 1°C by 2025 and 3°C by the end of the 21st century. Among all the abiotic factors, temperature is the most important one affecting insect distribution and abundance in time and space. Climate change will probably increase the spread of both the existing and emerging pests. Reduction in damage by insect pests is one of the potential areas for increasing food production. It is therefore, essential to focus our attention on the likely effects of global warming and climate change on the activity and abundance of insect pests and their implications for crop protection and food security.

In the state of Jammu and Kashmir, 70% population dependent on agriculture and allied sectors, engaging about 49% of the work force. Sixty percent area is rainfed and only 56% area is sown more than once in a year due either to non availability of irrigation facilities, water logging or climatic constraints. Horticultural crops are more susceptible to changing conditions than arable crops. Field vegetables will be particularly affected by temperature changes. Water deficits will directly affect fruit and vegetable production (Zarin, 2007). Similarly with highly temperatures and humidity, there could be an increased pressure from insects and disease vectors (Alexey *et al.*, 2007).

The state of Jammu & Kashmir (32-36^{0}N, 73-80^{0}E) has diverse agro-climatic conditions that have encouraged the cultivation of different types of vegetables, fruits, food grains besides medicinal and aromatic plants. Jammu and Kashmir State is known for quality fruit and vegetable production ranging from subtropical to intermediate and temperate areas. Jammu in addition to production of world known high quality basmati rice and rajmash, is known to produce variety of horticultural crops such as vegetables and sub-tropical fruits. However, the production of these crops is far below the national average. The major constraints in the production of crops is the ravages caused by various insect pests and diseases which not only result in the low production of these crops but also drastically impair the quality of these, rendering them unfit for human consumption, reducing marketable yield and thereby posing a serious threat to farming communities.

In particular climate changes could impact the ability of insects to overwinter, the geographic distribution or ranges of insects, and the number of generations and their abundance in agricultural systems. The situation will become more serious with impending shifts in fauna and flora associated with climate change. Therefore, there is a need to focus on technologies that will help in increasing crop productivity on the available arable lands with minimum adverse effects on the environment. This will have to be achieved through crop improvement, better use of natural resources, pest management, and improved productivity in developing countries in the future. Insect pests are major threat to agricultural production. *Helicoverpa armigera,* whitefly, fruit fly, mealy bugs and mites are some polyphagous insect pest of many agricultural and horticultural crops across the world (Fitt, 1989), because of their high fecundity, migratory behaviour, high adaptation to various climatic conditions and development of resistance towards wide range of insecticides. However it is a polyphagous insect pest but chickpea is its most preferred host (Tripathi and Sharma, 1985). Recent climatic changes have also influenced the population of *Helicoverpa armigera* in different crops (Srivastava, 2009).

Pheromone traps are the important component in integrated pest management programme, for monitoring its population.

15.1 Stresses caused by climatic changes

Due to climatic changes, extreme weather conditions such as high temperature, robust winds, hurricanes and tornadoes, etc., are predictable. Other abiotic stresses are also associated with climatic changes such as drought, floods, and increased levels of CO_2 or ozone. All of these can be harmful to plant growth and development (Ahmad and Prasad, 2012a,b; Ahmad *et al.*, 2013a,b; Ahmad and Wani, 2013a,b; Hakeem *et al.*, 2013). Furthermore, certain changes in climate lead to the proliferation of biotic pressures in farming organization including pests, weeds, bacteria, fungi, and viruses. It has been estimated that more than 50% of the crop yield is lost around the world due to environmental stresses (Bray *et al.*, 2000), and also these lead to decreased soil fertility and increased soil erosion (Bray, 2002). Following are the major expected effects on agriculture due to climate change.

Drought Farmers are already coping with drought and it has been responsible for many severe famines throughout history. It is estimated that due to drought, the yield from just over half of Earth's land will be decreased in 50 years' time (Cattivelli *et al.*, 2008). The competition for water among cultivated and urban areas will complicate the problem of water accessibility. The usage of saline water might also help to ease Earth's water difficulties, but this is only feasible with the development of crops that are salt tolerant (Sinclair *et al.*, 2004). In order to limit the problems of drought, there is an urgent need for crop diversity that can preserve water and maintain good harvest during phases of water shortage (Rosegrant *et al.*, 2009). Development of such crops is rather difficult, because of the interchanging crop reaction systems to drought at the genomic, biological, metabolic, and biochemical levels. In order to make varieties that are drought tolerant and easily accessible to growers, interdisciplinary teams of scientists must cooperate (Kelly *et al.*, 2005). Less crop yield and reduced grain size are the result of terminal drought (Vadez *et al.*, 2011). Temperature The development and growth of all crops is influenced by temperature. The present heat range in the Midwest and southern USA is best for crop production, resulting in increased profit. Temperature ranges that are higher than ordinary are predicted to decrease grain legume and cereal yields (Hatfield *et al.*, 2011). Raised temperatures decrease the pollen viability and grain-filling period (Boote and Sinclair, 2006). Furthermore, changes in temperature can lead to a warmer environment and less cold winters, which occasionally allow infections and pests to endure and hibernate, thus increasing the probability of subsequent reduced harvests (Ostberg *et al.*, 2013). In areas of high latitude and elevation, where temperature is the major constraint, high temperatures will increase the growing seasons, which lead to late maturation. However, adapting the latest technologies in agriculture and research on the development of new and resistant crop varieties can help to lessen these effects.

15.1.1 Increased carbon dioxide levels

Carbon dioxide (CO_2) levels increase consistently everywhere in the world, unlike ozone which varies locally. CO_2 is the major raw material of photosynthesis and its increased levels enhance photosynthesis resulting in efficient plant growth and yield (Battisti and Naylor, 2009). However, this is still debatable because plants may have low nutritional value including low protein content (Deressa and Hassan, 2009). It has also been found by researchers that poison ivy becomes more poisonous with higher levels of CO_2. Higher levels of atmospheric CO_2 make herbicides less effective, thus making plants more prone to damage caused by herbs (Taub, 2012).

15.2 Biotic factors

Biotic pressures on harvesting systems include pests, weeds, bacteria, fungi, and viruses. The chief element responsible for how pests and bugs affect crop manufacture and harvest is temperature (Coakley *et al.*, 1999). For example, some populations of insect species, such as flea beetles, are showing signs of overwintering because of warmer winter temperatures, high moisture, and elevated atmospheric CO_2. Bacterial, fungal, and viral agents and certain weeds also proliferate significantly due to temperature, rainfall, and humidity. It is quiet challenging to make plants with increased resistance to such biotic factors (Juroszek and Tiedemann, 2011).

15.3 Modern tools for enhancing crop adaptation

Not all plants' responses to stresses are the same. It depends upon the degree of severity of climate change. To maintain sustainable yields, it is important for plants to have greater adaptability to the changing climate (Capiati *et al.*, 2006). Research is ongoing in various domains such as plant physiology, cell physiology, plant breeding (both molecular breeding and conventional breeding), genetics, genomics and molecular biology in order to improve crop adaptations to climatic change for better yield and food quality (Bhatnagar-Mathur *et al.*, 2008; Ahmad and Prasad, 2012a,b; Ahmad *et al.*, 2013a,b; Ahmad and Wani 2013a,b). These approaches have led to the enhancement of plant genetic material with stress-tolerant and adaptive genes or alleles in the population (Redden, 2013).

Some of the methods to enhance crop adaptability are listed below-Farmers quite often apply insecticides and fungicides to combat these pests on horticultural ecosystem. Sometimes, they apply those pest control chemicals that are not at all recommended and quite often use large amounts of chemical pesticides by increasing the doses. This indiscriminate use of chemicals not only triggered the residues on food for consumers but also disrupts the ecological balance of Agri-Horti-Ecosystem. It is more dangerous in horticultural crops because most of them are being eaten raw.

Integrated Pest Management strategies are the only remedy to contain the use of such toxic chemicals on these horticultural crops. Development of regional specific bio-intensive and bio-rational IPM technology is one of the possible strategies to control these pest problems in different horticultural ecosystems of Jammu province. Various IPM tools are already known for these important pests. Integration of these tools that include use of various bioagents, pheromones, light traps, biological formulations (Bt and NPV), inter-cropping (either as trap crop or as deterrent crop) and also management of nursery by soil solarization, seed treatment with bioagents like Trichoderma etc. and covering the nursery beds with agro-nets are the strategies that will be validated in the field. Identification and use of insect/disease resistant crop varieties through screening, if available, will be prime strategy to control the pest populations. Use of environmentally degradable pesticides like plant extracts (NSKE or neem oil formulation, oil spray, pongamia soap formulation etc) will occupy main stay of spray schedules. If necessary, selective and eco-friendly chemicals will be identified and used for managing the pest populations as a last resort. This will be the elegant way of reducing the amount of pesticides residues on crops for sustainable production.

The studies envisage the improvement in pest management strategies by adapting ecologicallu sound pest management approaches for the selected crops *viz.*, tomato, brinjal, okra, cucurbits, mango, citrus, paddy and rajmash. The strategies will include proper identification, screening of varieties/germplasms, validation and promotion of pest management strategies for safe crop production through improving indigenous technologies for enhancing system productivity. It also help to predict or forecast the insect pests first appearance, incidence and damage potential in relation to climate change. The above said studies will provide an ample scope for management of various insect pest species for better production and to improve the quality of fruit produce. This type of research would help to develop sustainable pest management programs and provide job opportunities for the rural youth and extra source of income to the farming community to uplift their socio-economic status.

Increased temperatures have drastically affected rice (*Oryza sativa* L.) production because of reduced crop duration in Philippines (10% reduction in rice yield per 1°C rise in temperature) (Peng *et al.*, 2004). An increase of 6°C in temperature and a precipitation deficit of 300 mm reduced maize (*Zea mays* L.) yield by 36% in the European Union (Ciais *et al.*, 2005). The swarms of locust produced in the Middle East usually fly eastward into Pakistan and India during summer season and they lay eggs during monsoon period. The swarms as a result of this breeding, return during autumn to the area of winter rainfall, flying to all parts of India and influencing *kharif* crops (Rao and Rao 1996). Changes in rainfall, temperature and wind speed may influence the migratory behaviour of locust. Forecasting of aphids (*Lipaphis erysimi* Kalt) on mustard crop, grown during winter season in northern part of India based on the movement of western disturbances, has been established (Ramana Rao *et al.* 1994). Western disturbances bring in cold and humid air from the Mediterranean region, resulting in cloudy and favourable weather conditions for occurrence of aphids on mustard crop. It was observed that there was a sharp increase in the population of aphids when the mean daily temperature ranged from 10 to 14°C, with relative humidity of 67–85% and cloudiness greater than 5 octas.

15.4 Climate change scenario in Jammu and Kashmir

According to the recently released report by the Inter-governmental Panel on Climate Change (IPCC), global temperatures are likely to rise by 1.10 Celsius to 6.40 Celsius by 2100. Such a scenario means that the plains of Kashmir would no longer receive snow and whatever snow the upper reaches may receive would hardly last to feed the rivers of the region throughout the year, threatening livelihoods at a large scale. Temperature on an average in Kashmir region has shown a rise of 1.450 Celsius while in Jammu region the rise is 2.320 Celsius. The Indian Meteorological Department's monitoring reveals that temperatures are increasing in both Jammu and Kashmir valley, with significant increase in maximum temperature of 0.050 Celsius per year in Kashmir, and minimum of 0.08 degree Celsius in the Jammu region.

Effects of climate change are very much visible in the state, the glaciers in the valley are fast melting. Half a century back Kashmir's forest cover was 37 per cent, which has now reduced to a mere 11 per cent. Deficit in food production in Kashmir region has reached 40 per cent, while the deficit is 30 per cent in vegetable production and 69 per cent in oilseed production for a population of 6 million, putting food security at a greater risk.

15.4.1 Impact of climatic change

Amount of snowfall has reduced by over half to one-fourth of what has been witnessed in Kashmir about four decades ago. What is discernible is that the actual time period for snowfall has also undergone a change with months of December and January receiving scant or no snow while February and March witnessing heavy snowfall. What has also been observed is that the amount of rainfall during winter months of December, January and February has also increased in comparison to the past. This can be attributed to higher temperature prevalent closer to the earth's surface in the plains and the Karewas areas, thereby constraining snowfall in the Kashmir Valley. In most of the areas the snowfall months have reduced from four months a year to only two months a year. Actual time period for snowfall has also undergone a change with December and January receiving scant or no snow while February and March witness heavy snowfall. Quantity of snowfall in Kashmir has clearly reduced over the last few decades. It has been found that there has been an overall 21 per cent reduction in the glacier surface area in the Chenab basin. The mean area of glacial extent has also declined from 1 sq km to 0.32 sq km during 1962–2004.

Arthropods can serve as good indicators of environmental change such as pollution, habitat loss, and fragmentation (McGeoch, 1998; Scherm *et al.*, 2000; Gregory *et al.*, 2009) and monitoring of terrestrial arthropods can provide early warnings of ecological effects attributable to climate change. Further, insects are likely to be most affected by climate change because environmental factors have a strong influence on the development, reproduction, and survival of insect pests and their natural enemies (Bale *et al.*, 2002). Insects have short generation times and high reproductive rates, and hence they are more likely to respond quickly to climate change than plants and vertebrates. Potential responses include changes in phenological patterns, habitat selection and regulation of geographic distribution.

Temperature has a strong influence on the viability and incubation period of *H. armigera* eggs, which can be predicted on the basis of degree-days required for egg hatching (Dhillon and Sharma, 2007). An increase of 3°C in mean daily temperature would cause the carrot fly, *Delia radicum* (L.), to become active a month earlier than at present (Collier *et al.*, 1991) and temperature increases of 5°C–10°C would result in the completion of four generations each year compared to two to three generations presently (Biron *et al.*, 2000), necessitating the adoption of new pest control strategies. Global warming will lead to earlier infestation by insect pests, such as *Helicoverpa zea* (Boddie) in North America (EPA, 1989) and *Helicoverpa armigera* (Hub.) in north India (Sharma, 2010), resulting in increased crop losses. Rising temperatures are likely to result in the availability of new niches for insect pests. The emergence of some insect pests from diapause now occurs much earlier than their natural enemies, resulting in a mismatch between the interacting species. The legume pod borers *H. armigera* and *Maruca vitrata* (Geyer) (Figure 12.3), presently confined to tropical climates in Asia, Africa, and Latin America, are most likely to move to northern Europe and North America over the next 50 years as a result of global warming and climate change (Sharma, 2010). Increases of 1°C and 3°C in temperature would cause northward shifts in the potential distribution of the European corn borer, *Ostrinia nubilalis* (Hub.), up to

1220 km, with an additional generation in nearly all regions where it is currently known to occur (Porter *et al.*, 1991). In a six-decade long study, Gordo and Sanz (2005) found increasingly earlier time of first appearance for the honey bee; cabbage white butterfly, potato beetle and olive fly. Climate change could disrupt the synchrony between host plants and insect herbivores that in trophic chains and webs evolved to cope with expected range of fluctuations in their thermal and other requirements. Projected climate changes could modify extant relationships and interactions in food chains and webs (*i.e.,* plant-herbivore-natural enemy interactions) with unknown consequences in natural, and agricultural systems and diseases vectored by insects, and other arthropod pests (Gutierrez and Baumg¨artner, 1984; 2007; Patz *et al.*, 2003; Baylis and Githeco, 2006).

Large-scale changes in rainfall caused by global warming have a major bearing on the abundance and diversity of arthropods. Increased levels of tree stress are associated with an 8 to 10 fold decline in arthropod species richness and abundance. Arthropod richness and abundance on individual trees were positively correlated with a tree's radial growth during drought, suggesting that tree ring analysis could be used as a predictor of arthropod diversity (Stone *et al.*, 2010). Interactions that involve two or more trophic groups, such as plant–herbivore, plant–pollinator, and insect–host–natural enemy interaction are likely to undergo a major change (Harrington *et al.*, 1999). Some plant species may be unable to cope with climate change, resulting in the extinction of species that are specific to particular host plants (Thomas *et al.*, 2004).

Climate change will have a major effect on geographic distribution of insect pests, and low temperatures are often more important than high temperatures in determining the distribution (Hill, 1987; Thomas *et al.*, 2001). A common approach in predicting developmental dynamics and migration of insects in relation to weather conditions involves the use of degree-day models (Bryant *et al.*, 1998; Roltsch *et al.*, 1999). However, temperature does not act in isolation to influence pest status, and therefore it is important to consider interactions with other variables, such as rainfall, humidity, radiation, and CO_2 concentrations (Harrington *et al.*, 2001). Increasing temperatures may impart greater ability to insect species limited by low temperatures at higher latitudes to overwinter, extending their geographical range (EPA, 1989; Hill and Dymock, 1989; Elphinstone and Toth, 2008). Sudden outbreaks of insect pests can wipe out certain crop species and also encourage invasions by exotic species (Kannan and James, 2009). Spatial shifts in the distribution of crops under changing climatic conditions will also influence the distribution of insect pests in a geographical region (Parry and Carter, 1989). However, whether or not an insect pest would move with a crop into a new habitat will depend on other environmental conditions, such as the presence of overwintering sites, soil type, and moisture.

Climate change may prove catastrophic for some species and human populations unable to adjust to habitat change and unable to migrate to favourable areas. The increasing introduction of invasive species with globalization is a further important complication. Current losses due to invasive species in agricultural and natural systems are in excess of a $130 billion per year in the United States alone (Pimentel *et al.*, 2000), and hence they have become a major focus of much of the climate change research on biological systems (*e.g.,* alien weeds, insect pests (*e.g.*, aphids, mealybugs, fruit flies, bollworms, defoliators, forest bark beetles), and diseases). Global warming will lead to faster development of immature stages, and adults will emerge much earlier than before. Observed responses include early adult emergence and an increase in the length of the flight period. Changes in butterfly phenology have been reported in Europe (Roy and Sparks, 2000; Stefanescu *et al.*, 2003), and species have advanced their flight periods by 2–10 days for every 1°C increase in temperature. In Spain, butterflies appeared earlier by 1 and 7 weeks during the past 15 years, and by 8 days per decade in California (Forister and Shapiro, 2003). Several species of microlepidoptera have also shown temporal shifts in their phenology in Europe (Ellis *et al.*, 1997).

Many insects such as *H. armigera*, *M. separata*, and *Spodoptera litura* (F.), which are migratory, may be well able to exploit new opportunities by moving rapidly into new areas as a result of climate change (Sharma *et al.*, 2002b; Sharma, 2005, 2010). Pest outbreaks are more likely to occur with stressed plants as a result of weakening of the plants' defensive system, thus increasing their level of susceptibility to insect pests (Rhoades, 1985). However, many plant species in tropical regions have the capability to withstand the phenological changes that are a result of climate change (Corlett and LaFrankie, 1998).

Early adult emergence and early arrival of migratory aphids have been observed in the United Kingdom (Zhou *et al.*, 1995). A common phenological response has been observed in four unrelated species of insects (a butterfly, a bee, a fly, and a beetle) (Gordo and Sanz, 2005). In most cases, earlier appearance was correlated with an increase in spring temperatures. Global warming and changes in climate will influence–

- Activity, diversity, and abundance of insect pests.
- Geographical distribution of insect pests.
- Overwintering; development and population dynamics.
- Expression of host–plant resistance to insects.
- Pest outbreaks and pest invasion.
- Synchrony between plants and insect pollinators.
- Effectiveness of crop protection technologies.

As a result of global warming, insects may find suitable alternative habitats at greater latitudes and higher altitudes. Genetic variation and multifactor inheritance of innate recognition of environmental signals may mean that many insect species will have to adapt readily to such disruption. Many species may also have their diapause strategies disrupted as the linkages between temperature, moisture regimes, and day lengths will be altered. Climate change will also result in increased problems with insect-transmitted diseases. These changes will have major implications for food security, particularly in developing countries where the need to increase and sustain food production is most urgent. Long-term monitoring of population levels and insect behaviour, particularly in identifiably sensitive regions, may provide some of the first indications of a biological response to climate change. In addition, it is also desirable to keep ahead of undesirable pest adaptations, and it is important to consider global warming and climate change for planning research and development efforts for pest management and ensuring food security.

Host plant resistance to insects is one of the most environment-friendly components of pest management. However, climate change may alter the interactions between insect pests and their host plants (Bale *et al.*, 2002; Sharma *et al.*, 2005, 2010). There will be increased impact on insect pests that benefit from weakened host defenses as a result of the stress caused by the lack of adaptation to suboptimal climatic conditions. The chemical composition of some plant species changes in direct response to biotic and abiotic stresses, which renders the host plant less suitable for the growth and survival of insect pests (Sharma, 2002). Problems with new insect pests will occur if climatic changes favor the introduction of insect-susceptible cultivars or crops into new areas as one of the adaptive methods to mitigate the effects of climate change (Parry and Carter, 1989).

Tritrophic interactions between plants, herbivores, and parasitoids are the result of a long coevolutionary process, which is specific to a particular environment. Relationships between insect pests and their natural enemies will change as a result of global warming, resulting in both increases and decreases in the status of individual pest species. The higher trophic levels are more likely to be affected by climate change because they depend on the capacity of the lower trophic levels for development and survival. Biodiversity plays an important role in the relative abundance of insect pests and their natural enemies, and there is a need to increase functional diversity in agroecosystems vulnerable to climate change to improve system resilience and decrease the extent of losses caused by insect pests (Newton *et al.*, 2009). Changes in cropping patterns as a result of climate change will drastically affect the balance between insect pests and their natural enemies. Because climate change will lead to a shift in cultivation of crops in nontraditional areas, this may influence the prevalence and importance of specific insect pests and their natural enemies (Maiorano *et al.*, 2008). System diversity can be exploited to enhance the resilience of agroecosystems, improve resource utilization, and stabilize yields to cope with the effects of global warming and climate change on food security (Alteiri 1994; Sharma and Waliyar, 2003; Newton *et al.*, 2009).

Growing awareness about the hazardous effects of synthetic pesticides has led to increasing interest in the use of biopesticides in pest management. Natural plant products, entomopathogenic viruses, fungi, bacteria, and nematodes are highly sensitive to climate change. An increase in temperature and ultraviolet (UV) radiation and a decrease in relative humidity may render many of these control tactics less effective, and such an effect will be more pronounced on natural plant products and biopesticides (Isman, 1997). However, changes in temperature, relative humidity, CO_2, and UV radiation have shown differential effects on biopesticides (Lathika and Jacob, 1974; Verhaar *et al.*, 1998; Coviella and Trumble, 2000; Godoy *et al.*, 2007; Campos-Herrera *et al.*, 2010). Several studies have shown that the virulence of entomopathogenic bacteria on insect pests increases at high temperatures (van Frankenhuyzen, 1994). In case of entomopathogenic fungi and viruses, high relative humidity is associated with high mortality in insect hosts, but the effects on insect hosts infected with bacteria are quite variable (Walstad *et al.*, 1970; Thompson *et al.*, 2006). UV light is known to affect the biological activity of biopesticides adversely (Krieg *et al.*, 1980). Insect viruses are inactivated by extreme temperatures, sunlight, and UV radiation (Krieg *et al.*, 1980; Thompson *et al.*, 2006). Forecasting of aphids (*Lipaphis erysine* Kalt) on mustard crop, grown during winter season in northern part of India based on the movement of western disturbances, has been established (Ramana Rao *et al.* 1994). Western disturbances bring in cold and humid air from the Mediterranean region, resulting in cloudy and favourable weather conditions for occurrence of aphids on mustard crop. It was observed that there was a sharp increase in the population of aphids when the mean daily temperature ranged from 10 to 14°C, with relative humidity of 67–85% and cloudiness greater than 5 octas.

Temperature has shown a positive effect on organochlorines, organophosphates, and carbamates (although there are a few exceptions) but has shown a negative effect on synthetic pyrethroids (Lydy *et al.*, 1999; Li *et al.*, 2006; Uddin and Ara, 2006; Wang and Shen, 2007). Higher temperatures and lower humidity reduce the fumigation period of phosphine against *Liposcelis bostrychophila* (Badinnel) (Nayak and Collins, 2008). Soil texture, moisture, organic matter content, and organic acids influence toxicity and, in turn, affect the efficacy of insecticides for pest management (Achik *et al.*, 1989). Rainfall reduces insecticide toxicity; however, the effects vary with the intensity and amount of rainfall, formulation, and the adjuvants used. Carbon dioxide, pH, and sunlight have shown variable influences on insecticide toxicity (Warwick, 1985; Taylor and Matthews, 1986; Nayak and Collins, 2008). Therefore, there is a need to take a critical look at the effectiveness of various insecticides under global warming and climate change for sustainable crop production and food security.

References

Aerts, M., P. Cockrell, G. Nuessly, R. Raid, T. Schueneman, and D. Seal. 1999. Crop Profile for Corn (Sweet) in Florida.

Andrew, N.R. and L. Hughes. 2005. Diversity and assemblage structure of phytophagous agricultural insects. Ann Rev Entomol 48: 261-281.

Awmack, C.S., C.M. Woodcock and R. Harrington. 1997. Climate change may increase vulnerability of aphids to natural enemies. Ecological Entomology. 22:366-368.

Bale, J.S. G.J. Masters, I.D. Hodkinson, C. Awmack, T.M. Bezemer, V.K. Brown, J. Butterfield, A. Buse, J.C. Coulson, J. Farrar, J.E.G. Good, R. Harrington, S. Hartley, T.H. 2002. Herbivory in global climate change research: direct effects of rising temperature on insect herbivores. *Glob. Change Biol.* 8:1–16

Baniecki, J. and M. Dabaan. 2000a. Crop Profile for Corn (sweet) in West Virginia. http://www.ipmcenters.org/CropProfiles/docs/wvcorn-sweet.html

Baniecki, J. and M. Dabaan. 2000b. Crop Profile for Potatoes in West Virginia. http://www.ipmcenters.org/CropProfiles/docs/WVpotatoes.html

Bergthorsson, P., Bjornsson, H., Dyrmundsson, O., Gudmundsson B., Helgadottir, A., and Jonmundsson, J.V.1988. "The effects of climatic variations on agriculture in Iceland", in Parry, M.L., Carter, T.R., and Konijn, N.T., (eds), *The Impact of Climatic Variations onAgriculture, Volume 1, Assessments in Cool Temperate and Cold Regions* (Dordrecht, The Netherlands: Kluwer, 1988), pp.383-509.

Cannon, R.J. 1998. The implications of predicted climate change for insect pests in the UK, with emphasis on non-indigenous species. 4:785-796.

Castor, L.L., J.E. Ayers, A.A. McNabb, R.A. Krause. 1975. Computerized forecast system for Stewart's bacterial disease on corn. Plant Dis. Rep. 59:533-536.

Chakraborty, S., A.V. Tiedemann, P.S. Teng. 2000. Climate Change: Potential impact on plant diseases. Environ. Poll. 108:317-326

Coakley, S.M., H. Scherm, S. Chakraborty. 1999. Climate Change and Disease Management. Ann. Rev. Phyto. 37:399-426.

Coviella, C. and J. Trumble. 1999. Effects of elevated atmospheric carbon dioxide on insectplant interactions. Conserv. Biol. 13:700-712.

Das, D.K., Singh, J. and Vennila, S. 2011. Emerging Crop Pest Scenario under the Impact of Climate Change – A Brief Review. Journal of Agricultural Physics 11: 13-20.

Easterling DR, Meehl GA, Parmesan C, Changnon SA, Karl TR, Mearns LO. 2000. Climate extremes: observations, modeling, and impacts. *Science* 289:2068–74.

FAO (Food and Agriculture Organization), 2004: Data Base. Food andAgriculture Organization of the United Nations, Rome.

Gaston, K.J. and P.H. Williams. 1996. Spatial patterns in taxonomic diversity. *In: Biodiversity* 202-229. Blackwell Science, Oxford.

Georghiou, G.P. and Taylor, C.E. 1986. Factors influencing the evolution of resistance. *In:* Pesticide Resistance: Strategies and tactics for management. National Research Council, National Academy Press. Pages 143-157.

Godfray HCJ, Hassell MP, Holt RD. 1994. The population dynamic consequences of phenological asynchrony between parasitoids and their hosts. *J. Anim. Ecol.* 63:1–10

Hamilton, J.G., O. Dermody, M. Aldea, A.R. Zangerl, A. Rogers, M.R. Berenbaum, and E. Delucia. 2005. Anthropogenic Changes in Tropospheric Composition Increase Susceptibility of Soybean to Insect Herbivory. Envirn. Entomol. 34:2 479-485.

Harrington, R., R, Fleming, I. P. Woiwood. 2001. Climate change impacts on insect management and conservation in temperate regions: can they be predicted? Agricultural and Forest Entomology 3:233-240.

Harrington, R., S. and Clark, J. 2007. Environmental change and the phenology of European aphids. *Global Change Biology* 13(8): 1550-1564.

Hijmans, R.J., G.A. Forbes, and T.S. Walker. 2000. Estimating the global severity of potato late blight with GIS-linked disease forecast models. Plant Path. 49:697-705.

Hoffman, B. 1999. Crop Profile for Potatoes in Pennsylvania. http://www.ipmcenters.org /CropProfiles/docs/papotatoes.html

Houghton, J.T, Meiro Filho, L.G, Callander, B.A, Harris, N, Kattenberg A, Maskell K (eds) (1996) *Climate Change 1995: The Science of Climate Change. Contribution of Working Group I to the Second Assessment of the Intergovernmental Panel on Climate Change.* Cambridge University Press, Cambridge, 584pp.

http://www.ipmcenters.org/CropProfiles/docs/FLcorn-sweet.html

Hunter, M.D. 2001. Effects of elevated atmospheric carbon dioxide on insect-plant interactions. Ag. Forest. Entomol. 3:153-159.

Karl TR, Nichols N, Gregory J. 1997. The coming climate. *Scientific American*, 276, 54–59.

Kaukoranta, T. 1996. Impact of global warming on potato late blight: risk, yield loss, and control. Agric. Food Sci. Finl. 5:311-327.

Kenny, G.J, Harrison, P.A, Parry, M.L. 1993. *The Effect of Climate Change on Agricultural and Horticultural Potential in Europe.* Research Report no. 2, January 1993. Environmental Change Unit, University of Oxford.

Kunkel KE, Pielke RAJ, Changnon SA. 1999. Temporal fluctuations in weather and climate extremes that cause economic and human health impacts: a review. *Bull. Am. Meteorol. Soc.* 80:1077–98

Landsberg J, Smith MS. 1992. A functional scheme for predicting the outbreak potential of herbivorous insects under global atmospheric change. *Aust. J. Bot.* 40:565–77

Leemans R and Soloman A.M .1993. Modelling the potential change in yield and distribution of the earth's crops under a warmed climate. *Climate Research*, 3, 79–96. Lewis, T. 1997. Thrips as crop pests. CAB International, Cambridge: University Press. 740 pp.

McMichael, A. *et al.,* eds., *Climate Change and Human Health* (World Health Organization, Geneva, 1996), p. 218.

Mcvean, R. and Dixon, A.F.G. 2001. The effect of plant drought-stress on populations of the pea aphid *Acyrthosiphon pisum.* Ecol. Entomol. 26: 440-443.

Messenger, G. 1989. "Migratory Locusts in New Zealand?", *The Weta,* vol. 11-2: 29 (1988). Hill, M.G., and Dymock, J.J., *Impact of Climate Change: Agriculturall Horticultural Systems* (DSIR Entomology Division, submission to New Zealand Climate Programme, Department of Scientific and Industrial Research, Wellington, New Zealand, 1989).

Musser, F. P and Shelton, A. M. 2005. The influence of post-exposure temperature on the toxicity of insecticides to *Ostrinia nubilalis* (Lepidoptera:Crambidae). Pest Manag Sci. 61:508-510.

Parry, M.L, Porter, J.H, Carter, T.R. 1990. Agriculture: climate change and its implications. *Trends in Ecology and Evolution*, 5, 318–322.

Pimentel, D. *et al.*, 1993. Ethical Issues Concerning Potential Global Climate Change on Food Production, "*Journal of Agricultural and Environmental Ethics* 5"., pp. 113-46.

Pimentel, D. *et al.*,1991.Environmental and Economic Impacts of Reducing U.S. Agricultural Pesticide Use, D. Pimentel ed., in *Handbook on Pest Management in Agriculture*, Vol. I. (Boca Raton, FL: CRC Press), p. 679.

Porter, J.H 1990. IPCC, *The potential impacts of climate on agriculture* (Geneva and Nairobi: WMO and UNEP).

Reiners, S and C. Petzoldt (eds). 2005. Integrated Crop and Pest Management Guidelines for Commercial Vegetable Production. Cornell Cooperative Extension publication #124VG http://www.nysaes.cornell.edu/recommends/

Rosenzweig C, Parry, M.L, Fischer G, Frohberg, K. 1993. *Climate change and world food supply.* Research Report no. 3. Environmental Change Unit, University of Oxford, Oxford.

Rosenzweig, C. 1993. Recent global assessments of crop responses to climate change. *In: International Crop Science I* (eds Buxton DR *et al.*), pp. 265–272. Crop Science Society of America, Madison, WI.

Rosenzweig, C., A. Iglesias, X.B. Yang, P.R. Epstein and E. Chivian, 2001. Climate change and extreme weather events: implications for food production, plant diseases and pests. *Global Change and Human Health*, 2, 90-104

Karl TR, Trenbeth KE. 2003. Modern global climate change. *Science* 302:1719–23.

Shelton, A.M., W.R. Wilsey, and Soderlund, D.M.. 2001. Classification of insecticides and acaricides for resistance management. Dept. of Entomology, NYSAES, Geneva, NY 14456. 315-787-2352

Smith, J.B and Tirpak, D., The Potential Effects of Global Climate Change on the United States, Report to Congress (Washington, DC: US Environmental Protection Agency, 1989). A summary is given in Adams, R.M. (and 9 others), "Global climate change and US Agriculture" *Nature, vol.* 345: 219-224 (1990).

Sombroek WG, Gommes R (1996) The climate change – agricultural conundrum. *In: Global Climate Change and Agricultural Production: Direct and Indirect Effects of Changing Hydrological, Pedological and Plant Physiological processes* (eds. Bazzaz F, Sombroek W), pp. 1–14. John Wiley, Chichester.

Sparks TH, Yates TJ (1997) The effect of spring temperature on the appearance dates of British butterflies 1883–1993. *Ecography*, 20, 368–374.

van der Putten WH, de Ruiterb PC, Bezemera TM, Harvey JA, Wassen M, Wolters V. 2004. Trophic interactions in a changing world. *Basic Appl. Ecol.* 5:487–94

Wallin,J.R. and P.E. Waggoner. 1950. The influence of climate on the development and spread of *Phytophthora infestans* in artificially inoculated potato plots. Plant Dis. Reptr. Suppl. 190. pp 19-33.

Watson RT, Zinyowera MC, Moss RH (1996) *Climate Change 1995: Impacts, Adaptations and Mitigation of Climate Change. Contribution of Working Group II to the Second Assessment of the Intergovernmental Panel on Climate Change.* Cambridge University Press, Cambridge, 880pp.

Whitney, S, J. Whalen, M. VanGessel, B. Mulrooney. 2000. Crop Profile for Corn (sweet) in Delaware. http://www.ipmcenters.org/CropProfiles/docs/DEcorn-sweet.html

Yamamura, K. and K. Kiritani. 1998. A simple method to estimate the potential increase in the number of generations under global warming in temperate zones. Appl. Ent. and Zool. 33:289-298.

Ziska, L.H. 2003. Evaluation of the growth response of six invasive species to past, present and future atmospheric carbon dioxide. *J. Exp. Bot.*, 54, 395- 404.

INTEGRATED MANAGEMENT OF PESTS OF STORED GRAINS

16

Devinder Sharma, D. P. Abrol, V. Kaul, Hafeez Ahmad, Suheel Ahmad Ganie and Karandeep Singh Sasan

16.0 INTRODUCTION

The introduction of high yielding varieties in the early sixties which has ultimately blossomed into Green Revolution has helped India, to emerge as one of the leading developing countries in the front of agricultural economy resulting in increased food grain production, touching more than 230 million metric tonnes/year (Anonymous, 2009). The sustained increase in production brought with in the problems of storage. Millions of tonnes of valuable grains worth more than 2000-3000 crores of rupees are either damaged or last for want of knowledge of scientific methods of storage every year due to various agencies like insects, rats, moisture and birds. Insects are responsible for enormous spoilage in storage. Nearly one thousand species of insects have been found associated with stored products in various parts of the world. The majority of insects pests belong to order Coleoptera and Lepidoptera, which accounts for about 60 per cent and 10 per cent respectively of the total number of species of stored insect pests. They feeding grain bore the kernel; destroy the germ portion, cause heating and deterioration in stored produce, and loss in both qualitative and quantitative. Nutritional depletion, reduction in market value besides contaminating losses, mixing with their excretory products are hazardous to health of human beings who handle and eat infested grains. The loss is not merely in terms of quantity but also in quality of foodgrains. The qualitative loss is attributed to chemical changes in protein, carbohydrates, amino acids, fatty acids and vitamins which affect the nutritive value of the grains.

In most countries grains are among the most important staple foods. However they are produced on a seasonal basis, and in many places there is only one harvest a year, which itself may be subject to failure. This means that in order to feed the world's population, most of the global production of maize, wheat, rice, sorghum and millet must be held in storage for periods varying from one month up to more than a year. Protection of food grains through sound scientific storage practices occupies a vital place in the economies of developed and developing countries alike is thus, a matter warranting urgent attention.

16.1 Factors affecting the storage of foodgrains

The ability of man to grow his own food grains and to store them for future use as food, feed and seed is an important attribute to the socioeconomic evolution of human civilization. Storage of food grains acts as an insurance against fluctuation in production, stabilizes price, restores confidence among consumers, strengthens the public distribution system and replenishes depleted stocks. Safe storage of food grains is of paramount importance in the present day context of food problems in the country. During storage, various factors viz. Physical (Temperature and Moisture), Biological Properties of produce, Biotic factors (insects, rodents, mites, birds, microorganisms), Chemical (break-down of produce, pesticides), Engineering (Structural and mechanical) and Socio-economic problems like finance, farming methods, storage and marketing methods etc. play a crucial role in the upkeep and storage life of the agricultural produce.

16.2 Identification of stored grain insect-pests

Farmers are generally aware that an infestation of weevils and lesser grain borers can cause significant quality and monetary loss if left uncontrolled. These insects often are classified as "primary grain pests" because they attack and destroy whole, undamaged grain. The immature stages occur inside the kernels, thus a "hidden" infestation may develop. "Secondary grain pests," the so-called "bran bug groups," include most other grain-attacking beetles (Table 16.1). These insects frequently cause more serious losses where some type of kernel damage precedes their establishment.

Table 16.1. Stored grains pests

Primary pest	Secondary pest
These attack sound foodgrains	These attack the broken or grains already damaged by primary stored pest
Rice weevil, *Sitophilus oryzae*	Red rust flour beetle, *Tribolium castaneum*
Lesser grain borer, *Rhizopertha dominica*	Long headed flour beetle, *Latheticus oryzae*
Khapra beetle, *Trogoderma granarium*	Saw toothed grain beetle, *Oryzaephilus surinamensis*
Pulse beetle, *Callosobruchus spp.*	Red rust grain beetle, *Cryptolestes ferrugineus*
Grain moth, *Sitotroga cerealella* Rice moth, *Corcyra cephalonica*	Flat grain beetle, *Cryptolestes pusillus*
Indian meal moth, *Plodia interpunctella* Almond moth, *Ephestia cautella*	

16.2.1 Detection of infestation

Significance of infestations depends on the species, density, and ultimate plans for the grain. Proper sampling and identification during storage help the manager recognize problems early and thereby prevent further damage (Table 16.2, 16.3). Selecting the most appropriate curative or preventative action from among available alternatives is not easily accomplished, especially when the type of insect present is not known. For example, if insects that feed inside the kernels are already present in significant numbers, a surface protectant applied as a dust or spray would not be the best management option. Usually a combination of tactics will provide the most reliable protection.

16.2.2 Methods of detection of visible infestation foodgrains

Sieving, agitation of sacks, disturbance of stack or bulk surfaces, the feel of grain in bulk, artificial crevices, presence of dead insects, sprays which stimulate insect activity, determination of grain temperature and moisture content, presence of white spots of powdered material outside the grain bags and skin cast by the larvae also indicate the insect infestations in grain mass and use of probe trap and pit fall trap.

Table 16.2. Detection of infestation

Methods	Applicability
Physical methods	
Visual inspection	Whole grains, processed foods
Floatation method	Whole grains
Berlese method	Whole grains

Methods	Applicability
Sampling and sieving	Whole grains, milled products
X-ray technique	Whole grains
NMR and Near Infra Red Spectroscopy	Whole grains
Chemical method	
ELISAtest	Whole grains, milled products, processed foods
Uric acid analysis	Whole grains, milled products and spices
Analysis for CO_2	Whole grains
Specific gravity method	Whole grains except oats and maize
Fragment count	Whole grains, milled products, processed foods
Staining technique	
1. Egg plugs	Whole grains
2. Ninhydrin method	Whole grains

Table 16.3. Indicators of insect infestation in stored product commodities

Indication	Commodities	Insect
Exit holes	Wheat, rice, maize	*R. dominica*
		Sitophilus spp.
	Paddy	*S. cerelella*
	Pulses (whole)	*Callosobruchus* spp.
	Whole spices	*Stegobium paniceum*
Eggs on grain surface	Pulses (whole)	*Callosobrucus* spp.
Webbing or silken strands present	Cereals, whole and milled	*C. cephalonica* *P. interpunctella*
	Oilseeds/oilcakes/meals	*E. cautella* *P. interpunctella*
	Dry fruits	*E. cautella*
	tree nuts	*E. cadidella*
Pupal cases sticking to shells and gunny bags	Peanut in shell	*Caryedon serratus*

16.2.3 Methods of detection of hidden infestation

1. *Use of stains:* solutions of various dyes are used for staining egg plugs in 94 grain samples, this is a rapid method of detecting the presence of weevil infestation. 2. *Use of acid fuchsin:* the stain is prepared by mixing 50 ml glacial acetic acid in 950 ml of distilled water and adding 0.5 g acid fuchsin. Uniform samples of grains are prepared by soaking in warm water for 5 min and are immersed in the stain for 2-5 min after which

the excess stain is removed by washing in tap water. The egg plugs of weevils are stained bright cherry red and feeding punctures including mechanical injures in light pink. (Frankenfeld, 1948). 3. *Gentian violet:* in this method, the wheat sample is exposed for 2 min. In a solution containing 10 drops of 1 per cent aqueous stock solution of gentian violet in 50 ml of 95 per cent ethanol. The egg plugs are stained in purple colour (Goosen, 1949). 4. *Berberine sulphate:* in this method, kernels of wheat are immersed in dilute solution of the dye (20 parts per million) for one minute rinsed and examined under ultra violet light (wave length of 3660a). The egg plug stains show intense greenish yellow colour under ultra violet light. 5. *Floatation or density method:* the grain is immersed in the fluids consisting of sodium silicate in water (sp. Gravity 1.160 to 1.190) and methyl chloroform (sp. Gravity 1.30) with deobase oil. Sodium silicate remains on the top and a 3 layers separation occurs. The non infested kernels sink to the bottom, the infested one float and the light weight kernels including those infested by early stages of insect hang in the line of separation between the two fluids. 6. *Gelatinization method:* in this method the grain is boiled for 10 min. In a 10 per cent solution of sodium hydroxide. The treatment renders the kernels translucent, thereby, revealing the presence of internal infestation. 7. *Cracking floatation method:* cleaned grains are coarsely ground to release the internal insects and soaked in a water- alcohol solution or in boiling water and then mixed with gasoline or mineral oil. The insects are floated off with the oil layer in a flask and collected on a filter paper and counted. 8. *Spectrophotometric analysis:* this method is based on spectrophotometric analysis of dihydroxyphenol present in insect cuticle. These phenols produce certain dyes when these react with dichloroquinone chlorimide. 9. *Aural method:* the detection of the presence of insects by implication of 95 the sounds produced by the insects due to their movement or feeding has been utilized in detecting insect infestation quantitatively with the help of special instrument known as acoustic apparatus. The principle of this method in detecting grain infestation is to convert the mechanical vibration produced by the insects, picked up by a receiver into electric signals. After amplified several thousand times, the signal is conveyed to a transmitter or a head phone. It may be used to determine the grain infestation in laboratories as well as in storage places. The apparatus "acoustic" records only one indication of infestation. The species of the insect stage of development and dead insects cannot be determined on this apparatus. Besides, only active insects can be detected by thisapparatus, cold grain has to be heated to 20-25 °C to activate the insects. (Zaklandoi and Ratanova, 1987). 10. *Ninhydrin colour reaction:* this method which has been developed by dennis and decker (1962) is based on a chemical indicator technique in which the body fluid of the insects (free amino acids of coelomic fluids of insects) produce a colour reaction (purple spots) with ninhydrin impregnated fi1ter paper (0.7 % soln. In acetone). A machine has also been manufactured in which wheat kernels are crushed on a treated strip of paper tape and the same observed for stain while the machine is in operation. The machine is known as 'ashman simon hidden infestation detector'. 11. *Carbon dioxide determination method:* howe and oxley (1944) suggested that by measuring the quantity of carbon dioxide produced by a given sample of grain in 24 h, the extent of internal infestation can be estimated. Representative grain sample free from moving insects is incubated for 24h at 25°C. The concentration of CO_2 is measured with the help of gasometric method. Figure of 0.3 per cent CO_2 at 14 per cent moisture content indicates that the grain is insect free. If moisture content were below 14 per cent such a reading would indicate a slight infestation. In case of slight insect infestation or moisture content higher than the CO_2 concentration will range between 0.3 per cent and 0.5 per cent. A figure between 0.5 per cent to 1.0 per cent indicates that the grain is unfit for prolonged storage. This method fails to indicate the presence of dead insects and requires long period. 12. *X-ray radiographic method:* this method is described by milner *et al.* (1950). This method requires dark room facilities for processing the mm 96 but recently by the use of polaroid radiographic media, dark room and wet processing of film are eliminated. This method is now widely used all over the world for detection of insect infestation within the grain.

16.3 Damage caused by insects

Worldwide annual loss 8-10% (13 mt of grains lost due to insect and 100 mt due to failure to store properly).

- Insects damage stored grain in many ways leading to:
- Reduction in grain weight due to consumption by adults and immature stages.
- Contamination of grains by dead bodies, fragments of insects, moulting, excreta, excretions etc.
- Effect on chemical composition.
- Change in nutritional quality.
- Effect on end use products.
- Health hazards.
    - Inhalant allergy.
    - Ulcerative colitis.
    - Carcinogenic quinones.
    - Carrier of toxicogenic fungi.

- Change in humidity by
  - Re-distribution of water in the bulk as a result of heating.
  - Production of water by the insects as a result of oxidation of hydrogen in their food and the composition of the inert-granular atmosphere is changed by production of CO_2 and consumption of O_2.
- Damage to receptacles (storage) and mills.
- Loss due to screening.
- Loss of goodwill.

16.4 Sources of infestation

- Field infestation (Hidden infestation): *e.g.* pulse beetle and Paddy moth infest crop in the field.
- Cracks and crevices : Insects from old stock infest fresh stock (Cross infestation).
- Left over seed from bins/stores or spilled seeds under the stacks.
- Old Containers and bags.
- Trucks, trollies and bullock carts.
- Entry of insects from neighboring stores.
- Carry over of field infestation.
- Old infested stores.
- Cleanings of the processing plants.
- Human beings.

16.5 Requirement for successful storage

The following points should be considered for safe storage of food grains:

- Selection of site (location)
- Cleaning of storage structures
- Cleaning of Bags
- Cleaning of vehicles
- Proper aeration
- Selection of storage structure.
- Cleaning and Drying of grains.
- Separate storage of new and old stock.
- Use of dunnage.
- Regular Inspection.

Indoor bins	Outdoor bins
Domestic metal bins	Flat and hopper metal bin
Gharelu theka	Partly above ground and underground bin
Pucca kothi	Ferro-cement bins
Welded mesh wire bin	Pusa bin
Reinforced cement ring bin	Vaccum process storage
Paddy straw mud bin	Improved go-downs
	Seed storage bins

16.6 Traditional farm/village storage methods

Traditional methods of storing grain are the product of decades, if not centuries of development, perhaps by trial and error, but certainly as a result of experience of the users and their ancestors. Farm folk of the tropics and subtropics mainly follow Indigenous traditional (ITK) storage methods (Table 16.4). Resource poor farmers in developing countries use different Physicochemical and biological methods to protect stored grains against pest infestation by mixing grains with protectant made up of plant products. Though these methods have a strong scientific basis, it is seldom known to the users. The traditional methods have been used for many years with little or no modification and are successful because of the application of scientific principles, though unawares. Most of the traditional methods, at least unawares, minimize economic loss and damage to the environment. People of J&K follow time-tested customs and practices in various areas like food, medicines and agriculture. In recent years, documenting of traditional wisdom gained significant attention world over, because of its importance in developing high potential, environment friendly and sustainable management. Identification and utilization of such indigenous knowledge from the elderly people from rural and tribal belts will surely bridge the gap between the current scientific and age-old practices.

Table 16.4. Ethnic practices followed by farmers for managing stored pests

Methods	Effective against
Biological methods	
Neem and pongamia leaves	Rice stored pests
Neem, Vitex *negundo leaves*	Maize stored pests
Leaves of *Annona squamosa, Cymbopogon citratus, C. nardus, Erythrina indica*	Pulse pests
Physiochemical methods	
Sun drying	Most of the food grains
Fly ashes and diatomaceous soil	Pulses, maize
Different coloured polyethene bags	Pulse pest
Match boxes (Match sticks have phosphorus, which is anti repellent in nature	Wheat stored pests
Garlic, Turmeric powder has pesticidal insecticidal, and antifungal properties.	Rice pestss
Clove The strong flavour and bitterness of cloves protects the grain from spoilage.	Rice pests
Edible oil and salt (Oil is anti-feedent or feeding repellent and lubricating. Salt is hygroscopic and insecticidal and thus protects the pulses from spoilage.	Pulse pests
Red chillies (repellent)	Rice pests
Soil top uping and mixing	Many stored pests

Management Guidelines for Stored-Product Insects

Pest management measures, in general, have to be integrated into an operational system, be it large or small in scale, if they are to be effectively applied. This is a basic principle, not a novel concept, but it connects well with the modern idea of Integrated Pest Management (IPM). Integrated pest control can be defined as the acceptable use of practicable measures to minimize, cost-effectively. The losses caused by pests in a particular management system. For the measures to be cost-effective they must be appropriate to and acceptable into that system. They may be simple or complex but they must suit the system objectives and its technical capabilities. Furthermore, in this context, cost-effectiveness requires that all costs and benefits, including sociological and environmental effects, should have been taken into account (Table 16.5).

Surveillance, Inspection and Detection

Inspection and detection are necessary parts of a stored-product pest management program. They provide information that allows to evaluate control methods and to monitor for re-infestation. A complete and thorough inspection of the premises helps to locate potential infestation sources. Use pheromone traps inside a building or structure to monitor pest activity; pheromones are available for most of the insects that damage stored food. When using pheromones or food attractants for monitoring, place one trap per 250 to 500 square feet of storage space. For monitoring flying insects, locate traps near storage containers. Put traps inside containers for insects that do not normally fly. Pheromones or attractants can sometimes be used in traps for control of stored-product insects. Trapping may be preferable to insecticides because foods are not exposed to insecticide residues. Put traps close to the infestation source for maximum control, and increase the density of traps to about 1 to each 25 to 50 square feet of storage space. For stored-bulk grains, use pheromones with specially designed probes positioned at different levels inside storage bins. Check probes periodically for insect pests. Use this monitoring also to evaluate the effectiveness of other control measures.

Legal method

Entry of insect which is not found in the particular area can be prevented by the imposition of Destructive Insect pest Act (DIPA) of 1914.

Exclusion

Prevent insect entry into the storage facility by inspecting grains, cereals, flour, and other bulk and packaged products as they arrive. Check packages for holes, webbing, insect frass, eggs, living insects, and insect parts. Even new, unused packaging material, may be an insect source. Immediately dis- infest or destroy them. Prevent contamination of flour, grains, cereals, and dried fruit by keeping them in insect-proof containers. Promptly remove empty boxes and bags from the building or store them separately.

Keep insects out of buildings by using screens over doors and windows. Close off all other openings with wire screening or caulking. If it is not possible to exclude pests from the entire building, at least make sure the storage area is protected. Locate and close rodent holes where stored product insects can enter. If rodent baits are in the area, check them for infestation; even stored or unused bait may harbour insects. To keep from attracting insects into buildings, locate outdoor lighting away from doorways. Use sodium vapour lights rather than mercury vapour lights for outdoor lighting around warehouses and grocery stores, because insects are less attracted to yellow lights.

Sanitation

Sanitation is a critical part of controlling stored-product insects in homes, stores, warehouses, and processing facilities. Clean up spilled materials to eliminate food sources for pests. The cracks should be sealed to eliminate places where pests can hide and to keep grains, flour, or other food from accumulating. Raise shelving in warehouse and other storage areas off the floor to make cleaning underneath possible. The storage areas should be well lighted for ease in cleaning and spotting pest infestations. Moths may be detected more easily during evening hours when they are active. Clean conveyors, augers, and food-processing machinery thoroughly and regularly to prevent them from harboring pests. Drying of food grains of moisture content of below 10% is recommended. Dry the grains thoroughly and stored them in rat and moisture proof new gunny bags. Clean threshing floor/yard, machines, trucks/trollies/storage structures/go downs. All cracks, crevices, rat burrows be plugged. All dirt, rubbish, sweepings, webbings etc should be dumped/destroyed. Use improved storage structures. Disinfest stores/receptacles by malathion 0.5%, primiphos methyl 0.5% or pyrethrin 0.02% @ 3litre/100m2 with frequency treatment of 15 days. The old gunny bags should be treated by dipping in cypermethrin 0.01% or malathion 0.10% for 10 minutes and dry in shade before filling the grains.

Grain drying

Grain moisture regulates biological activity of insects in storages. Atmospheric humidity and grain moisture are closely related factors. Even though stored grains are dried to safe moisture level before storage, moisture content in due course comes to equilibrium with humidity in the air. Paddy at 13 per cent moisture does not allow insects to multiply in storage. (Hall, 1980). Grains stored at around 10 per cent moisture content escape from the attack of insects (except 'khapra' beetle).

Use of improved storage structures

The traditional receptacles used for storage of food grains are not resistant to insects, rodents, fire hazards, moisture and sub-soil water and are also not suitable for fumigation/ disinfestations during storage. Improved traditional storage structures: improved structures developed by different agencies are: pusa bin, 'puri', 'gade', 'patara', 'kothi', 'pucca kothi'. Modern storage structures: food storage receptacles constructed on scientific basis and techniques can store food grains in safe and sound conditions for long periods. Such scientific structures are: i) indoor bins such as (i) domestic designs (ii) 'gharelu thekka' (iii) high moisture hollow bamboo bins. It outdoor bins such as (i) flat bottom metal bins (ii) hopper bottom metal bins - module i to iv (iii) composite bins (iv) r.b. Bins (v) high moisture grain storage bins. Urban bins: circular or square in shape. Polyethylene lined gunny bags or polyethylene bags of 300-400 gauge thickness also protect paddy from cross infestation of insects for up to 2 years as well as stop moisture migration to the paddy stored in these bags. However, R. *dominica* and *S. oryzae* can penetrate polyethylene lining. Sometimes the loss of seed viability was also greater in polyethylene lined bags than in jute bags without polyethylene.

Environmental Modification

Heat treatment kills some pests outright; cold usually blocks their development. For adequate control, it may be necessary to subject products to a prescribed period of high temperatures followed by cold, after which they should be stored at a constant, lowered temperature. In general, a temperature of 60°F prevents insect feeding; 40°F kills insects over a period of time. Some products can be frozen to protect them from insect damage.

i. Cold treatment

Temperature below 14°C result in death particularly immature stages of almost all insect pests. Death rapidly occurs at freezing point (-10 to

20°C). *Tribloium castaneum* and *O. Mercator* are most cold susceptible whereas *Trogoderma* spp., *Ephestia* spp. and *Plodia interpunctella* are most cold tolerant species. In most field applications of cold temperature to control insect pests, the insects are exposed to 10 to 20°C for some days before being exposed to lethal cold temperature. Producers and grain handlers can use ambient cold air to prevent and control infestations. Stored product insect pests generally do not feed or reproduce at temperatures below 18°C. Lower temperatures can also be used to cause mortality. For example, grain kept at -5°C for 12 weeks will control stored insect pests at all life stages. You can lower the temperature of stored grain by: Mixing and transferring infested grain from one bin or pile to another. Transferring part of the crop to a truck or small pile to expose it to low air temperature and leaving it to cool before returning it to the bin. Aerating the bin. Aeration systems are very effective at reducing grain temperatures, as well as reducing moisture migration. Time required at low temperatures to control stored grain insect pest infestations. A Constant grain temperature of -5°C for 12 weeks, -10°C for 8 weeks, -15°C for 4 weeks, and 20°C for 1 week is to be maintained. Time required at low temperatures to control stored grain insect pest infestations: Immature stages of almost all insect pests die below 140°C. Death occurs rapidly at freezing point (-100°C to -200°C). Aeration cool ambient air results in low temperature. Means of low temperature: It can be obtained by aeration with cool ambient air. Bulk grains can be also be cooled by refrigeration in cold stores. *T. casteneum, T. confusum* and *Oryzaephilus mercator* most cold and susceptible and *T. granium, S. granarius, E. elutella* and *P. interpunctella* most cold resistant.

ii. High temperature

In the past, grain dryers have been used to help with disinfestation. There are a number of variables that are difficult to manage and may affect end use quality of grain, including temperature, time, and type of dryer. As a result, insect control using high temperature may produce variable results. Caution: After you apply heat treatment, grain must be immediately cooled so that it does not overheat, spoil and cause a new insect infestation. Energy costs for this method of control are high. Therefore, using a grain dryer is 98 recommended only when you need to rapidly dry grain. High temperature from 50°C–60°C for 10-20 minutes lethal to all insects. Pupal stage of *R. dominica* is heat tolerant. Muthu and Majumdar (1974) found that most of the insects attacking stored seed spices die when the body temperature reaches upto 60°C but in case of infrared, when the temperature attained is 65-68°C in the rotating aluminium or stainless tubes give 100 per cent mortality of S. paniceum. SOLAR BED: Three times exposures on solar bed on sunny days for 15 minutes before storage (first exposure) and at regular intervals of 60 (second exposure) 120 (third exposure) during entire storage period of 180 days respectively proved to be better as compared to exposure of 5 minutes and 10 minutes for control of pulse better i.e. subsequent increase in 10% and 20% level of pulse infestation green-gram infestation. Grain heating is done by hot air fluidized bed, infra red, high frequency dielectric and microwave heating for achieving uniform grain temperature throughout the storage structure.

iii. Controlled atmosphere (CAS)/ Hermetic storage

The CAS involves the use of high CO_2 (9.0 to 9.5%) and low O_2 (2-4%) in storage which is lethal to all insects. This technology for control of stored pests has been extensively used in the field.

Desiccants

Dusts, such as earth, silica gel, non silica and diatomaceous earth, can be combined with certain stored grains to provide protection against insect damage. These dusts kill target insects by desiccation. Dusts are removed from grain and other stored food before processing by a cleaning operation that also removes other debris. Because sorptive dusts are inert, they do not leave any potentially harmful residues on the food if traces of the desiccant remain. The rate for bulk grain is 100-300g/ton, depending on the insect species and the grain moisture content. The rate for surface treatment is 0.3 mg/g to the top 45.7 cm of the grain mass (300 ppm, 0.6 lb/ton to the top 1.5 ft).

Alternative dusting preparations includes ash, laterite dust clay dust or very fine sand. The quantites commonly applied vary considerably, and can reach up to 50 percent by volume. Depending on the type of dust however, it is also possible to achieve an acceptable protective effect with considerably smaller quantities.

Sorting

The food grains displaying insect infestation, mould, mechanical damage or any other inferior qualities, must be removed and processed as soon as possible. This will prevent contamination of the healthy grains, and maintain overall quality of the produce.

Dividing the harvest

The harvest should be divided into two parts. The first is for short-term daily requirements, whilst the other is for long-term storage. Studies have shown that, during the first three to four months after storage, as a rule no high losses are caused by insects, and hence no special protective measures are necessary during this period. Food grains intended for consumption during this period therefore need not be treated. Only those grains intended for longer-term storage need to be treated. This should be performed in a separate store or in a suitable storage area. In this way the farmer can save working time and money. This aspect is also of significance to market-oriented farmers.

Treatment With lime dust

A proven and inexpensive dusting preparation for foodgrains stored in their husks is quicklime which is distributed evenly in a fine coat over the stored products. This lime dust has a dehydrating effect on insects, blocks their respiratory orifices and renders

The integrated stored-food protection package

It has proved to be the case that individual measures of stored-food protection often do not lead to the desired result, as storage constitutes merely one element of the complex post-harvest system Each stage of this system influences the subsequent stages in the simplest case (subsistence farming), the post-harvest system of food grains is characterized by the following stages:

Sowing - harvest - transport - drying - sorting - measures of traditional stored-food protection - storage - processing or consumption.

The package of integrated measures to protect food grains against stored-food pests comprises the following components:

i. Selection of varieties

The storage properties of different varieties should already be taken into account when panning cultivation. For e.g. in Maize, local varieties with long husks tightly enclosing the cob are more suitable for traditional storage than high-yielding varieties with short husks Stored product pests infesting maize m the store have very great difficulty penetrating the protecting enclosing leaves.

ii. Harvest date

It is important to harmonise the framing calendar such that the required labour is available on the best date for harvest. If the harvest is brought forward, there is a risk that the crop will be too moist, and later mouldy in the store. On the other hand, the longer the crop remains on the field, the greater the risk that stored-food pests might already infest the crop in the field, and continue to feed on the crop in the store.

Botanicals

Botanicals such as neem possess repellent, anti feedant and feeding deterrent properties against storage insect pests. Neem seed Kernel powder @ 4.0%, neem seed oil @ 1.0% and mahua oil @ 1% proved repulsive, potent oviposition inhibitor in checking damage by pulse beetle, *C. chinensis* up to 8 months in pigeonpea. Wheat grains mixed with neem and dharek *(Melia azederach)* leaves @ 4% (w/w) were found to be less damaged by insects. The rice grains treated with turmeric powder @ 3.25% (w/w) were found to be least infested by rice weevil (Singal, 2005). Kirubal *et al.* (2008) have reported that 0.2 percent (v/w) ginger grass oil on red gram prevented oviposition and F1 emergence of *Callosobruchus chinensis* (L) for a long period after initial release of adults. Paranagama *et al.* (2003) reported that damage to grain was lower in *C. citratus* Stapf and *C. nardus* Rendle treated to *Oryza sativa* than in the control. *C. citratus and C. nardus* showed deleterious effects on oviposition and F_1 adult emergence of cow-pea bruchid, *C. maculatus* (F) compared to the control during no-choice tests. Use of edible oils mustard and groundnut oils @ 7.5 ml/kg of stored grains are effective for protection of stored pulse grains against pulse beetle 'dhora' upto 9 months. Pulse grain protectants namely groundnut oil @ 3.75 ml/kg + turmeric powder @ 1.75 g/kg, mustard oil @ 3.75 ml + turmeric powder @ 1.75 g/kg, neem oil @ 10 ml/kg, 7 cm sand and dung cake ash covering on gram proved effective in protecting the treated gram grain from the infestation of pulse beetle, Callosobruchus spp for a period of 180 days.

Microbial pesticides

Spinosad is a commercial bacterial insecticide derived from metabolites of the actinomycete bacterium *Saccharopolyspora spinosa* Mertz and Yao. It is highly effective in controlling insects associated with stored wheat (Flinn *et al.*, 2004). In field crops, spinosad loses activity after a week due to breakdown caused by UV radiation from sunlight. In farm bins, where most of the wheat is not exposed to sunlight, spinosad degraded very little during 12 month of storage without appreciable loss of insecticidal activity against the lesser grain borer, *Rhyzopertha dominica* (F.), and red flour beetle, *Tribolium castaneum* (Herbst) (Fang *et al.*, 2002b). Laboratory and field tests using stored wheat have shown spinosad to be effective against several stored-product insects. Spinosad applied to wheat at 0.1 and 1.0mg kg^{-1} was effective in killing all adults and preventing population growth of the lesser grain borer, *Rhyzopertha dominica* (F). A rate of 1.0mg kg^{-1} was necessary for complete control and progeny suppression of the rusty grain beetle, *Cryptolestes ferrugineus* (Stephens), flat grain beetle, *Cryptolestes pusillus* (Sch¨onherr) and confused flour beetle, *Tribolium confusum* (Toews and Subramanyam, 2003). *Bacillus thuringiensis* (Bt) is a registered protectant for use in stored grains in USA. The larva of *P. interpunctella* and *E. cautella* shows high susceptibility to *Bt* (Singal, 2003). NPV, GV and CPV isolated mainly from lepidopterous insects have demonstrated in several cases and may be included to provide long term protection.

Biological control

Controlling insect pests in stored grain and grain products can be very difficult because of the variety of species that can infest grain. Insect parasitoids have been shown to be effective in suppressing a limited number of pest species both in bulk grain storages and in food processing facilities and warehouses. One of the more effective parasitoids is *Theocolax elegans* (Westwood), a small pteromalid wasp (1–2mm) that attacks primary grain pests, whose immature stages develop inside the grain kernels, including the weevils, *Sitophilus* spp., lesser grain borer, *Rhyzopertha dominica* (F.), drugstore beetle, *Stegobium paniceum* (L.), cowpea weevil, *Callosobruchus* spp., and Angoumois grain moth, *Sitotroga cerealella* (Flinn *et al.,* 2006). *Trichogramma* spp. also have been evaluated against a variety of stored product moths in bulk peanut storage, bulk wheat storage and bakeries as well as in warehouses and retail stores in Europe (Grieshop *et al.*, 2007). Stored-product moths commonly oviposit on packages, and on shelves holding stored-product packages. *Trichogramma* spp. are especially promising as biological control agents on finished products because they attack the egg stage of the pests, thereby preventing the invasion of products by first instars. *Dinarmus* spp. is a larval pupal parasitiod of *Callosobruchus* spp., *Bruchus* spp., *Bruchidius atrolineatus* and *Acanthoscelides obtectus* in legume seed. Biological control has a limited scope in the stored pests management but is becoming an increasingly important part of IPM approach.

Insect Growth Regulators (IGRs)

Insect Growth Regulators (IGRs) have a low toxicity to humans as compared with organophosphate, carbamate, and synthetic pyrethroid insecticides. Several IGRs, including hydroprene (1.9 3 1023 mg(AI)/per cm2), have been evaluated for efficacy toward stored-product beetle species (Oberlander *et al.*, 1997). Because of the low toxicity of IGRs, they are usually safe to spray directly onto raw products. Use an IGR where fumigation is not possible or desirable. An IGR is only effective if it contacts the targeted insect pest; therefore, thorough coverage is necessary. Apply a spray of a labelled IGR to grains, nuts, or other foodstuffs during the filling of storage bins. Use enough spray to protect the entire stored product. Spray when insects are at the correct stage of development as described on the IGR label. Occasionally, the application of an IGR extends the larval period, and, therefore, larvae may feed more before they are destroyed.

Fumigants

Fumigants are used to control stored product insects in bulk containers, truck trailers, railroad cars, warehouses, and other large storage areas. Fumigants are effective because they penetrate areas where pests occur or might become problems. To be effective, fumigation must take place in a well-sealed area. This allows the fumigant to reach a lethal concentration of gas, and once this concentration is reached, it must be maintained for a specified period of time. Small quantities of cereals and similar products can be fumigated in containers such as plastic pails or glass jars using dry ice (frozen carbon dioxide); however, if containers are tightly closed immediately after treatment, a vacuum will form that may cause them to implode. Tighten down the lid after the container warms to room temperature.

Residual Insecticides

Short-term residual insecticides, such as pyrethrins can be used for rapid knockdown of some types of stored-product insects. These materials can be applied to bulk containers before adding foodstuffs. They are also used in cupboards and on shelves and areas close to where products are stored, but if infestations are high, frequent re-applications are usually necessary. Selectively use residual insecticides, including some persistent pyrethroids. Residuals are generally applied to surfaces of empty containers to prevent infestation, but rarely applied directly to foodstuffs.

The desirable properties of a grain fumigant, notably efficient penetration of the commodity, toxicity to target insects and lack of harmful residues, make it unlikely that new chemical compounds will become available as fumigants (Taylor, 1991). Carbon dioxide can be used as a conventional fumigant but low toxicity to insects and the consequent high degree of gas tightness necessary for effective insect control makes it unlikely that this gas will find widespread use except in controlled atmosphere (CA) storage systems. Where *Trogoderma* spp. are present, dosage should be increased by 50 percent.

If a 48hr exposure period is reduced to 24 hours the dosage rate should be increased by 50 per cent. If a 24hr exposure period is increased to 48 hours the dosage rate should be reduced by not more than 30 per cent. Where stacks of less than 30m3 (approximately 20 tonnes) are treated under sheets, dosages should be calculated as if the volume were 30m3 (20 tonnes). Phosphine, because of its availability in solid formulations of metal phosphides which are relatively easy to apply, compared with the pressurized gas fumigant methyl bromide, has become the most popular and widely used fumigant in most tropical countries. Methyl bromide, which is in some ways more versatile, retains its place as the fumigant of choice wherever circumstances do not easily accommodate the protracted fumigation period, of several days duration, that is required for the effective use of phosphine.

Table 16.5. Methods for managing stored grain insect pests

Ecological method	Temperature control	Temperature below 1 5 °C and above 42 °C retards growth and development. Heating of grains to 50-60°Cfor 10-20 minutes kill almost all the pests.
	Moisture control	Grains stored at around 10% moisture content escape from the attack of insect-pests except Khapra beetle
	Oxygen control	Oxygen level > 1 % is lethal to all insect-pests.
Mechanical method	Screening of grains	Regular screening and destruction of screening and infested bags
Physical method	Heat treatment	Infra red heating of grains
	Controlled atmosphere	Use of Low O_2 (2-4%) and high CO_2 (9-9.50%) is lethal to all insects
	Inert dust	Mixing of sand, clay, ash, silica aerosols, activated charcoal with food grains
	Edible oils	Coconut, groundnut or mustard oil @ 0.25-0.50% for pulses
Biological method	Bio-agents	Use of Bt, parasitoids, predators etc.
Cultural method	Air tight storage	Use of air tight sealed structure do not allow insects to survive
	Drying of grains	
	Splitting of pulses	Split pulses escape attack of pulse beetle
Chemical method	Prophylactic treatment	Malathion 0.50%, Pyrethrum with 2% pyrethrin EC, Primiphos methyl 0.50% in 1:100 ratio @ 3litre/100m^3 at 15 days interval
	Curative treatment	
	a. Knockdown chemicals	Pyrethrum sprays, lindane smoke generators or fumigation stripes against flying insects
	b. Grain protectants	Pyrethrum dust or malathion 5%@250g/q of seed.
	c. Fumigants	Aluminium phosphide @2 tablets (3g each)/ tonne
		EDB @ 3ml/q for wheat and pulses and3ml/q for rice and paddy
		EDCT @ 55 ml/q stored grains.

Conclusions

It is high time that both the Centre and state governments should think about the severity of the problem of storage losses as it is estimated that the country might be losing every year Rs. 5000 crores to 7000 crores worth of food grains due to defective storage. As a beginning the farm level storage practices. One strategy could also be that the FCI can keep the stocks with the farmers instead of purchasing the grain from them. As per the present categorization in vogue that if a farmer stores the grain without infestation then he should be given some incentive apart from normal government price. This would encourage the farmers to store the grains properly and ultimately it would lead to zero loss level and the pressure on storage space with FCI can also be reduced. Systems approach rather than piece meal must be adopted. The existing post harvest system has to be improved to cut post harvest losses at farm level where about 70% grains are stored and consumed as food, feed & seed.

References

Annonymous 1996. A Practical Guide to Save Grain Campaign. Ministry of Agriculture & Irrigation, Deptt. Of Food. Govt of India.

Annonymous 1996. Training Manual on Scientific Methods Of Storage of Foodgrains, Govt of India. Indian Grain Storage Management & Research Institute Hapur (UP).

Bhargava M. C., Kumawat, K. C. 2010. Pests of Stored Grains and Their Management. New India Publishing, New Delhi. p264.

Hall, D.W. 1970. Handling and storage of food grains in tropical and subtropical areas.

FAO. Agricultural Development Paper No. 90 and FAO, Plant Production and Protection Series No. 19, FAQ, Rome.

Jayas, D.V.,White N.D.G. and Muir, W.E. 1995. Stored Grain Ecosystem. Marcel Dekker.

Khader, V. 2004. Textbook on Food Storage and Preservation. Kalyani Publishers.

Khare, B.P.1994. Stored Grain Pests and Their Management. Kalyani Publishers.

Pruthi, Hem Singh and Mohan Singh 1948. Pests of stored grain and their management.

The Special Number replaces Vol. XVIII. Part 4 of Indian Journal of Agricultural Sciences.

Subramanyam, B. and Hangstrum, D.W. 1995. Integrated Management of Insects in Stored Products. Marcel Dekker.

INTEGRATED PEST MANAGEMENT IN HORTICULTURAL CROPS 17

Uma Shankar

17.0 INTRODUCTION

Jammu and Kashmir State is known as the bowl of horticultural crops and has endowed with the varied habitats and climatic conditions which not only provide the congenial condition but also encouraged the cultivation of a number of fruit and vegetable crops. Horticulture is the backbone of the State's economy with a yearly turnover of Rs. 5000 crores. Presently, it provides direct and indirect employment to about 23.00 lakh people and most of them earn their livelihood from this sector. According to an estimate about 9.10 lakh tons of fruits are produced annually, out of which more than 7.10 lakh tons are exported to the foreign countries. Presently, an area of 1.21 lakh ha is under fruit cultivation in Jammu Province with an annual production of 1.87 lakh M.T. (2014-15). Horticulture has been declared as a thrust area by the state. The major constraint in the production of these crops is the ravages caused by insect-pest which not only result in the low production of fruits and vegetables but also drastically impair their quality by reducing marketable value and thereby posing a serious threat to fruit and vegetable for the export potential.

17.1 Integrated Pest Management

Integrated pest management is a systems approach to pest control that combines biological, cultural, and other alternatives to chemical control with the judicious use of pesticides. The objective of IPM is to maintain pest levels below economically damaging levels while minimizing harmful effects of pest control on human health and environmental resources. IPM is a dynamic and constantly evolving system in which all the suitable control tactics and available surveillance and forecasting information are combined into a holistic management programme delivered to the farmer at requisite interval as part of the sustainable crop production technologies. In another words, IPM aims at combining all available methods or tools of insect pest control in a judicious manner that minimizes insecticide use and disturbance to the ecosystem but becomes a multidisciplinary one.

17.1.1 Prerequisites in Integrated Pest Management

17.1.1.1 Understanding of an Agro-ecosystem

IPM technology manages the pest population in such a manner that economic loss is avoided and adverse side effects of chemical pesticides are minimized. The IPM packages encompass various management strategies for containing the pest and disease problems. Pest monitoring is one of the important components of IPM to take proper decision to manage pest problem well in time. It can be done through agro-ecosystem analysis (AESA), field scouting, light, sticky and pheromone traps. The economic threshold level (ETL) of important pests and disease are determined to take appropriate control measures when pest population crosses ETL.

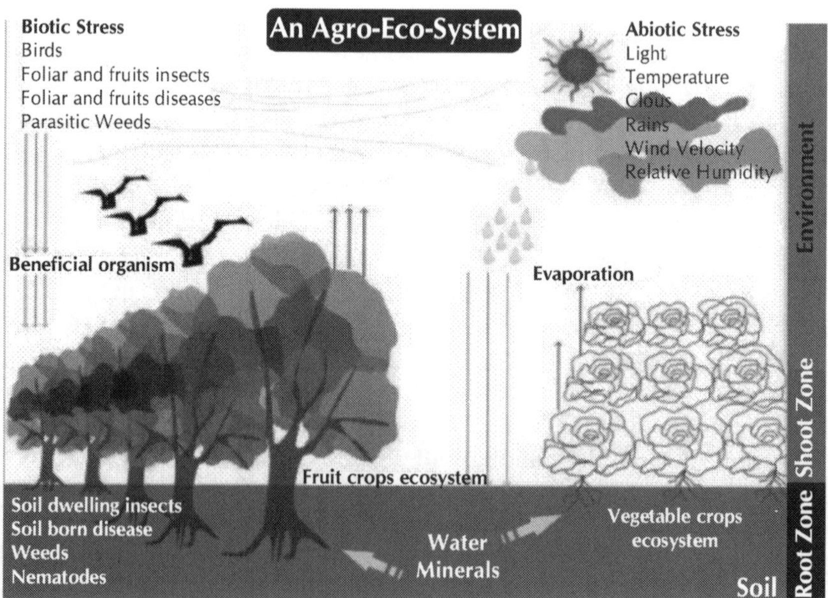

Fig. 17.1

Certain pests are recorded by (DPPQS, 2003) as the pests of national significance. AESA is an approach in which extension functionaries and farmers to analyze the field situations with regard to pests, defenders, soil conditions, plant health, the influence of climatic factors and their relationship for growing healthy crop. Such a critical analysis of the field situation will help in taking appropriate decision on management practices. An agro-ecosystem is a set of model having uniform pattern or cultivation of crops and exposed for the biotic and abiotic components to act upon (Figure 17.1). Many agroecosystems are unfavorable environments for natural enemies due to high levels of disturbance. Habitat management, a form of conservation biological control, is an ecologically based approach aimed at favouring natural enemies and enhancing biological control in agricultural systems (Landis *et al.*, 2000). Understanding the existing eco-system and their relationship in an environment is the fundamental aspects in IPM programme. Although continuous flowering not only encourages but also conserve the natural enemies fauna in short duration vegetable crop. Shennan (2008) discussed the biotic interactions in agro-ecosystems and how they may be manipulated to support crop productivity and environmental health by provision of ecosystem services such as weed, pest and disease management, nutrient cycling and biodiversity conservation. Important elements for understanding biotic interactions include consideration of the effects of diversity, species composition and food web structure on ecosystem processes; the impacts of timing, frequency and intensity of disturbance; and the importance of multitrophic interactions.

17.1.2 Understanding of the ecology of the cropping system

IPM requires an understanding of the ecology of the cropping system including that of the pests and their environment. For example, temperature is the primary factor determining the rate at which insects develop; higher temperatures increase the rate of development. Temperature can be important when determining the frequency of insecticide applications. In sweet corn for example, the interval between insecticide applications for the European corn borer can be increased when the weather is cool because eggs are not hatching as rapidly and the rate of larval development is slower; with warmer temperatures the situation is reversed. Degree-day models can aid in determining how fast insects are developing and the timing of applications. Weather conditions can also influence insect pest populations. For example, wind-driven rain is a major source of mortality to nymphs of the potato leafhopper on potatoes and thrips on onions and frame leaves of cabbage.

17.1.2.1 Sampling and Monitoring

To estimate the severity of pest infestations, the crop must be sampled at least a week. Sampling may involve examining plants and recording the number of insect pests or the amount of damage observed, or traps may be used to capture the adult pest species to estimate pest abundance. Sampling is conducted at regular intervals throughout the season or during critical stages of crop growth. Monitoring includes detecting, identifying, and determining the level of pest populations on a timely basis. Insect traps can often be used to detect pests and identify them when scouting should be intensified or control measures should be taken. Scientifically based, accurate, and efficient monitoring methods are available for many pests on vegetable crops.

17.1.2.2 Forecasting

Weather data and other information help predict when specific pests will most likely occur. Weather-based pest forecast models for diseases and insects of many crops have been developed.

17.1.2.3 Economic threshold level

Use thresholds to determine when pest populations have reached a level that could cause economic damage. The decision to use an insecticide, or take some other action, against an insect infestation requires an understanding of the level of damage or insect infestation a crop can tolerate without an unacceptable economic loss. The level of infestation or damage at which some action must be taken to prevent an economic loss is referred to as the 'action threshold'. Thresholds levels should be adjusted based on market, environmental conditions, variety, etc.

17.1.2.4 Record keeping

Records kept from year to year on insect pest occurrence in fields can be valuable tools for avoiding insect pests in the future.

The insect pests of horticultural crops are discussed in two broad categories i.e., insect pests vegetable crops and insect pests fruit crops for easy understanding.

(A) Important insect pests of Vegetable crops and their Management.

17.2 Insect pests of Brinjal

Eggplant (*Solanum melongena*) is one of the most important vegetables in South Asia, which accounts for almost 50% of the world's area under eggplant cultivation (Alam *et al.* 2003). Eggplant fruit and shoot borer (EFSB), *Leucinodes orbonalis* Guenee (Lepidoptera: Pyralidae) is one of the most serious insect pests of eggplant (Alam *et al.* 2003). Larvae of this pest bore inside tender shoots. This can result in reduction of plant growth and subsequently the yield (Mall *et al.* 1992). Farmers in South Asia rely primarily on the application of chemical insecticides to combat EFSB. This has resulted in tremendous misuse of pesticides. Yield losses of up to 90% have been reported in India (Kalloo, 1988) and similar losses are also common in neighboring Bangladesh (Ali *et al.*, 1980).

1. Brinjal Shoot and Fruit Borers (BSFB), *Leucinodes orbonalis* (Lepidoptera: Pyralidae)

Brinjal shoot and fruit borer inflicts a serious damage to the fruit causing drastic reduction in the marketable quality fruit yield. The damage starts soon after transplanting the crop and continues till harvest of the fruits. The adults are small moths having brown to black spots on the wings. The adult female lays eggs on the ventral surface of the leaves, flower buds and on young fruits. After hatching within the week, small pinkish larva initially bores into the terminal shoots resulting in wilting and drying of the shoot and at later stage, it damages the young fruits by making holes and feeds inside which makes the fruits unfit for consumption and marketing. Wilting reduces plant growth, which causes a reduction in fruit number and size. Over 90% of fruits can be infested in high pest pressure years (Dhandapani *et al.* 2003). Regular pruning and prompt disposal of pest damaged eggplant shoots, especially up to the first harvest, is an important component in the BSFB IPM strategy (Rahman *et al.*, 2002; Talekar, 2002; Miller *et al.*, 2003; Satpathy *et al.*, 2005). Often the entry hole encourages the secondary bacterial infection and rotting of fruits. As its larvae are concealed within the twigs as well as in fruits, it is very difficult to control.

Management

- Monoculture or continuous cropping of brinjal on the same piece of land should be avoided.
- The affected twigs should be regularly clipped along with the insect and destroyed. Fruits showing any boring symptoms should also be picked and destroyed.
- The moths can be mass trapped and killed by installation of water pheromone traps @ 100 per ha at 10 m spacing (Cork *et al.*, 2001, 2003)
- Apply ash dust and avoid the excessive dose of nitrogenous fertilizers in brinjal crops to reduce the infestation of BSFB.
- Long and oblong varieties with bunchy fruiting/ flowering behavior are less susceptible to BSFB.
- Before spraying, all the affected twigs and fruits should be clipped off or removed from the field (Shankar *et al.*, 2010; Alam *et al.*, 2003, 2006).

- Spraying the crops alternatively with profenophos 50 EC @ 2ml/ litre of water at 15 days interval and Cartap hydrochloride @ 1ml/lit or flubendiamide @ 0.5 ml/lit of water starting from 20 days after transplanting to control the pest.
- The safer insecticides like malathion, endosulfan, deltamethrin etc. may be applied in alternate or in association with *Bt* formulations as per need (Raju *et al.*, 2007).
- The larval parasitoids such as *Trathala flavor-orbitalis* and *Goryphus nursei* should be encouraged through the strips of flowering plants.

2. Jassids, *Amrasca biguttula biguttula* (Hemiptera:Cicadellidae)

Both nymphs and adults suck the sap from the lower surface of the leaves. The infested leaf curl upward along the margins, which may turn yellowish and show, burnt up patches. They also transmit mycoplasma disease like little leaf and virus disease like mosaic or leathery leaves. Fruit setting is adversely affected by the infestation.

Management

- Seed treatment with thiamethoxam 70 WS @ 3g/kg seed at the time of sowing.
- Foliar spray of imidacloprid (0.006 %) during the vegetative stage of the crop.

3. Red Spidermite, *Tetranychus cinnabarinus* and *Paratetranychus indicus* (Acari: Tetranychidae)

Nymphs and adults of phytophagous mite suck cell sap and leaves exhibiting yellowish white symptoms leading to their wilting, burning and shedding. It also infests floral buds and tender fruits. The loss in yield is related to the population level and stage of infestation. Castanera and Arroyo (1979) found that initial infestation of 20 or 60 mites / plant significantly reduces the yield. The red spider mite infests brinjal particularly in low humid dry condition. Different stages of mites are found in colonies covered by white-silky webs on lower surface of leaves.

Management

- During egg stage and the resting stages, most acaricides are ineffective. At high temperatures, it may be necessary to apply these at an interval of two days.
- Acaricides like dicofol (0.05%) and wettable sulphur (0.3%) gives effective control of mites. Avoid use of synthetic pyrethroids for borer control.
- Collection and burning of severely infested plant parts reduces further multiplication of mites.
- Proper irrigation and clean cultivation are essential to reduce the damage caused by mites.

- Insect pests of Tomato is a very popular constituent of Indian diet. It is devastated by an array of insect pests, however the major damage is caused by fruit borer, *Helicoverpa armigera*, serpentine leaf miner, *Lariomyza trifoli* and white flies, *Bemesia tabaci* (Shankar *et al.*, 2009b).

1. Tomato fruit borer, *Helicoverpa armigera* Hubner

Helicoverpa has been recorded on more than 200 hosts in India (Pawar, 1998) and is most damaging to cotton, pigeonpea, chickpea, groundnut, sorghum, pearl millet, tomato and many other crops of economic importance. Caterpillars are the immature stage of the insect which causes the considerable damage to the crops in various stages. In the pre-fruiting stage, the caterpillar feeds on the tender foliage including leaves, flowers, buds and even sometimes they make holes in the tender twigs which reflects the perforated look to the crops. After fruiting, the larvae bore large, clear, circular holes into fruits and feed on the pulp. Its name has been changed according to its host. Its larvae are stoutly built and robust. In tomato, it is responsible to destroy many fruits and make them unfit for the human consumption. Its larvae have peculiar feeding habit on host crop. The larvae thrusts its half of the body into the fruits and feeds the inner contents and rendering them unfit for the human consumption. It causes damage up to the tune of 50 to 80 per cent if not intervened at right time.

Management

- Soil solarization in nursery beds as well as in the main field.
- Plant one row of marigold as trap crop for every 16 rows of tomato.
- Monitor and clip top three tender leaves for *Helicoverpa* eggs destruction.
- Early maturing varieties should be planted to escape the damage by borers.
- Installation of pheromone traps @ 5-7 per ha for early detection and 12-15 per ha for trapping and mass destruction for mating disruption.
- Release *Trichogramma chilonis* @ 50,000 adults /ha per release (6 times) at weekly interval when eggs of *Helicoverpa* are noticed.
- Conservation of *Campoletus chloridae*, a potential parasitoid of *H. armigera.*
- Spraying of EPN *Steinernema carpocapsae* against *Helicoverpa* gave an excellent result in killing the larvae within 24 to 48 hrs.
- Spray Ha-NPV 250 to 500 LE/ha (freshly prepared) 2-3 times at 10 days interval at evening hours (Sivaprakasam, 1998; Satpathy *et al.*, 1999; Satpathy and Rai, 2000).
- Spraying of bacterial formulation *Bt* @ 500 g /ha also may proved beneficial in regulating the *Helicoverpa* larvae in the field condition.

2. Serpentine leaf miner, *Liriomyza trifoli*

Leaf miners lay eggs below the epidermal layer of leaves and cause heavy leaf mining on upper or lower side of the leaves. Mines starts from the margins of leaves and progress towards the base/centre of leaf. Mines are initially narrow and gradually enlarge often twisting through leaf which causes stippling of young seedlings. Severe leaf mining slows down the plant growth and accelerates the leaf drop.

Management

- Nuvon / DDVP or dimethoate @ 2 ml per lit of water are found to be effective against leaf miner.
- Hymenoptera (Eulophidae) an endo-larval parasiotid of leafminer, Hemiptarsenus varicornis is found to be very effective.
- Neem seed kernel extract (NSKE) @ 5 per cent concentration found promising in controlling leaf miner followed by Karanj, Pongamia pinnata oil next in the order of efficacy.

17.3 Insect pests of chillies

1. Chilli mites, *Polyphagotarsonemus latus*

Chilli crops are attacked by yellow mite, *Polyphagotarsonemus latus* which has become a serious problem in the entire country. Mite affected leaves of the chilli plant turn towards underside and look like an inverted boat. As young and adult mites attack the tender leaves, almost all new leaves curled downward and lower surface appears silvery whereas, the upper surface of leaves look more dark green as compared to healthy ones. Further the plant growth is checked and the internodes become shorter.

Management

- Since last few decades, farmers have normally been using different kinds of insecticides for the control of mites. Being insecticides these were not able to give satisfactory control of mites, due to which mites also developed resistance.
- Some insecticides have been reported to be effective for mite control (chilli pest complex) - triazophos 40 EC @ 2ml per lit of water or Dicofol 18.5 EC @ 2ml per lit of water or propargite 57 EC @ 3-4 ml per lit of water at the beginning of mite attack followed by another spray at an interval of 12 days.
- Seedling root dip treatment with imidacloprid 200 SL @ 1ml per lit of water before planting and 3 sprays with the same at 15 days interval.
- The following lines/varieties have been reported resistant/ tolerant to mites viz., Pusa Jwala.

2. Thrips, *Scirtothrips dorsalis*

Thrips cause severe damage to the plants in the nursery and in transplanted crop and is the main constraint in the chilli and *Capcicum* cultivation. Both the nymphs and adults of chilli thrips lacerate the leaf and imbibe the oozing sap. Sometimes even buds and flowers are also attacked. Tender leaves and growing shoots are preferred for feeding, while the older leaves are rarely attacked. The infested leaves start curling and crumbling and ultimately shed while, buds become brittle and drop down. Main visible symptom are curling of leaves upward (normal boat shaped), small leaf size and shorter internodal space with stunting growth of plants, no fruiting or small deformed fruiting in infected plants takes place.

Management

- Rouge out the infected plants and burn them.
- Grow resistant varieties.
- Seed treatment of imidacloprid 70 WS at 10, 20 and 30g per kg protected the seedlings in the nursery up to 45 days and foliar treatment of imidacloprid 200 SL at 250, 375 and 500 ml /ha reduced the thrips population significantly and also recorded higher yield of chilli fruits (Santharam *et al.*, 2003).
- Periodic spray of metasystox @ 2 ml per lit of water or imidacloprid @ 0.5 ml per lit water at 10 days interval for vector (*Bemesia tabaci*) Management.

17.4 Insect Pests of Cucurbits

1. Cutworms in cucurbits: *Agrotis ipsilon*

It is a polyphagous insect pest which feeds mainly in seedling stages of many vegetables, potato, maize and other crops. Caterpillar is the only damaging stage of this pest. The short stout larvae are highly destructive. At night, larvae cut the plant at ground level at an early stage of their growth. A few caterpillars in the field kill many small plants during one night. They fell/cut more plants rather consume.

Management

- The population of cut worm larvae can be considerably suppressed by applying 1.3 % lindane dust @ 20 kg /ha or drench the soil with chlorpyriphos @ 2.5 ml per lit of water or carbaryl @ 4 g per lit of water.

2. Red Pumpkin Beetle, *Aulacophora foveicollis*

- The beetles eat leaf lamina, cause perforation, irregular holes or complete defoliation and damage is severe, particularly at cotyledonary stage of crop. The grubs feed on the underground stem and root portion of the host plant forming holes/ galleries leading to dry up of branches or whole plant. They may also attack the fruits touching the ground surface by making holes.

Management

- Plant the crop in November to avoid the damage from this insect.
- After harvesting the cucurbits, the field should be ploughed to exposed and kill the grubs in soil.
- Spray of the following insecticides are recommended against red pumpkin beetles- carbaryl @ 2g per lit of water or endosulfan @ 2ml per lit of water.

3. Fruit flies, *Bactrocera cucurbitae.*

Cucurbit fruit fly is a major pest of all cucurbitaceous vegetables. It is serious and destructive pest of cucurbits like melon, pumpkin, cucumber and gourds. Adult flies are reddish brown with lemon yellow markings on thorax. Female fly lacerates or puncture the fruits by its pointed ovipositor and laid eggs on soft fruits. Female secrets the gummy secretion which cements the punctured tissue and make the entrance water proof. The most damaging stage is the maggots. Maggots feed on the pulp of fruit and cause premature dropping of fruits and also make them unfit for the human consumption. Sometimes secondary bacterial infection also takes place. Oozing of brown resinous fluid from the fruits and fruits become distorted and malformed. In severe condition or in rainy season 50-100 % damages are reported.

Management

- In endemic areas, change the sowing dates, as the fly population is low in hot dry conditions and its peak during rainy season.
- Collect infested fruits and dried leaves and burn it in deep pits. Expose the pupae by deep ploughing and turning over soil after harvest.
- Use ribbed gourd as trap crop and apply carbaryl (0.15 %) on the congregating adult flies on the undersurface of leaves.
- Use poison baiting in severe infestation: Saturated sugar solution 5 ml + Malathion 50EC 5ml+ gur or jaggary 100 ml are kept in the container and put it to the four corners of the field as a poison baits.
- Use fly traps @ 50 traps /ha : Use 5 g wet fish meal in a polythin bag (20×15 cm in size and having 6 holes of 3 mm in diameter) and add 0.1 ml diclorovos (1 drop) in the cooton plug and put it in the bag. For better performance, dichlorovos is added every week and fishmeal should be renewed after 20 days.

17.5 Insect pests of Okra

1. Okra Fruit borer, *Earias vittella* - pest of humid areas and *E. insulana* - pest of dry regions.

Eggs are laid singly on buds and flowers and occasionally on fruits. During the early stage of crop's growth, the eggs are laid on shoot tips. When the crop is only a few weeks old, the freshly hatched caterpillars bore into tender shoots and tunnel downwards. These shoots wither, droop and ultimately the growing points are destroyed, and side shoots may arise giving the plant a bushy appearance. With the formation of buds, flowers and fruits, the caterpillars bore inside these and feed on inner tissues.

Management

- Use of resistant or tolerant varieties.
- Intallation of pheromone traps @ 25 per ha is quite effective.
- As soon as flowering starts give 3 sprays at fortnightly interval with any of the following insecticides carbaryl @ 2g or endosulfan@ 3ml or fenvalerate @1 ml per lit of water.

2. Caterpillar pests of crucifers' vegetables

The cruciferous crops (cole), which includes economically important crops like Cabbage, *Brassica oleracea* var. *capitata*; Cauliflower, *B. oleracea* var. *botrytis*; Knol khol, *B. caulorapa* (L.) throughout the world. However, poor productivity is mainly attributed to the incidence of diamondback moth (DBM), *Plutella xylostella*; cabbage caterpillar, *Pieris brassicae*; leaf webber, *Crocidolomia binotalis;* cabbage head borer, *Hellula undalis* and aphid *Brevicoryne brassicae* infesting the crop in various seasons (Shankar *et al.*, 2009a).

1. Cabbage butterfly, *Pieris brassicae* and *P. rapae*)

It is also known as imported cabbage worm. Two species of cabbage butterflies i.e. large butterfly, *Pieris brassicae* and small butterfly, *P. rapae* have been recorded and found damaging the cole crops in Jammu province. They are serious threats to the crops grown for seed purpose especially in hills as well as in plain areas of Jammu. Adults are small to large sized butterflies, pale white with black markings. Female butterflies have two black circular dots on the dorsal side of each forewing. After mating, female butterfly lay about 100-150 conical, yellowish eggs in clusters on lower or upper side of the leaf. The eggs hatch in 3-15 days depending upon the environmental temperature. Just after hatching, the caterpillars feed gregariously on leaves in early stages. They passes from five larval instars and become full grown in 2-3 weeks. Their incidence has been recorded in fields from October to April- May except in cool winter months during December-January. They completely skeletonize the leaves leaving only mid ribs. Due to complete loss of photosynthetic area crops failed to produce the economic yields.

Management

- Hand picking and mechanical destruction of caterpillars during early stage insect pest attack is beneficial.
- *Cotesia glomeratus* are recorded as potential parasitoids against cabbage caterpillar larva.
- *Phryxe vulgaris* (dipteran fly) are also observed abundantly in cole crops fields to parasitize the caterpillars.
- Foliar spray of *Bt* formulations @ 500 g/ ha along with sticker (0.5 ml/lit of water) is found promising to control all lepidopteran insect pests.
- Need based application of cypermethrin @ 1 ml per lit or malathion @ 2ml per lit or endosulfan @ 3ml per lit of water are beneficial in case of severe infestation.

2. Cabbage head borer *(Hellula undalis and Helicoverpa armigera)*

Cabbage head borer is seen sporadically but its attack coincides with the head/curd initiation stage and a single larva is responsible to damage the entire head/curd. The early instar may feed on the growing point, and bore inside the head and damaging the head and excrements which prevents head formation. When fully grown, larvae pupate inside the head, or on stems, or in soil. *Helicoverpa armigera* larvae have also been recorded at several places in Jammu region as head borer of cabbage. It may be due to absence of preferred host in winter months or some environmental and behavioural change.

Management

- Monitoring and mechanical destruction of larvae at the time head/curd initiation may proved to be good.
- Bracon spp. are important larval parasitoids against head borer.
- Fungal formulations like *Beauvaria bassiana* and entomopathogenic nematodes (EPN) may be beneficial to suppress the attack of head borer.
- Spraying of synthetic pyrethroids like cypermethrin, fenvalerate, deltamethrin @ 1 ml per lit of water is effective to control head borer.

3. Diamondback moth (*Plutella xylostella*)

Diamondback moth is the universal pest of crucifers' crops and remains one of the most serious problems for the intensive cole crops growing areas. Plants at all stages of growth may be attacked by and completely devastated by this pest. In Jammu province, diamondback moth is recorded only on early and on late season cole crops and on crucifers i.e., during September-November and from February onwards.

Management

- Use of pheromone traps are useful in monitoring and detection of the DBM pest popualtion in early stages.
- Practice of trap or intercropping of mustard, coriander and marigold with cole crops may reduce the DBM pest infestation.
- Release of *Cotesia plutellae*, natural bio-agent of DBM larvae is effective to suppress its population in field condition (1000 adults per release every 2 week interval up to harvest).
- Use of resistant varieties are also effective against DBM.
- Some safer insecticides and neem products are recorded to be effective against DBM.
- Spray endosulfan@ 3ml, cypermethrin@ 1ml, cartap hydro-chloride@ 2ml, spinosad@ 1ml and *Bt* formulation@ 500 g/ha may be promising in controlling the DBM pest population.

4. Tobacco catepillar : *Spodoptera Litura*

Spodoptera litura has also wide host range and economic insect pest of cole crops and tomato but it is usually kept under damaging levels by management practices targeted to *Helicoverpa.* They are voracious feeders and scrape the leaves and later on it may completely defoliate the leaves. The insect is active during the entire year except severe cold months and females can lay up to 300-600 eggs each, usually in groups of about 100. Eggs are laid on the underside of lower leaves and are covered with fuzzy, white creamy scales. Under warm conditions, eggs hatch within three to four days. The neonate larvae feed from one to three weeks, in groups at younger stage and scattered on the plants when larger.

Management

- Summer ploughing is beneficial to expose the hibernating pupae to natural enemies and for bird predation.
- Hand picking and mechanical destruction of egg masses, caterpillars and spraying of NSKE 5% during early stage may proved beneficial.
- Installation of pheromone traps @ 25 to 30 /ha may be the useful tool for early detection and also for mass trapping and destruction of tobacco caterpillar.
- Spraying of Splt NPV @ 250 LE with gur or jaggary (10 g/lit) and sticker during evening hours may be the promising to control tobacco caterpillar (Monobrullah and Shankar, 2008).
- Foliar spray of Bt formulations @ 500 g/ ha is the good for controlling the caterpillars.
- EPN may be quite promising to suppress the Spodoptera larval population in the field condition.

- Chemical sprays recommended for the DBM control will also work against this pest but in this case, spot application may be enough if the attack is not wide spread in the field.

17.6 Onion

Onion thrips, *Thrips tabaci*

Besides onion, several other species of thrips infest tomato, brinjal, capsicum, and on cucurbits. They are tiny, slender insects that may vary in colour from yellow to dark brown or black. Both nymphs and adults rasp the epidermis of tender leaves and imbibe the oozing sap. As a result of which leaves develop numerous white specks. Heavily infested leaves curl up and get distorted and ultimately the top of attacked plants wither away. The symptoms develop by feeding on foliage may cause a silvering of foliage.

Management

- In the case of onion, the only reliable method of controlling onion thrips is through the use of tolerant varieties. Advantages of pest-resistant crop varieties include ease of use; compatibility with other integrated pest management tactics; low cost; cumulative impact on the pest (each subsequent generation of the pest is further reduced); and reduced negative impact on the environment.
- The following lines/varieties have been reported resistant/ tolerant to onion thrips: White Persian, Guano, Sweet Spanish, Crystal Wax, Spanish White, Bombay White, Pusa Red Udaipur 103, N 780-1, Pbr-3, PBR-4, VL-1, No. 18, No. 19, Pusa Ratnar.

  Some insecticides and neem products have been reported to be effective against thrips.
- Biological control agents like minute pirate bugs, lacewing larvae and ladybirds prey on thrips.
- Application of systemic insecticides like imidacloprid or thimethoxam or dimethoate after 10 days of transplanting may help to suppress the thrips population.

(B) Insect pests of fruit crops and their Management

17.7 Conclusion

Lack of knowledge about biocontrol agents and lack of availability of IPM devices and other inputs is the major constraints for farming communities to adopt the IPM practices on their farm. By enhancing the knowledge base of the farmers with modern technology and farmer's friendly policies, improving the skills of farmers, availability of inputs at various levels can bring a significant change in vegetable crop production. Extra attention is to be given in case of the chemical control of insect pests which leave toxic and hazardous residue on vegetables. As most of the vegetables are being eaten row, so, proper care and precautions are to be taken care of while recommending the use of pesticide. Sole reliance on chemical control should be avoided. The preference should be given to the other effective methods of control like cultural practices, host plant resistance, biocontrol, use of biopathogen/botanicals to minimize the use of pesticides. The present technology should embrace conservation of land and water resources, adoption of integrated nutrient and pest management, and application for hybridization and biotechnology towards food quality and high yield production for domestic and export market of vegetable crops. Observations for forecasting the possible pest or disease appearance, scouting the pests and beneficial insects in farm unit, Proper diagnosis and Economic threshold level determination. ETL for all major pests should be standardized and this scientific information should be disseminated among farmers. But, due to more complexity in determining the ET level, it has been replaced with Agro-Ecosystem Analysis in modern agriculture. Preventive methods comprises cultural practices like summer and pre monsoon ploughing, suitable crop rotation, incorporation of intercrops, use of resistant / tolerant varieties, proper water management by judicious irrigation while Intervention methods mainly focused on mechanical, biological and chemical control.

References

Adams, W.M. 1990. Green Development: Environment and Sustainability in the Third World. London: Routledge.

Aluja, M. 1994. Bionomics and management of *Anastrepha. Annu. Rev. Entomol.* 39:155-178.

Angeles, D.E. 1991. *Mangifera altissima. in:* Coronel, R.E. and Verheij, E.W.M. [Eds.] Edible Fruits and Nuts. *Plant Res. S.E. Asia* 2 (PROSEA), Pudoc, Wageningen, the Netherlands. pp. 206-207.

Arneson, P.A. and Losey, J.E. 1997. Integrated Pest Mangement-Lecture Notes, Cornell University.

Bailey, J., Hoffman, M. and Olsen, K. 1988. Black light monitoring of two avocado insect pests. *California Agriculture*. 42: 26-27.

Bajwa, W.I. and Kogan, M. 1998. "Compendium of IPM Definitions (CID): A Collection of IPM Definitions and their Citations in Worldwide IPM Literature." Integrated Plant Protection Center (IPPC), Oregon State University.

Barnes, B.N. 1990. FFTRI monitoring Manual for Orchard Research Institute, Stellenbosch, 21 pp.

Barnes, B.N. 1999. Fruit fly baiting in deciduous fruit doing the right things right. Deciduous Fruit Grower. 49: 5-7.

Barnes, B.N. and Blomefield, T.L. 1997. Goading growers towards mating disruption: the South African experience with *Graphotita molesta* and *Cydia pomonella* (Lepidoptera: Tortricidae). *IOBC WPRS Bulletin*. 20: 45-46.

Barnes, B.N. and Eyles, D.K. 2000. Feasibility of eradicating *Ceratitis* spp. fruit flies from the Western Cape of South Africa by the sterile insect technique. In: *Tan, K. H. (ed.) Area wide control of fruit flies and other insect pests.* Penerbit Universiti sains, Pulau Pinang, Malaysia, pp. 449-455.

Beaument, P. 1993. Pesticides, Policy and People. *London: Pesticide Trust.*

Beers, E., Hull, L. and Jones, V. 1994. Sampling pest and beneficial arthropods of apple. In: *Pedigo, L. and Buntin, D.(eds.) Handbook of sampling methods for arthropods in agriculture.* CRC Press, Boca Raton, Florida, pp. 383-415.

Bentley, J.W. and O'Neil, R.J. 1997. On the ethics of biological control of insect pests. *Agriculture and Human Values.* 14(3):283-289.

Bhagat, K.C., Masoodi, M.A. and Koul, V.K. 1988. A note on the occurrence of a coccinellid predator in Kashmir. *Current Research, University of Agricultural Sciences, Bangalore.* 17: 49-51.

Blomefield, T.L. 1997. Managing resistance of codling moth, *Cydia pomonella* (L.) in South African pome fruit orchards. In: *Combating Resistance, IARC-Rothamsted,* Harpenden, UK. pp. 14-16.

Boavida, C., Neuenschwander, P. and Schulthess, F. 1992. Spatial distribution of *Rastrococcus invadens* Williams (Homoptera: Pseudococcidae) in mango trees. *J. Appl. Entomol.* 114:381- 391.

Bull, B.C.; Raupp, M.J.; Hardin, M.R. and Sadof, C.S. 1993. Suitability of five horticulturally important armored scale insects as hosts for an exotic predaceous lady beetle. *Journal of Environmental Horticulture.* 11: 28-30.

Campbell, C.W. and Knight, A.J. 1983. Production de litchis. Communication at XIII Congreso de N.O.R.C.O.F.E.L Puerto de la Cruz. Ministerio de Agricultura Espana (En Prensa).

Chadha, K.L. and Choudhary, M.L. 2007. *Report of the* Working Group on Horticulture, Plantation Crops and Organic Farming for the XI Five Year Plan (2007-12), GOI, Planning Commission, January, 2007.

Chang, K.P. 1970. A study of Anoplophora macularia, aserious insect pest of litchi. Taiwan Agriculture Quarterly. 6(3):133-143.

Cochran, W.G. 1977. Sampling Techniques. New York: Wiley 3rd ed.

Cunningham, I.C. 1984. Mango insect pests. Australian Mango Research Workshop. CSIRO, Melbourne, Australia. pp. 211-224.

Daku, L., Norton, G.W., Pfeiffer, D.G., Luther, G.C., Pitts, C.W., Taylor, D.B., Tedeschini, J. and Uka, R. 2000. "Farmers' knowledge and attitudes towards pesticide use and olive pest management practices in Vlora, Albania: A Baseline Survey." IPM CRSP Working Paper 001-1, Virginia Polytechnic Institute and State University.

de Laroussilhe, F. 1980. Le Manguier. G.P. Masoneuve & Larose, Paris, France.

DeBach, P.1964. Biological control of insect pests and weeds. Chapman & Hall, London. 844 pp.

Dent, D. 2000. Insect pest management. 2nd edition, Ascot, CABI Publishing, 399p.

DPPQS, Directorate of Plant Protection, Quarantine and Storage. 2001. *Integrated Pest Management Package,* Government of India. *(IPM Package No: 28, 31, 32, 34, 37)*

Drew, R.A.I. and Hancock, D.L. 1994. The *Bactrocera dorsalis* complex of fruitflies (Diptera: Tephritidae: Dacinae) in Asia. *Bulletin of Entomological Research.* Supplement Series. 2:68.

Falcon, L.A. and Huber, J. 1991. Biological control of codling moth. In: Tortricid Pests, their Biology, Natural Enemies and Control (van der Geest L.P.S. and Evenhuis, H.H. eds.). *World Crop Pests.* 5: 355-369.

Farahbakhch, G. and Moini, M. 1975. Olive pests in Iran. In: *Report of Plant Pests and Diseases Research Institute.* National Agricultural Research Organization, Ministry of Agriculture and Natural Resources, Tehran. pp.1-14.

Flint, M.L. and van den Bosch, R. 1981. Introduction to Pest Management. Plenum Press, New York, NY.

Galan-Sauco, V. 1990. Los frutales tropicales en los subtropicos. Ed. Mundi Prensa, Madrid, Spain. pp. 83-84.

Gold, C.S., Nankinga, C., Niere, B. and Godonou, I. 2003. IPM of banana weevil in Africa with emphasis on microbial control, p. 243-257. *In: Neuenschwander, P., Borgemeister, C. and Langewald, J. (eds.), Biological control in IPM systems in Africa.* Wallingford, UK, CABI Publishing, 400p.

Golez, H.G. 1991. Bionomics and control of the mango seed borer, *Noorda albizonalis* Hampson (Pyralidae: Lepidoptera). *Acta Hortic.* 291:418-424.

Haji, F.N.P., Barbosa, F.R., Lopes, P.R.C., Moreira, A.N., de Alencar, J.A. and Ferreira, R.C.F. 2004. Monitoring mango pests within an integrated production programme in Brazil. *Acta Hort.* 645: 163-165.

Hameed, S.F., Sharma, D.D. and Agarwal, M.L. 1992. Integrated pest management in litchi. National seminar on recent developments in litchi production held at Rajindra Agril. Uni., Bihar, Pusa, May 30-31, Abstract of papers, pp 38.

Hansen, J. D. and Amstrong, J. W. 1990. The failure of field sanitation to reduce infestation by mango nut weevil, Cryto mangiferae in Hawaii. *Tropical pest management.* 36: 359-361.

Hare, D.J. 1994. Sampling arthropod pest in citrus. In: Pedigo, L. and Buntin, D. (eds.) Handbook of sampling methods for arthropods in agriculture, CRC press, Boca Raton, Florida, pp. 417-432.

Haseeb, M. 2007. Current status of insect pest problems in guava. *Acta Hort.* 735:453-467.

He, D.P. 2001. An overview of integrated management of insect pests in litchi orchards of Guangdong. *Acta Hort.* 558: 401-405.

Hill, D. 1975. Agricultural Insect Pests of the Tropics and Their Control. Cambridge Univ. Press, New York, NY.

Howard, P.A., Michalenko, E.M. Sage, G.W. Jarvis, W.F. Meylan, W.M. Basu, D.K. Beauman, J.A. and Gray, D.A. 1991. Handbook of Environmental fate and Exposure Data for Organic Chemicals, Lewis Publishers: Chelsea, Michigan.

Hoyt, S.C. and Burts, E.C. 1974. Integrated control of fruit pests. Ann. Rev. of Ento. 19: 231-252.

Huang, M.D. and Li, S.X. 1989. The damage and economic threshold of citrus leafminer, *Phyllocnistis* citrella Stainton to citrus. In: *Studies on the IPM of Citrus Insect Pests.* Guangzhou, Guangdong, China. Academic Book and Periodical Press.

Huda, A.K.S. and Luck, J. 2008. Early warning of pest/diseases for selected crops using climate information and crop simulation modeling approach under climate change scenarios. *Proceeding of International Symposium on Agrometeorology and Food Security, 18-21 Febuary 2008, Hyderabad, India:* pp 11.

Jothi D.B., Tandon, P.L. and Verghese, A. 1994. Hot water immersion as a quarantine treatment for Indian mangoes infested with the oriental fruitfly, *Bactrocera dorsalis* (Hendel) (Diptera:Tephritidae). *FAO Plant Protection Bulletin*. 42(3):158-159.

Kaul, V. and Kesar, Y. K. 2003. Incidence and management of lepidopteran fruit borers of guava (*Psidium guajava* L.) in Jammu, India. *J. Asia-Pacific Entomol.* 6(2): 201-205.

Kaul, V., Malik, G.H., Shankar, U. and Monobrullah, Md. 2007. Incidence and management of olive psylla, *Euphyllura pakistanica*. *Indian Journal of Entomology*. 69(4): 331-340.

Kaul, V., Tiku, A.K., Shankar, U. and Monobrullah, Md. 2007. Green stink bug (Hemiptera: Pentatomidae) recorded as a new pest of olive in India. *Journal of Asia-Pacific Entomology*. 10(1): 81-83.

Keenan, M. 1997. *In:* Smith, D., Beattie, G. and Broadley, R. (eds.) Citrus pests and their natural enemies. Deptt. of Primary Industries, Q 197030, Brisbane, Australia, 272 pp.

Khangura, J. S., Kanwar, J. S., Singh, H. and Singh, D. 1992. Pest management in litchi (*Litchi chinensis* Sonn.). National seminar on recent developments in litchi production held at Rajindra Agril. Uni., Bihar, Pusa, May 30-31, Abstract of papers, pp 40-41.

Kozar, F., Brown, M.W. and Lightner, G. 1994. Spatial distribution of homopteran pests and beneficial insects in an orchard and its connection with ecological plant protection. *Journal of Applied Entomology*. 117: 519-529.

Kumar, N., Suresh, J. and Anbu, S. 2007. Problems and prospects of guava cultivation under Tamil Nadu conditions. *Acta Hort.* 735:329-334.

Lichtenberg, E., Zilberman, D. and Archibald, S.O. 1990. "Economics and Pesticides." Chapter 5 in D. Zilberman and E. Lichtenberg (Eds) "Economic Perspectives on Pesticide Use in California: a Collection of Papers." Berkley University, California, November, 1990.

Malo, S.E. and Campbell, C.W. 1994. The guava. Horticultural Sciences Department Fact Sheet HS-4. Florida Cooperative Extension Service, Institute of Food and Agricultural Sciences, University of Florida. April 1994

Mani, M. and Krishnamoorthy, A. 1990. Evaluation of the exotic predator *Cryptolaemus montrouzieri* Muls. (Coccinellidae, Coleoptera) in the suppression of green shield scale *Chloropulvinaria psidii* (Maskell) (Coccidae, Homoptera) on guava. *Entomon*. 15: 45-48.

Mani, M., Krishnamoorthy, A. and Singh, S.P. 1989. The impact of the predator, *Cryptolaemus montrouzieri* Muls. on the pesticide resistant populations of the striped mealybug *Ferrisia virgata* (Ckll.) on guava in India. *Insect Science and its Application*. 11: 167-170.

Masoodi, M.A. 1985. Growth response of *Lymantria obfuscate* Walker in relation to tannin content in different host foliage. *Indian Journal of Entomology*. 47(4): 422-426.

Mathur, A.C. and Tondon, P.L. 1974. Litchi mite can be controlled. *Indian Horticulture*. 19:11- 12.

Mathur, R.N. 1975: Psyllidae of the Indian subcontinent. Indian Council of Agricultural Research, New Delhi, 429 pp.

Miller, J.C., Hanson, P.E. and Dowell, R.V. 1987. The potential of gypsy moth as a pest of fruit and nut crops. *California Agriculture*. 41: 10-12.

Mohyuddin, A.I. and Mahmood, R. 1993. Integrated control of mango pests in Pakistan. *Acta Hortic.* 341:467-483.

Morin, C. 1967. Cultivo de Frutales Tropicales. Librerias ABC, Lima, Per´u. pp. 225-229.

Mullen, J.D. 1995. "Estimating Environmental and Human Health Benefits of Reducing Pesticide Use through Integrated Pest Management Programs." Master Thesis. Virginia Polytechnic Institute and State University, USA, 1995.

Murray, R.C. 1991. Arthropod pests of mango in the Caribbean Islands. *Rev. Crop Prot. Univ. West Indies, Kingston.*

Navon, A., Cohen, M., Keren, S., Levski, S., Harel, M. and Dunkelblum, E. 1997. Trapping of the leopard moth, *Zeuzera pyrina* L. (Cossidae: Lepidoptera) with sex pheromone traps for monitoring the pest experiments in apple and pear orchards. *Alon Hanotea.* 51 (8): 358-366.

Nelson, K.C. 1994. "Participation, Empowerment, and Farmer Evaluation: A Comparative Analysis of IPM Technology Generation in Nicaragua." *Agriculture and Human Values*, (Spring-Summer 1994): 109-125.

Nijjar, G. S. 1972. Litchi cultivation. Punjab Agriculture University, Ludhiana, India.

Norton G.W. and Mullen, J. 1994. Economic Evaluation of IPM Programs: A Literature Review. Report 448-120. Blacksburg, Virginia: Cooperative Extension.

Norton, G.W., Rajotte, E.G. and Gapud, V. 1999. "Participatory Research in Integrated Pest Management: Lessons from the IPM CRSP." *Agriculture and Human Values*, 6: 431-439.

Pandey, V., Patel, M.G., Chaudhari, G.B., Patel, J.R., Bhatt, B.K., Vadodaria, R.P. and Shekh, A.M. 2003. Influence of weather parameters on the population dynamics of mango hopper. *Journal of Agrometeorology.* 5(1): 51-59.

Pedigo, L.P. 2001. *Entomology and Pest Management,* 4th ed. Prentice Hall, New Jersey.

Pena, J. and Johnson, F. 2001. Insect management in guava. Document ENY- 412. Florida Cooperative Extension Service, Institute of Food and Agricultural Sciences, University of Florida. November 2001

Peña, J. E. 2004. Integrated pest management and monitoring techniques for mango pests. *Acta Hort.* 645: 151-161.

Peng, R. K. and Christian K. 2005. Integrated pest management in mango orchards in the Northern Territory Australia, using the weaver ant, *Oecophylla smaragdina*, (Hymenoptera: Formicidae) as a key element. *International Journal of Pest Management.* 51(2): 149 – 155.

Pimental, D., H. Acquay, M. Biltonen, P. Rice, M. Silva, J. Nelson, V. Lipner, S. Giordano, A. Horowitz, and M. D'Amore. 1992. "Environmental and Economic Costs of Pesticide Use." *BioScience.* 42(10): 750-760.

Prasada Rao, G.S.L.H.V and Beevi, S. P. 2008. Forewarning Tea Mosquito Bug *Helopeltis antonii* Sign. (Miridae:Hemiptera) in Cashew. *Proceeding of International Symposium on Agrometeorology and Food Security, 18-21 Febuary 2008, Hyderabad, India,* pp 203.

Rao, V.P., Ghani, M.A., Sankaran, T. and Mathur, K.C. 1971. A review of the biological control of insects and other pests in South East Asia and the pacific region. *CIBC Tech. communication.* 6: 149.

Roling, N. and E. van de Fliert. 1994. "Transforming Extension for Sustainable Agriculture: The Case of Integrated Pest Management in Rice in Indonesia." *Agriculture and Human Values*, Spring-Summer, 96- 108.

Roy, R.S. and De, B.N. 1950. Control of litchi mite. *Indian Journal of Horticulture.* 7(1): 16.

Sauco, G. V. and Menini, U.G. 1989. Litchi cultivation. FAO Plant production and protection paper. 83.

Shankar, U., Kaul, V. and Rai, S. 2007. New recode of anar fruit borer, Deudorix isocrates (Fabricius) on aonla. *Journal of Research, SKUAST-J.* 6(2): 292-295.

Sharma, D.D. 1985. Major pests of litchi in Bihar. *Indian Farming.* 35(2): 25-26.

Sharma, D.R. 2006. Management of insect pests of tropical and sub-tropical fruit crops. *In:* Proceedings of advance training course on Emerging trends in economic entomology, February 21st - 13th March 2006, Hisar, India, pp 123-135.

Sharma, D.R. and Singh, S. 2006. Management of insect pest of temperate fruits. *In: Proceedings of advance training course on Emerging trends in economic entomology,* February 21st - 13th March 2006, Hisar, India, pp 111-122.

Singh, A., Trivedi, T.P., Sardana, H.R., Dhandapani, A. and Sabir, N. 2001. Resource Manual on validation and promotion of IPM. National Centre for IPM, New Delhi, pp.119.

Singh, G. (1991) Loss assessment, ecology and management of mango fruitfly, *Dacus* sp. *Acta Hortic.* 291:425-436.

Singh, S.P. 1989. Achievements of AICRP on Biological Control. *Tech.Bull.* No. 2. Biological Control Centre, NCIPM, Faridabad, 20 pp.

Singh, S.P., Rao, N.S., Ramani, S.and Poorani, J. (*eds*) (2001) Annual Report 2000-01, Project Directorate of Biological Control, Bangalore. Bangalore, India; PDBC, 218 pp.

Smith, D, Beattie, G. and Broadley, R. 1997. Citrus pests and their natural enemies: IPM in Australia. Queensland, Department of Primary Industries, Brisbane, Australia, 272 pp.

Srivastava, A.S. and Masoodi, M.A. 1985. Influence of host plants on the development and survival of *Lymantria obfuscate. FAO Plant Protection Bulletin.* 33 (2): 67-69.

Syed, R.A., Ghani, M.A. and Murtaza, M. 1970. Studies on trypetids and their natural enemies in West Pakistan. III. *Dacus (Strumeta) zonatus* (Saunders). *Tech. Bull. Commonw. Inst. Biol. Control.* 13:1-16.

Tandon, P.L. and Verghese, A. 1985. World list of insect, mite and other pests of mango. *Tech. Doc. Indian Inst. Hortic. Res.* 5.

Thakur, J.N.; Rawat, U.S. and Pawar, A.D. 1989. Investigations on the occurrence of natural enemies of San Jose scale, *Quadraspidiotus perniciosus* Comstock (Hemiptera: Coccidae) in Jammu & Kashmir and Himachal Pradesh. *Entomon.* 14: 143-146.

Thakur, J.R., Sharma, P.C. and Gupta, P.R. 1989. Occurrence of olive psylla, *Euphyllura pakistanica* on olive in India: A new record. *Tropical Pest Management.* 35 (3): 331.

Veerish, G.K. 1989. Pest problems in mango – world situation. *Acta Hortic.* 231:551-565.

Villeirs, E.A.D.E and Matthee, W.A. 1973. The occurrence of bark borers on litchi and other fruit tree. *Citrus Grower and sub-tropical Fruit Journal.* 477:17-19.

Wearing, C.H. 1975. Integrated control of apple pests in New Zealand. 2. Field estimation of fifth-instar larval and pupal mortalities of codling moth by tagging with cobalt-58. *New Zealand Journal of Zoology.* 2: 135-149.

Whalon, M.E. and Croft, B.A. 1984. Apple IPM implementation in North America. *Annu. Rev. Ent.* 29: 435-70.

Whitehead, V.B. and Rust, D.J. 1972. Control of fruit piercing moth, *Sorrodes partita* (Fabr.) (Lepidoptera: Noctuidae). *Phytophylactica.* 4:9-12.

Zouiten, N. and Hadrami. I.E. 2001. Olive tree psyllid: State of knowledge and perspective of control. *Agricultures.* 4(10): 225-232.

18

CANOPY AND NUTRIENT MANAGEMENT IN LITCHI

S.D. Pandey, Amrendra Kumar
Kuldeep Srivastava and R.K. Patel

18.0 INTRODUCTION

In India about 83000 ha area is under this crop and more than 20 cultivars are cultivated. Among them Shahi, China, Bedana, Rose scented, Bombai, Purbi, Saharnpur are commercial cultivars. Bihar is the premier state in litchi production and marketing. Litchi is being cultivated in large area in Bihar (31000 ha) with an annual production of about 2,54,600 MT of fruits having productivity of around 8.0 tonnes/ha (Table 18.1). The harvesting of litchi fruits starts from last week of April in Tripura to end of July in Himachal and foot hills of Uttarakhand. In Bihar the harvesting period lies between 2nd week of May to 3rd week of June for different cultivars like Early Bedana, Shahi, Mnadraji, China, Purbi, Rose Scented, Longia, etc. There has been slow increase in the area and production of litchi in the country because of its specificity to the climatic requirements. It is restricted to northern parts particularly in the foot hills of Himalayas from Tripura to Jammu and Gangetic plains. Frost free, cool, dry winters and humid summers free from hot and dry wind, hails and storms characterize as favourable rowing zone for litchi.

In developing countries like India, sustainable development of horticulture is a biggest challenge. There is need to formulate strategies for their management through "state of art" technologies for harnessing the naturally available resources in a sustainable manner to meet the growing demands of the society. The emerging concerns in the area of canopy architecture and its management are the existence of large number of old senile orchards among the fruits crops and the development of a rejuvenation technology to restructure their canopy and enhanced productive and remunerative plant life. However, the adoption of rejuvenation technology is affected by certain socio-economic constraints prevailing in the society. We should know what had been a problem with earlier system of orchard management and was lack of focus upon regulating the development of tree and canopy architecture since the establishment stage. Thus canopy management can be defined or stated as "Designing the plant as per need, using inherent plant characteristics in accordance with given set of conditions and resources to perform the plant maximum". The productivity of fruit crops depends on several agronomical/ horticultural factors among them poor management of canopy architecture being the most important one. In fact fruiting potential of the trees is largely governed by their architecture, canopy density and photosynthetic efficiency. Canopy management in fruit crops deals with the development and maintenance of their structure in relation to the size and shape for maximum productivity with quality fruits. The tree architecture or canopy management, especially size control, has become a priority for reducing production cost and increasing yield and quality. Canopy design and shape influence light interception with assured higher monetary returns to fruit growers. Therefore, early height control and trees canopy management are important techniques which should be practiced in fruit crops for higher returns to growers.

Table 18.1. State wise area and production of litchi

States	Area in 000' ha					Production in 000' MT				
Bihar	30.5	30.60	31.10	31.10	31.28	216.90	215.10	227.00	236.40	254.60
West Bengal	8.40	8.40	8.60	8.90	9.19	81.20	82.40	85.10	85.30	89.00
Assam	4.80	5.10	5.20	5.30	5.63	34.80	39.20	40.50	41.50	49.64
Punjab	1.50	1.60	1.60	1.70	1.74	21.30	22.40	23.90	24.50	25.49
Jhar-khand	4.30	4.30	4.30	4.80	5.26	20.30	51.10	35.80	57.50	58.21
Uttra-khand	8.80	9.10	9.30	9.50	9.49	15.30	15.70	18.70	19.00	19.16
Tripura	2.80	2.80	2.90	3.20	3.20	14.40	17.00	16.60	16.60	16.56
Orissa	4.20	4.30	4.40	4.50	4.46	13.00	17.10	19.20	20.10	20.26
Others	6.60	8.20	10.2	11.40	13.09	6.20	23.30	30.50	37.20	43.60
Total	**71.90**	**74.40**	**77.60**	**80.40**	**83.34**	**423.40**	**483.30**	**497.30**	**538.10**	**576.52**

18.1 Canopy architecture management

A series of operations on the above ground portion of the plant, aiming towards maximized production of quality fruits per unit canopy area.

An appropriate canopy architecture management encompasses the development of :

- Strong frame of primary branches
- Wider crotches in scaffold branches
- Healthy and well distributed secondary and tertiary branches
- Sufficient fruiting terminals in most productive areas
- Healthy foliage with high photosynthetic efficiency to maximize the solar radiation use efficiency
- Enough space for air circulation inside canopy

18.2 Need of canopy architecture management

Canopy architecture management is one of the most important production factors confronting to the burgeoning fruit industry in India. Indeed, canopy management is the manipulation of trees canopy to optimize its production potential with excellent quality fruits. In many fruit crops, increase in production with enhanced fruit quality is obtained by managing canopies for short-statured trees. Small trees capture and convert sunlight into fruit production in a better way than larger ones. Fruit production involves the capturing and conversion of sunlight into production of fruit biomass (dry-matter content). The main controlling factors in this context are amount of incoming radiation and percentage of that radiation which is intercepted by energy into fruits. For economical fruit growing, the natural form and shape of fruit trees have to be modified through the practices of pruning to achieve the target of more yield with better quality fruits. The fruit growers desire to regulate the growth and development of their trees in a specific manner. The removal of unwanted plant parts and having a balance between vegetative and reproductive growth is thus known as pruning. The practice of pruning, as and when performed in an appropriate manner, keeps the plant in such a shape and condition so as to yield the fruits of desired quality. Climate has significant bearing on pruning. The fruit crops grown in different climate zones (tropical, subtropical and temperate zones) are required to be trained and pruned accordingly depending upon the type of plant and specific objectives of the fruit growers. Canopy management depends on the choice of trees size, number of trees/ ha and pruning techniques to get optimum yield. Pruning should be done when trees are young to establish their basic shapes and sizes and continued as regular process as and when required.

Principles of Canopy Architecture Management

- Facilitating aeration inside the canopy.
- Maximizing light penetration in and outside of canopy.
- Facilitating conversion of sunlight into chemical energy by plant leaf.

- Controlled plant structure including stature.
- Improved fruit colour, quality and yield.

An ideal training strategy centres around the arrangement of plant parts especially to develop a structures that optimizes the utilization of sunlight and promotes productivity.

Steps of Canopy management

a. Selecting dwarf cultivars.
b. Selecting appropriate planting system.
c. Use of dwarfing rootstock and inter-stocks.
d. Initial frame development.
e. Training the plants with open centre.
f. Allow the canopy to grow horizontally.
g. Centre opening of grownup trees.
h. Pruning of the shoots and excessive growth.

18.2.1 Pruning and training

Pruning refers to removal of parts of a tree, especially shoots, roots, limbs, buds or nipping away of the terminal parts. It is done to make a tree more productive and bear quality fruit, increase longevity of the tree, make it into more manageable shape and to get maximum returns from the orchard. In ornamentals, pruning gives a more aesthetic look to plant, but it is being given the lowest priority in commercial production. Pruning is as important as plant protection, weeding and irrigation.

18.2.2 Different types of pruning

1. There are 2 main types of pruning i.e. heading back and thinning out. In heading back, the terminal part of the shoot is removed to the desired level, for example, one-fourth or two-third of the shoots may be headed back. Such pruning encourages development of secondary branches. In thinning out, the entire branch is cut right from the base. It allows more of sunlight to enter the interior of the tree because of its opening effect.
2. Another type of pruning is dehorning. It consists of removal of all the wood after leaving 7-10cm thick stub all over the tree. Such severe treatment should normally not be given except when it is the last resort it into commercial bearing.

18.2.3 Nutrient Management

Litchi require good amount of mineral nutrients to sustain good yield and quality produce in commercial plantations. These can be supplied either by growing the crop in very fertile soils or by supplementing moderate soil fertility with applied fertilizers. Sooner or later even the most fertile soils become depleted and require additional fertilization. However, of the total acreage of litchi, only a small portion is intensively fertilized. The remaining is grown with few applied nutrients, apart from some organic matter and house hold refuse. This undoubtedly contributes to the gradual depletion of soil fertility and the phenomenon of yield decline, which is so prevalent in traditional plantations.

Litchi crop have precise requirement of nitrogen, phosphorus, potassium and other micro nutrients. The early stages of vegetative growth are critical for later quality production of fruits. Thus nutrients must be available to the plants from establishment to the stabilized fruiting stage of plants. Large quantity of nutrients removed by the litchi plants and these must be effectively replaced to sustain the yield and quality of fruits. One of the major factors limiting fruit production in litchi is lack of a judicious nutrient programme.

Nutrient Standards

Tentative standard leaf nutrient levels for Australian litchi orchards after panicle emergence during May to August are: 1.5-1.80 percent for N; 0.14 – 0.22 percent for P; 0.70 – 1.10 percent for K; 0.66 – 1.00 percent for Ca and 0.30 – 0.50 percent for Mg. The suitable leaf nutrient levels for Guangdong province of China are suggested as 0.93 – 2.10 percent for n, 0.08 – 0.21 percent for p and 0.12 -0.33 percent for K. In India optimum nutrient content of leaf N, P and K are in range of 1.5-1.8, 0.15 -0.30 and 0.70 – 1.10 percent of dry matter. Ca content in the leaves varies from 0.60 – 3.0 percent of the dry matter basis. It has been observed that a yield of 100 kg fruits removes about 90-250g N, 35-50g P, 240-320g K 20-60g Ca, 2.0- 2.50 g Cl, 1.40g Na.

18.2.4 Nutrient doses for bearing plants

Field experiments conducted on different cultivars at different locations have clearly demonstrated the effect of a graded dose of NPK on growth, yield and quantity of fruits. Application of 600-800 g N, 200-300 g P_2O_5 and 400-600 g K_2O per plant is recommended for 12-15 year old trees. Nitrogen and Potassium should be applied in 2-3 splits and P_2O_5 intwo splits. Excessive application of nitrogenous fertilizer before flowering should be avoided. Phosphorus application at the time of flower bud differentiation improves flowering and fruiting. Application of cakes and manure is generally practiced to get better quality fruits. In general, lychee orchards, maintained with higher doses of organic manure, have better yield and quality as compared to orchards maintained with chemical fertilizers.

Table 18.2. Fertilizer recommendation for litchi in calcareous soil of Bihar

Age of trees (Years)	Organic man (Kg/plant/year)		Nutrients (g/tree/year)		
	F.Y.M.	Cake	N	P_2O_5	K_2O
1	10	1.0	50	25	50
2	15	1.5	100	50	75
3	20	2.0	150	75	100
4	25	2.5	200	100	125
5	30	3.0	250	125	150
6	35	3.5	300	150	175
7	40	4.0	350	175	200
8	45	4.5	400	200	225
9	50	5.0	450	225	250
10	55	5.0	500	250	275
11 or more	60	5.0	500-600	250	300

Table 18.3. Recommendations for correcting micronutrient deficiencies in litchi

Nutrient	Concentration	Compound used
Zn	0.50-2.70 g Zn/1	$ZnSO_4\ 7H_2O$ or ZnO
B	0.10-0.50 g B/1	H_3BO_3 Or $Na_2B_4O_7$
Cu	0.50-3.00 g Cu/1	$CuSo_4\ 4H_2O$
Mn	0.60-0.70 g Mn/1	$MnSO_4\ 5H_2O$
Fe	0.40-0.50 g Fe/1	$FeSO_4\ 7H_2O$

18.2.5 Time of Application

The fertilizer programme for young trees begins when the plants begin to flush. The rate of nutrients is increased each year with increase in age of plant. Fertilizer application is usually followed by irrigation unless rain falls within a day of application. On poor soils, application of organic manures can be used to great advantage in the first few years. The newly planted trees are not likely to need feeding in the very first year. The fertilization programme may start from the second year. In case of fully developed bearing trees, nutrition management is directed firstly to gain maximum vegetative growth immediately following harvest in May. The second requirement is to maintain the tree in state of dormancy for 3-4 months before flowering. The third requirement is to maintain nutrition at a high rate once fruit have set. To achieve this, nutrients are applied twice in the year. The first application is made 4 weeks after fruit set to encourage fruit development and the second two weeks after harvest to encourage one growth flush. Nitrogen application during autumn and winter may encourage excessive flushing prior to the normal period of flower initiation. Application in early spring can promote growth flushing which may compete with the flowers and reduce shedding.

Mycorrhizae

Litchi plants make better growth ifmycorrhizal fungi grows on its root. The mycorrhizae (*Glomus litchi*) belong to VAM (Vasicular- arbuscular mycorrhizae group) since there is symbiotic relationship between VAM and litchi plants. It is suggested that new plants must be grown in the vicinity of soil taken from old litchi orchards. It has been observed that fungi grow better in acid soil compared to alkaline soils. The poor growth of litchi under alkaline pH has been attributed to micro nutrient toxicity/deficiency or lack of mycorrhizae. It is also observed that VAM increases the uptake of P and K in to the shoot and increased root weight, trunk diameter and fruit yield of litchi.

Management of biotic stress

Major pests of litchi include fruit borer, shoot borer, leaf roller, mite, bark eating caterpillar, leaf eating weevil and litchi bug. *Trichogram machilonis* @ 50,000/ha and application of Nimbicidine @0.5% and Cypermethrin @ 0.005% have been found effective in minimizing the damage of fruit borer. Less damaged litchi fruits at harvest has been found by use of pheromone trap + Tricho. Card @50000/ ha and *Kamdhenu Keet Niyantrak* (5%) sprays before panicle immergence and flowering time which opens up scope for safe crop management. Integrated management of litchi mite where mechanical and chemical measures are applied to control the mite. General cleaning and pruning of infested twigs/leaves in the month of June, September and January followed by spraying of dicofol or propargite (3 ml/lit.) or dimethoate (0.05%) or profenofos (0.05%) twice at weekly interval has successfully control the mite.

18.3 Conclusion

Among the several factors associated with production of litchi, balanced nutrition is considered to be the most important which determines productivity and quality. Canopy architecture management is another essential tree management operation starts from the first year of plant establishment which of-course varies with species and environments but enables the plant to produce high quality and quantity by providing proper framework, fruiting area and utilization of available resources and reduction in biotic stress.

References

Bose, T.K., Mitra, S.K. and Chattopadhyay, P.K. 1992 Optimum plant density for some tropical fruit crops. *ActaHorticulturae* No. 296: 171-176.

Galan – Sauco, V. and Menini, U.G.1989. Litchi cultivation. FAO plant production and protection paper-83, Rome.

Goren, M., Gazit, S., and Erez, A. 1993. Small-statured litchi orchards: a new approach to the growing of litchi. *ActaHorticulturae* No. 349: 69-72.

Menzel, C.M., Banks, A.G. and Simpson, D.R. 1986. Effect of aerial and soil environment on leaf water potential of lychee: implications for irrigation scheduling and productivity. *ActaHorticulturae* No. 175: 363-370.

Rai, M., Nath, V. and Dey, P. 2000. *Chota Nagpur Kshtera men litchi kikethi.* CHES, Ranchi.

Singh, S.N. and Pathak, R.A. 1983. Effect of irrigation intensity, its frequency and manures on the growth of litchi, cv. Calcuttia. *Punjab Horticultural Journal* 23 (3-4): 197-202.

Stern, R.A., Adato, I., Goren, M., Eisenstein, D. and Gazit, S. 1993. Effects of autumnal water stress on litchi flowering and yield in Israel. *ScientiaHorticulturae* 54 (4): 295-302.

Stern, R.A., Gazit, S. and Erez, A. 1993. Autumnal water stress checks vegetative growth and increases flowering and yield in litchi (*Litchi chinensis* Sonn.). *ActaHorticulturae* No. 349: 209-212.

INSECT PESTS MANAGEMENT IN LITCHI ECOSYSTEM

19

Kuldeep Srivastava, R. K. Patel and Devinder Sharma

19.0 INTRODUCTION

Litchi (*Litchi chinensis Sonn*) is an important subtropical evergreen fruit crop belongs to family *Sapindaceae.* It has high nutritive value and refreshing taste. Litchi is consumed as fresh fruit, pulp and various processed products like squash, RTS, wine etc (Singh *et al.,* 2012). In India litchi is being grown in an area of 82700 ha with a total production of 580 000 tons annually (NHB, 2013). It is commercially grown in Bihar, Uttarakhand, West Bengal and Jharkhand. Due to its high economic returns and ever increasing demand in the domestic markets, the crop is also gaining momentum in Punjab, J&K, Himachal Pradesh, Assam, Tripura, and Orissa.

Among biotic stress, insect pests are the major constraints in healthy litchi production. About 40 insects and mite species reported attacking trees and fruits of litchi at different stages of growth (Singh and Singh, 1954; Hameed *et al.*, 1992; Waite and Hwang, 2002). Earlier, only two species, namely, erineum mite and bark eating caterpillar were reported causing serious damage to litchi trees (Mathur and Tandon, 1974; Butani, 1977). Recently, litchi fruit borer and litchi leaf roller have acquired the status of major pests and now, litchi looper, litchi bug and bag worm are emerging pests of litchi (Hameed *et al.*, 2001; Kumar *et al.*, 2011; Choudhary *et al.*, 2013; Kumar *et al.*, 2014). Besides, few other insect pests *viz. Lepropus lateralis* (Coleoptera: Curculionidae), *Thalassodes* sp. (Lepidoptera: Geometridae), *Statherotis* sp. (Lepidoptera: Tortricidae) were also identified as minor pests in litchi ecosystem (Table 19.1).

Table: 19.1. Taxonomic position and economic status of insect pests in litchi

S. No.	Common name	Scientific name	Order: Family	Economic status
1.	Litchi fruit & shoot borer	*Conopomorpha cramerella Snellen*	Lepidoptera: Gracillariidae	Major
2.	Bark eating caterpillar	*Indarbela tetraonis Moore*	Lepidoptera: Metarbelidae	Major
3.	Leaf roller	*Pletypeplus aprobola Meyer*	Lepidoptera: Tortricidae	Major
4.	Leaf roller	*Statherotis* sp.	Lepidoptera: Tortricidae	Minor
5.	Ash weevil	*Myllocerus undecimpustulatusn Faust*	Coleoptera: Curculionidae	Major
6.	Red weevil	*Apoderus blandus*	Coleoptera: Curculionidae	Major
7.	Ash weevil	*Lepropus lateralis*	Coleoptera: Curculionidae	Minor
8.	Bag worm	*Eumeta crameri Westwood*	Lepidoptera: Psychidae	Major
9.	Looper	*Perixera illepidaria*	Lepidoptera: Geometridae	Major
10.	Looper	*Thalassodes* sp.	Lepidoptera: Geometridae	Minor
11.	Litchi bug	*Tessaratoma javanica Thunberg*	Hemiptera: Pentatomidae	Major
12.	Litchi mite	*Acerya litchi Keifer*	Acari: Eriophyidae	Major
13.	Tobacco caterpillar	*Spodoptera litura*	Lepidoptera: Noctuidae	Minor/ Sporadic
14.	Hairy caterpillar	*Spilarctia sp.*	Lepidoptera: Arctidae	Minor/ Sporadic
15.	Mango mealy bug	*Drosicha mangiferae*	Hemiptera: Coccidae	Minor

19.1 Litchi fruit & shoot borer: *Conopomorpha cramerella* Snellen (Lepidoptera: Gracillariidae)

Litchi fruit borer known as the litchi stem-end borer in China and the Litchi fruit borer in Thailand, is one of the major pest of litchi. Besides India, pests occur all litchi growing countries *viz.* China, Taiwan, Thailand etc. The pest shows a preference for litchis over longan but in both crops, damaged fruit may fall from the tree. Fruit loss and leaf infestation has been estimated from 24-32% and 7-70%, respectively by this pest (Lal and Sharma, 1976; Hameed *et al.*, 2001). In Taiwan, 96.1-

100% litchi fruit get damaged by *C. sinensis* Bradley, a closely related species in untreated orchards (Huang *et al.*, 1994). Bhatia *et al.* (2000) observed 13.6-64.9% incidence of *C. cramerella* in Himanchal Pradesh.

19.2 Litchi mite *Aceria litchi* (Acari: Eriophyidae)

Litchi mite has been reported from all the litchi growing countries of the world including India, where litchi mite has been found in almost all litchi growing areas. Litchi mite is a destructive pest, causing curling of leaves and chocolate-brown growth on the ventral surface of leaves at later stage (Singh and Raghuraman, 2011). Both nymphs and adults damage the leaves, inflorescence and young developing fruits by puncturing and lacerate the tissues with their stout rostrum and suck the cell sap (Lal and Rahman, 1975). The infestation generally begins from lower portion of the trees and gradually extends upwards. The incidence of litchi mite can be observed round the year on litchi trees; however, severe infestation has been noticed from March to September and goes under hibernation in extreme cold (Mishra, 1912; Mishra, 1980). However, Lal and Rahman (1975) reported active period of mite is January to October and hibernation in adult stage during November to December under surface of leaf. The highest population (18.1 per cm2 leaf surface) of mite was observed during March, while lowest population (1.4 per cm2 leaf surface) during November. Sharma *et al.* (1986) also reported litchi mite during April-May and November-December but mentioned that the population was significantly influenced by temperature and relative humidity.

19.3 Leaf roller : *Platypeplus aprobola* (Meyrick) (Lepidoptera: Tortricidae)

The leaf roller is a cosmopolitan pest and has been reported in various fruit crops in Hawaii, South Africa, Australia and China. Leaf roller is a very serious litchi pest in Bihar and surrounding areas. The incidence of leaf roller can be observed on litchi trees throughout the year, however more population has been observed from July-February (Singh, 1971). The symptoms of leaf injury by the larvae are manifested through rolling of tender leaves and feeding inside. According to Lal and Mallik (1976) leaf injured by the leaf roller varied between 16.70 to 71.60% while trees infestation varied 12.88 to 53.54% during August to February.

19.4 Litchi looper: *Perixera illepidaria* (Lepidoptera: Geometridae)

Looper attacks tender leaves in mass and defoliate the new shoots. In severe attack, it completely defoliates the newly emerged flush. Besides litchi so many host plants *viz.* longan, rambutan, mango, and castor have been reported for this pest (Schreiner and Nafus, 1992; Kumar *et al.*, 2014). Incidence of this pest has been observed from July to December however, highest population recorded during September- October.

19.5 Leaf cutting weevils: *Myllocerus* Sp and *Apoderus blandus* (Coleoptera: Curculionidae)

Grey weevil (*Myllocerus* Sp.) and red weevil (*Apoderus blandus*) are polyphagous pest therefore, besides litchi its damage other important crops (Singh, 1974; Singh, 1978). Adult has long snout with grey colour, though poor flier but very active feeder on the leaves of litchi. It attacks leaves, shoot and flower. Adult weevils congregate on the tender leaves and nibble irregular holes on the leaves and sometimes consume the entire leaf leaving the midrib only (Singh *et al.*, 2012). The damage of red weevil is more severe at the time of shoot emergence as it prefers newly leaves therefore; newly established orchard/ nurseries are more vulnerable for pest attack (Kumar *et al.,* 2011).

19.6 Litchi bug : *Tessaratoma* sp (Hemiptera: Pentatomidae)

Litchi bug has been emerged as major pest in all litchi growing countries (Choudhary *et al.*, 2013). There are many species of bugs *viz. Tessaratoma javanica* Thunberg, *T. papillosa* Dury, *T. quadrata* Distant, *T. nigripes* Dallas and *T. malaya* Stal that attack litchi plant, out of which, *T. javanica* Thunberg is the most destructive in Australia, China, India, Myanmar and Thailand (Han *et al.* 1999; Leksawasdi and Kumchu 1991; Lu *et al.* 2006). In India, pest is also noticed in all litchi growing areas particularly, Jharkhand, Bihar, U.P., Punjab, H.P. and J. & K. Both adults and nymph suck cell sap mostly on tender plant parts such as growing buds, leaf petioles, fruit stalks and tender branches of litchi tree. In case of severe infestation drying of growing buds and tender shoots has been observed resulting into poor fruit set. The bugs when crowd on the developing fruit, it causes the fruits to fall a couple of week later. Longan, rambutan, kusum, pummelo, castor, pomegranate, eucalyptus, loquat and rose are major host plants of this bug (Singh *et al.*, 2009). In the north-western part of India, the insect appears on litchi from the last week of April and disappears from the orchard after the last week of August and undergoes hibernation in adult stages (Kumar *et al.*, 2008), however, in eastern India incidence of this bug has been observed from February to September (Choudhary *et al.,* 2013).

19.7 Bag worm: *Eumeta crameri* Westwood (Lepidoptera: Psychidae)

Bag worm larvae infest the older leaves and bark. Young caterpillars construct silken bags covering with bark and dry twigs and by living inside scrap the leaf surface (Agrawal and Pati, 2002). Coffee, tea, maize, pomegranate, tamarind, castor, sandal, Casuarina, Cinnamon, *Shorea robusta*, are the alternate host of bagworm (Agrawal and Pati, 2003). Incidence of this pest has been noticed throughout the year on litchi with maximum population during the month of September-October.

19.8 Bark eating caterpillar: *Indarbela tetraonis* Moore and *Indarbela quadrinotata* Walker (Metarbelidae: Lepidoptera)

There are two species of bark eating caterpillars, (*Indarbela tetraonis* Moore and *Indarbela quadrinotata* Walker) and both are polyphagous in nature with wide range of host plants including, longan, rambutan, mango, mulberry aonla, ber, citrus, phalsa guava, jack-fruit, jamun, loquat, pomegranate and rose (Verma and Khurana, 1974; Khurana and Gupta, 1972.). The pest is distributed throughout Indian subcontinent, mainly U.P., M.P., Bihar, Rajasthan, Haryana, Orissa, A.P., T.N. and Maharashtra. Particularly in Bihar, *Indarbela tetraonis* Moore is a very serious pest of litchi, guava, mango and ber (Sharma and Kumar, 1986). Damage is caused by caterpillar, which bore into trunk, main stems and thick branches of litchi tree and destroying xylum tissues resulting into poor growth and fruiting of the tree. Caterpillars remain within the tunnel inside the stem during day, come out in night and feed upon the bark. Older and uncared trees are more affected by the pest. Infestation may be noticed by the presence of ribbons of wood chips, frass and silken thread over the bark surface (Hameed *et al.*, 2001; Atwal and Dhaliwal, 2008).

19.9 Integrated Management of litchi Pests

Various established and also innovative methods can be applied to keep the litchi pest under control. Some of the methods have been described here. However, for effective management the smart blend of all methods in befitting manner can be tried.

19.9.1 Cultural and mechanical methods

The infested litchi shoots and leaves due to litchi mite, leaf roller, fruit borer and bark eating caterpillar are pruned twice in a year during June and August and may be utilized in making compost. Infested samplings with mites should not be used. Area around the trees is ploughed during summer to kill the eggs and hibernating stages of soil insects and litchi mealy bug by exposing to sun heat, ants and birds. Banding the tree trunk with 300 mm wide alkathine sheet (400 guage) prevents the nymphs of mealy bug from climbing the tree. Larvae of bark eating caterpillar can be easily destroyed by pouring kerosene oil in the hole or by using a mixture of ethyleneglycol and kerosene (3:1) (Singh *et al.*, 2012). Manuring the litchi trees with 4 kg of castor and 1 kg of neem cake, when applied at root zone after the first shower monsoon have been proved effective in reducing mites, leaf roller and fruit borer infestation (Lal and Rahman, 1975). The alternate host such as *kath, jamun* and *chhota amaltas* which harboured the pest during off-season be collected and destroyed in order to minimize the chance of heavy infestation of the leaf roller in subsequent year.

19.9.2 Host plant resistance

Sufficient information is not available on this aspect. Bhatia *et al.* (2000) studied the response of different cultivars against *C. cramerella* and observed minimum fruit damage with Seedless Early (33.3%) followed by Deharadun (61.1%), Large Red (73.7%), Muzaffarpur (75%), Saharanpur (76.5%), Rose scented (78.6%), Calcuttia (78.8%), Seedless Late (81%) and McLean(85.7%). Litchi cultivar Bedana showed the highest infestation from litchi mite, followed by Shahi and China cultivars (Hameed *et al.*, 1992).

19.9.3 Biological control

Natural enemies play important role in reducing the pests population of litchi. Hameed *et al.* (1992) found eight species of predatory mites, namely *Amblyseius caecineae* Gupta, *A finlandicus* Oudemans, *A largoensis* Muma, *A paraacriallis* Muma, *A purni* gupta, *Phytoseius intermedius* Evans and Macfarhane, *Typhlodromus fleschneri* Chant and *T homalii* Gupta were associated with litchi mite. Population of these predatory mites were remain high during summer months (April to September) with peak in June. With the onset of winter the population of predatory mites starts decreasing. Three species of hymenopterous parasitoids belonging to genera *Apanteles* species, *Bracon* species and *Mesochorus* species have been reported to parasitizing the larvae of *Cnopomorpha cramerella* during July to September (Hameed *et al.*, 1992). The classical biological control of litchi stink bug utilizing *Anastatus japonicas* (Hymenoptera: Eupelmidae) and O. *phongi* (Hymenoptera: Encyrtidae) has resulted in effective management of the pest in China, Hong Kong and Thailand (Han *et al.* 1999; Leksawasdi & Kumchu 1991). Choudhary *et al.* (2013) also noticed *Anastatus bangalorensis* Mani & Kurian (Hymenoptera: Eupelmidae) and *Ooencyrtus* sp. (Hymenoptera: Encyrtidae) infesting eggs of *T. javanica.* Obviously, these findings would help in the development of effective integrated pest management modules.

19.9.4 Chemical control

The main objective of IPM is to reduce the pesticide load, however, pesticide could be recommended when its use, becomes necessary. Spraying the orchard with Deltamethrin 2.8 EC (0.0028%) 1ml/l or Fipronil 5 SC (0.01%) 2 ml/l or Flubendiamide 39.35 SC (0.008%) 1ml/ 5l of water after fruit set and second spray may be given before 15 days fruit harvest for the management of litchi fruit borer. Further, *Trichogramma chilonis* cards @50000 eggs/ha in combination with Nimbicidine 0.5% or Kamdhenu Keet Niyantrak 5% can be used for effective management of fruit borer infestation.

Earlier, litchi mite had been effectively control by using sulpur (0.05%), Kelthane (0.05%) (Mishra, 1912; Hameed *et al.*, 1992). Pruning of infested twigs in June after harvest followed by spraying of kelthane or monocrotophos (0.05%) twice at 7 days interval at the onset of flushing during September-October. Pruning of affected twigs in June + Pruning in October followed by spraying of Profenofos 0.05% is effective combination against litchi mite (Kumar *et al.*, 2014). In case of litchi bug if chemicals are used, the timing of sprays is critical because the bugs vary in their susceptibility to insecticide at different times of the year, depending on body fat content and its nature. These bugs may be controlled with any systemic insecticide. Litchi looper can be effectively managed by spraying of deltamethrin 2.8 EC (0.0028%) or Fipronil 5 SC (0.01%).

Litchi leaf roller has been successfully controlled by spraying of dimethoate (0.03%). The second best treatment is methyl-demeton (0.025%). Spraying must be given in the month of January when the population remains very low. Subsequent spraying may be done at fortnightly intervals in September if the infestation persists. For effective control of bark eating caterpillar dichlorvos (0.05%) may be sprayed on the tree and at forking of branches as soon as caterpillars start nibbling (Sharma and Kumar, 1986). At tunneling stage of the caterpillar holes are treated by dissolving 1 part of fumigant i. e. carbon disulphide or bisulphide, kerosene oil, petrol, formalin in 100 part of water. Cotton wool is then soaked well in the solution and inserted into the holes and sealed with mud.

19.10 Conclusions

Various insect pests have been reported to occur on litchi at various growth stages, but a few are a real menace to the cultivation of this crop. The major constraints in the production of litchi is the ravages caused by various insect pests which not only result in the low production but also drastically impair the quality, rendering them unfit for human consumption, reducing marketable yield and thereby posing a serious threat to fruit industry. Natural enemies constitute an important mortal factor in the serious insect pest on litchi. Integrated Pest Management is a strategy to manage pests on the basis of a systems approach that looks at the whole orchard ecosystem. This includes understanding how the pests interact with their plant hosts, with the general climatic conditions, plant health and nutrition and with each other. When implementing an IPM system, growers should select ways to reduce overall pest levels in their orchard and ensure that the management of pests is compatible with their other crop management strategies. It is important that growers realize that IPM system is updated from time to time in response to biological changes that occur in their orchard and new techniques or technologies are introduced as soon as additional relevant information becomes available.

References

Agrawal, A. and Pati, A. K. 2002. Increment in the size of an inanimate object in the case of bagworm moth *Eumeta crameri* Westwood (Lepidoptera: Psychidae). *Current Science* 83:71–73

Agrawal, A. and Pati, A. K. 2003. Larval case renovation-a unique behavior in bagworm moth, *Eumeta crameri* Westwood (Lepidoptera: Psychidae). *Current Science* 85:1674–1679.

Atwal, A. S. and Dhaliwal, G. S. 2010. Agricultural pests of south asia and their management. Kalyani Publishers, Ludhiana. 616 pp.

Bhatia, R., Sharma, R. and Agnitotri, R. P. 2000. Incidence, varietal preference and control of fruit borer, *Conopomorpha cramerella* (Lepidoptera: Gracillariidae) on litchi (*Litchi chinensis*) in Himachal Pradesh. *Indian Journal ofAgricultural Sciences*, 70 (5): 301-4.

Butani, D.K. 1977. Pests of litchi in India and their control. *Fruits,* 32(4): 269-273. Choudhary, J. S., Prabhakar S. C., Moanaro, Das, B., and Kumar, S.2013. Litchi stink bug *(Tessaratoma javanica*) outbreak in Jharkhand, India, on litchi. *Phytoparasitica*, 41: 73-77.

Hameed, S. F., Sharma, D.D. and Agarwal, M.L. 1992. Integrated Pest management in litchi. *Proceedings of National Seminar on Recent Developments in Litchi Production* held at R.A.U., Pusa. pp. 38.

Hameed, S. F., Singh, P. P. and Singh, S. P. 2001. pests. *In*: Chauhan, K. S. (ed). *Litchi: Botany, Production, Utilization.* Kalyani Publishers, Ludhiana. 228 pp.

Han, S., Liu, W., & Chen, Q. (1999). Mass releasing *Anastatus japonicas* to control *Tessaratoma papillosa* in Hong Kong. *Chinese Journal of Biological Control,* 15(2):54–56.

Huang, C. C. Chang, K., Sand, and Chu, Y. I. 1994. Damage and population fluctuation of the litchi fruit borer, *Conopomorpha sinensis* Bradley, in Chia-Nan district, Taiwan. *Plant Protection Bulletin*, Taipei 36 (2) : 85-95.

Khurana, A. D., and Gupta, O. P. 1972. Bark eating caterpillars pose a serious threat to fruit trees. *Indian Farmers Digest* 5: 51-52.

Kumar, A. Srivastava, K., Patel, R. K. and Nath, V. 2014. Management of litchi fruit borer and litchi mite using bio-rational approaches under subtropics of bihar. *The Ecoscan, Special issue,* Vol. VI: 285-289.

Kumar, H., Mittal, V., Gupta, A. and Singh, C. P. 2008. Population dynamics and seasonal occurrence of *Tessaratoma javanica* Thunberg in Litchi Orchards. *Annals of Plant Protection Sciences,* 16: 70-73.

Kumar, V, Venkataramireddy, P., Anal, AKD and Nath, V. 2014. Outbreak of the looper, *Perixera illepidaria* (Lepidoptera: Geometridae) on litchi, *Litchi chinensis* (Sapindales: Sapindaceae) - a new pest record from India. *Florida Entomologist,* 97(1): 22-29.

Kumar, V., Kumar, A. and Nath, V. 2011. Emerging pests and diseases of litchi (*Litchi chinensis* Sonn.). *Pest Management in Horticultural Ecosystems.* 17:11-13.

Lall, B.S. and Mallik, S.N. 1976. Binomics and control of litchi leaf roller *(Platypepla aprobola* Meyer). Lepidoptera :Tortricidae. *Proceedings of the National Academy of Sciences,* India, pp. 22-23.

Lall, B.S. and Rahman, M.F. 1975. Studies on the bionomics and control of the Erinose mite, *Eriophyes litchi* Keifer, Acarina: Eriophydae. *Pesticides,* 9(11): 49-54.

Lall, B.S. and Sharma, D.D. 1976. Studies on the bionomics and control of cacao moth *(Acrocercops cramerella* Sn. Gracillariidae: Lepidoptera). *Proceedings of the National Academy of Sciences,* India, pp. 22-23.

Leksawasdi, P. and Kumchu, C. 1991. Mass rearing and releasing of the parasitoid *Anastatus* sp. *Kasetsart Journal (Natural Science),* 25: 47–53.

Lu, F. P., Zhao, D. X., Liu, Y. P., Wang, A. P., & Chen, Q.(2006). Toxicity of neem seed extract to *Tessaratoma papillosa* (Drury) relative to its allozyme genotypes. *Acta Entomologica Sinica,* 49:241–246.

Mathur, A.C. and Tandon, P.L. 1974. Litchi mite can be controlled. *Indian Horticulture,* 19(1): 11-12.

Mishra, C.B. 1912. Litchi leaf curl. *Indian Agricultural Journal,* 7: 93.

Mishra, R.K. 1980. Gall formation and control of the erinose mite *Eriophyes litchi. Proceedings of International Symposium of TOBCIWPRS on intergated control in agriculture and forestry, Vienna,* 436 pp.

N.H.B. 2013. Indian Horticulture Database. Published by National Horticulture Board, Gurgaon, India.

Schreiner, I. H., and Nafus, D. M.1992. Changes in a moth community mediated by biological control of the dominant species. *Environ. Entomol,* 21: 664-668.

Sharma, D.D. and Kumar, H. 1986. How to control bark-eating caterpillars. *Indian Horticulture,* 31(1): 25.

Sharma, D.D., Singh, S.P. and Akhauri, R.K. 1986. Relationship between the population *of Aceria litchi* Keifer on litchi and weather factors. *Indian Journal of Agricultural Sciences,* 56(1): 59-63.

Singh Gorakh, Nath Vishal, Pandey, S.D., Ray, P.K. and Singh. H.S. (2012). The Litchi. Food and Agricultural Organization of the United Nations, New Delhi. 219 pp.

Singh, H. 1974. *Apoderus blandus* Faust (Coleoptera: Curculionidae)- A new pest of Litchi (*Litchi Chinensis* Sonn). *Indian Journal of Entomology,* 36: 239-40.

Singh, H. 1978. Pest complex of litchi in Dehradun and Saharanpur district of Uttar Pradesh. *Indian J. Entomol* 40(4): 464.

Singh, H.S., Nath Vishal, Singh, A. And Pandey, S.D. 2012. Litchi: Preventive practices and curative measures. Satish Serial Publishing House, Delhi. 479 pp.

Singh, J., and Raghuraman, M. 2011. Emerging scenario of important mite pests in North India, Zoosymposiim 6: 170-179 *In:* G. J. de Moraes and H. Proctor [eds.]. Zoosymposium 6: 1-304, Acarology XIII, Proc. Intl. Congress.

Singh, J.P., Jaiswal, A. K., Monobrullah, M. D., & Patamajhi, P.2009. Bioefficacy of insecticides against pentatomid bugs, *Tessaratoma javanica* Thunb.: A sporadic pest of Kusum, *Schleichera oleosa. Indian Journal of Entomology*,71: 259–273.

Singh, L.B. and Singh, U.P. 1954. *The litchi.* Superintendent, Printing and Stationary, Uttar Pradesh, Allahabad.

Singh, M. P. 1971. Occurrence of *Platypepla aprobola* Meyr. (Tortricidae: Lepidoptera) on litchi (*Litchi chinensis*) in North Bihar. *Indian Journal of Entomology*. 33 (1): 98.

Verma, A.N. and Khurana, A.D. 1974. Further new host records of *Indarbela* species (Lepidoptera : Metarbelidae). *Haryana Agricultural University, Journal of Research,* 4(3): 253-254.

Waite, G.K. and Hwang, J.S. (2002). Pests of litchi and longan. *In:* Peña, J.E., Sharp, J.L. and Wysoki, M. (eds). *Tropical Fruit Pests and Pollinators: Biology, Economic Importance, Natural Enemies and Control.* Wallingford, UK. CABI Publishing, pp. 331-359.

20

BIO-SECURITY ISSUES IN PLANT HEALTH MANAGEMENT

D. P. Abrol and Suheel Ahmad

20.0 INTRODUCTION

Bio-security is defined differently according to various disciplines. The original definition of bio-security started out as a set of preventive measures designed to reduce the risk of transmission of infectious diseases in crops and livestock, quarantined pests, invasive alien species, and living modified organisms (Koblentz, 2010). The term was first used by the agricultural and environmental communities. Starting from the late 1990s in response to the threat of biological terrorism, bio-security encompasses the prevention of the intentional removal (theft) of biological materials from research laboratories. These preventative measures are a combination of systems and practices put into its place at legitimate bioscience laboratories to prevent the use of dangerous pathogens and toxins for malicious use, as well as by customs agents and agricultural and natural resource managers to prevent the spread of these biological agents in natural and managed environments. More recent definition with advances in technology also involves oversight of dual-use research (Koblentz, 2010). Dual-use research has the potential to create new health security issues. The most inclusive definition of all would be bio-security as "security against the inadvertent, inappropriate, or intentional malicious or malevolent use of potentially dangerous biological agents or biotechnology, including the development, production, stockpiling, or use of biological weapons as well as outbreaks of newly emergent and epidemic disease" as defined by The National Academies of Science (Koblentz, 2010). This definition is analogous to that of health security. Although security is usually thought of in terms of "Guards, Gates, and Guns", bio-security encompasses much more than that and requires the cooperation of scientists, technicians, policy makers, security engineers, and law enforcement officials.

In today's scenario main focus on to gain maximum food production by modifying different agricultural practices or by adopting new technologies, the introduction and release of living modified organisms (LMOs) and their products (e.g. genetically modified organisms or GMOs) as are resulting in greater incidence of diseases, spread of insect pests and hazards to public, animal, plant health, economy and food security. To prevent or manage the risks to plant health and to protect the environment, there is requirement for the development of a tool to manage this risk. In doing so, bio-security is an essential element to overcome this problem and direct relevance to sustainable agricultural development. In other words, bio security is a set of precautions or an approach that encompasses the policy and regulatory frameworks to prevent the introduction and spread of harmful organisms have relevant risks to human, animal, plant life/health and environment. The strategic and integrated approach of bio security provides a regulatory framework in association with higher authorities and policies to provide safety of food supply, prevent the zoonotic aspects of public health, ensure the sustainability of agriculture, and safeguard the terrestrial, freshwater and marine environments and biodiversity. Globalization in the world, ever increasing requirement of food production, adoption of new technologies, legal obligations for signatories of relevant international agreements, movement of people across borders increase the incidence of pest and disease across the world, dependence of some countries on food, import/ export are the important factors to meet the requirement of bio security at global level as well as for Plant Health Management. The increasing demand to gain more production by adopting new technologies and changing agricultural practices results in new hazards to health that are readily able to cross borders. Changing human ecology and behaviour also contribute to the greater incidence and spread of hazards. Due to which public awareness is increasing towards impact of adverse bio-security events and interventions (Hawkes and Ruel, 2006).

National programs are already in place to prevent, control and manage associated risks to life, health and environment but bio-security work as a new concept by encompassing polices, planning, sciences and all the operations in a strategic way to safeguarding the natural resources from biological threats. The strengthening of policy and regulatory frameworks for bio-security in food and agriculture must be the main priority for their development. These policies and legislative frameworks need to include bio-safety, risk analysis and management which will provide an opportunity to harmonize terminology and methodology. Several commonalities in risk assessment and management across the sub-sectors and organisms which must ideally be addressed by the National Agricultural Bio-security System (Singh, 2008). Risk Analysis of commodities and their management constitute bulk of the animal and plant health management, thus bio-security would automatically occupy the centre stage in Agriculture and Plant health management.

Bio-security is a strategic and integrated approach that encompasses the policy and regulatory framework to analyse and manage risks in the sectors of food safety, animal life and health, and plant life and health, including associated environmental risk. It covers the following:

- The introduction of plant pests, animal pests and diseases, and zoonoses.
- The introduction and management of invasive alien species and genotypes.
- The introduction and release of genetically modified organisms (GMOs) and their products. (http://www.fao.org)

Thus, bio-security is a holistic concept of direct relevance to the sustainability of agriculture, food safety, and the protection of the environment, including biodiversity. It encompasses the concept of bio-safety which is the term used to describe policies and procedures adopted to ensure the environmentally safe application of modern biotechnology. The recent advancements in biotechnological products derived from the use of recombinant DNA technology has led to global concern on bio-safety issues. It mainly deals with the phyto-sanitary aspects of bio-security with a focus on dangers from exotic pests, including those which are invasive, the status of preparedness and the need for a holistic approach to address the issues.

20.1 What is Agricultural Bio-security?

The rebranding of the centuries-old practice of battling agricultural pest and disease introductions as 'bio-security' is itself interesting and reflects how current societal concerns about globalization and terrorism influence agriculture in new ways. Josling *et al.* (2003) observed 'since the terrorist attacks on the World Trade Centre and the Pentagon on September 11, 2001 bio-security has taken on new dimensions and products that move across borders are treated more suspiciously, [creating] uncertainty and transaction costs that impinge particularly on trade that could put domestic animal, plant or human populations at risk'. Following a short survey of recent bio-security threats, we examine and compare bio-security systems for crops and livestock. We then explore whether threats to bio-security are increasing and examine developments in the measurement of risk and in prevention and eradication of new threats. Finally, we explore major changes in today's bio-security systems and make recommendations about building a more bio-secure future for agriculture.

20.2 Regulatory Framework for bio-security

Codex Alimentarius Commission (CAC), the World Organisation for Animal Health (OIE) and the Commission on Phyto-sanitary Measures (CPM) are the International standard setting organizations and bodies, develop standards for different bio-security sectors in accordance with their mandates.

Responsibilities for sectors of bio-security at the international level are shared among a number of organizations and bodies. FAO plays a leading role in normative work and technical assistance, at the both national and international levels. FAO hosts the Secretariat for the Codex Alimentarius Commission, under the Joint FAO/WHO Food Standards Programme, as well as the Secretariat for the International Plant Protection Convention (IPPC). In addition, FAO's participation in the Standards and Trade Development Facility (STDF) aims to enhance collaboration between the three SPS-recognized standard-setting bodies and FAO, the World Bank, the World Health Organization (WHO) and WTO. In addition to the standards and related texts developed by the Codex Alimentarius Commission (CAC), the World Organization for Animal Health (OIE) and the Commission on Phyto-sanitary Measures (CPM), several other international legal authorities, agreements and texts are relevant to bio security. These include the Sanitary and Phyto-sanitary agreement (SPS) and to some extent, the Agreement on Technical Barriers to Trade (TBT Agreement), the Convention on Biological Diversity (CBD) its Cartagena Protocol on Bio-safety, and the International Health Regulations. At national level different stakeholders are involved in bio-security. Government agencies, scientific research institutes, specialist interest groups, nongovernmental organizations (NGOs), several branches of government, both the national and sub-national levels, general public, food safety, public health, agriculture, forestry, fisheries and the environment all have a vital role to play in Bio-security. However, other parts of government responsible for sectors such as trade, customs, transport, finance and tourism are also important in bio security, depending on national circumstances (Khetarpal and Gupta, 2006).

20.3 Areas of bio-security

i. Human life and health (including food safety),

ii. Animal life and health (including fish),

iii. Plant life and health (including forests),
iv. Environmental protection.

Bio-security approach in a holistic manner used to manage biological risks associated with food and agriculture and also focus to enhance productivity, sustainability and profitability. It also covers food safety, zoonoses, the introduction of animal/plant diseases and pests, introduction/release of living modified organisms and their products (e.g. genetically modified organisms or GMOs), their introduction and management of invasive alien species. Thus, bio-security is a wide-ranging aspects to the public health, plant health and protection of the environment. The overarching goal of bio security is to prevent, control and/or manage risks to life and health. In accordance with, bio-security is an essential element of sustainable agricultural development and food safety at global level.

20.4 Need of bio-security

In Agro ecosystem humans, plant, animal life and environment are associated to each other. If any harm taken at one stage then it will affect whole ecosystem. If homeostasis of environment disturb then it will affect all associated entities. This is the fundamental rationale for an integrated approach to bio security development. Bio security hazards of various types exist in each sector and have high potential to move and establish. In respect of food chains, hazards can be introduced anywhere from production to consumption level. If any pest enters though import/export or any other mean then it can affect trophic levels. At the national and international level, there are likely to be significant benefits in integrating bio security activities to overcome these problems.

20.5 Process of bio-security

Bio-security is a strategic and integrated approach require collaboration, high level support across different agencies, their agreement on purpose, scope and process, profile the bio-security context at the country level, assess existing bio-security capacity, describe the desired future situation of bio security, identify capacity needed to reach desired future and generate options to address identified needs are the essential components for bio security. Successful bio security program require conveying the information, needing to convince policy and decision-makers, to ensure transparency and best use of available resources, examine the context for bio-security at the national level. Bio-security issues, general needs, prevailing and opportunities, context shape bio security goals, programs and activities for essentiality to establish bio-security as a national priority and ensure cross-sectoral collaboration and participation. Assess existing bio-security capacity by examine current situation of bio-security capacity and performance, ensure capacity building activities and provide broad framework for the successful development of bio-security system. Next steps for Bio-security program are to describe the desired future situation of bio-security (developing vision and goals across the bio security system, enhancement of the future outcome, effectiveness and maximized potential cross-sectoral gains), identify capacity needed to reach desired future and generate options (Bio security policy framework, legal and Institutional framework).

20.6 Bio-security laws in India

India has a plethora of laws which deal with Bio-security but it needs to be noted that they do not stem from an understanding of the term. The pieces of legislation have been enacted with differing objectives and public concerns in mind. Though disparate and scattered, these pieces of legislation serve an essential function in specifically addressing the sectoral concerns, and they carry forth the intent contained in the preambles. Likewise, the institutions, though numerous, have been established to serve the purposes of the original enactments.

20.6.1 Constitution of India

Though there is no specific reference or use of the term Bio security in the Constitution of India, a number of its provisions are of relevance to understanding the legal framework dealing with bio security in the country. The Constitution is also the key to understanding how the general legal set-up works.

20.6.2 Fundamental rights

Part III of the Constitution of India contains the fundamental rights. Among these is the right to life, which is enshrined in article 21, and which has the most relevance for the legal framework for Bio-security. Since the late 1970s, the Supreme Court, which is the highest court of the country, has progressively widened the scope of the rights granted under this article. This has been achieved by giving an expansive interpretation of the term "life". As a result of judicial interpretation, the right to life has become a sort of repository of various human rights. Some of the pertinent rights thus included are:

i. The right to health,

ii. The right to a healthy environment,

iii. The right to pollution-free water and air; and

iv. Protection against hazardous industries.

20.6.3 Federal scheme

Since India has a federal Constitution, it necessarily provides for a division of power and functions between the centre and the federal units (states). The Indian federal system leans slightly in favour of the centre while keeping a federal pattern and framework. The Constitution has created three functional areas regarding law-making by the two components of the federal system. These are:

i. An exclusive area for the centre called the Union List,
ii. An exclusive area for the states called the State List; and
iii. A common or concurrent area in which both the centre and the states may operate simultaneously, though with the centre having overall supremacy, called the Concurrent List.

20.7 Plant bio-security programme and its need

"Plant bio-security is a set of measures designed to protect a crop or a subgroup of crops from plant pests at national, regional and individual farm levels."

Several control measures has been developed to control or manage exotic pests and diseases. Plant Bio-security program develops quarantine policies that protect the plant health from exotic pests. Threat or quarantine pest identification and pest risk analysis are significant components of plant bio-security. Plant Bio-security policies are based on our national and international obligations under the World Trade Organisation (WTO) and in particular the Sanitary and Phytosanitary (SPS) measures. Globalisation and movement of people across world are increasing the chances of spreading the diseases and exotic pests among different forests and woodlands. Pests are mostly transported in soil or organic material, such as plant debris, that can be carried on footwear or by the wheels of vehicles and forest machinery. Diseases may also be spread via the equipment. Some pathogens are dispersed in water and so the risk of these being spread increases when conditions are wet (Khetarpal and Gupta, 2006). Several insect pests and diseases were introduced into country and now become a serious threats to our plant health and environment viz., *Agrobacterium tumefaciens, Corynebacterium michiganense, bacterial pathogens,* Ficus mosaic virus, Grape fan leaf virus, *Puccinia helianthi* from Europe, North America, North America, South Pasific Asia, Africa respectively. Cottony cushion scale, *Icerya purchasi*, Citrus Pest introduced from Australia, San Jose scale, *Quadraspidiotus perniciosus*, apple pest from China and Wooly apple aphid, *Eriosoma lanigerum*, a pest of apple entered from European countries are now become a serious threat for our crops and plant health So there is essential to secure plant health by adopting Plant Bio-security.

20.7.1 Plant health

With increase in population, agricultural production has to be increased in order to feed billions of people. So for this we are adopting new technologies to enhance the production. But on the other hand, these technologies are affecting the plant health. So the main focus is plant health management.

20.7.2 Threats affecting plant health

Insects, nematodes, bacteria, virus and fungal pathogens are the serious threat for plant health. They affect the plant growth, development and the associated environment through which food production collapse, chances of diseases transmission and insect pest movement all over world increase. Some of the important threats are:

20.7.3 Bacterial and Fungal Pathogens

Bacterial wilt and ring rot of potato, Fire blight of apple and pear, Black pod of cocoa, Powdery rust of coffee, Sudden death of oak, South American leaf blight of rubber, Vascular wilt of oil palm, Soybean downy mildew, Blue mold of tobacco, Tropical rust of maize.

20.7.4 Nematodes

Pine wood nematode, Red ring nematode of coconut. Barley stripe mosaic virus, Coconut cadang cadang (Viroid), Palm lethal yellowing (Phytoplasma)

20.7.5 Insect biotypes

"A population of a pest species that differs from other populations of the species in its ability to attack a particular cultivar is known as biotype". The continuous growing of insect-resistant varieties may lead to certain physiological and behavioural changes in insect pests so that they are capable of feeding and developing on the resistant varieties. (i) Hessian fly has developed 16 biotype, these are maximum in all insects but in insect groups maximum no. of biotypes developed in Aphids. (ii) Biotypes in Rice brown plant hopper is 5 while in Rice gall midge no. is 13.

20.7.6 Invasive alien species

Alien species are not of native origin; they introduced from another country or continent and become serious threat in the introduced country or region. Introduction, establishment, spread and naturalization are the important process for invasion of alien species (Table 20.1).

Table 20.1. Invasive alien insect species in India (Rana et *al.,* 2004 and Raghubanshi et *al.,* 2005)

S.No.	Common Name	Scientific Name	Host	Year
1.	Woolly apple aphid	*Eriosoma lanigerum* (Hausmann)	Apple, Pear	1889
2.	San Jose Scale	*Quadraspidiotus perniciousus* Comstock	Apple	1911
3.	Lantana bug	*Orthezia insignis* (Browne)	Lantana, coffee, citrus, potato	1915
4.	Cottony cushion scale	*Icerya purchasi* (Maskell)	Cotton, *Acacia sp.*	1921
5.	Potato tuber moth	*Phthorimaea operculella* (Zeller)	Potato, tobacco, tomato, brinjal and beat	1937
6.	Diamond-back moth	*Plutellc xylostella* (Linnaeus)	Cabbage, cauliflower, radish	1941
7.	Pine woolly aphid	*Pineus pini* (Macquart)	*Pinus sp.*	1970
8.	Subabul psyllid	*Heteropsylla cubana*	Crawford Subabul	1988
9.	Serpentine leaf miner	*Liriomyza trifolii*	Burgess cucurbits, tomato, castor	1990
10.	Coffee berry borer	*Hypothenemus hampei* Ferrari	Coffee	1990
11.	Spiraling whitefly	*Aleurodicus disperses* Russell	Polyphagous	1994
12.	Silver leaf whitefly	*Bemisia argentifolii* Bellows	Tomato	1999
13.	Blue gum chalcid	*Leptocybe invasa* (Fisher and LaSalle)	Eucalyptus	2006

20.8 Plant Protection Organisation in India

Plant Protection has significant importance in plant health management and sustainable agriculture. Plant protection aims to minimize crop losses or to manage the incidence of insect pests, diseases, weeds, nematodes, rodents etc. Enforcement of Destructive Insects and Pests Act (DIP Act, 1914), promotion of Integrated Pest Management, implementation of Insecticide Act, 1968, monitoring and control of desert locust in the Scheduled Desert Area, providing the information through different trainings are the key role in Plant Protection (Khetarpal and Gupta, 2005). The Central Directorate of Plant Protection is functioning with the support of 87 Sub offices: 29 Central Integrated Pest Management Centres, 35 Plant Quarantine Stations, 2 Regional Pesticides Testing Laboratories, 29 Locust Control Stations, 1 National Plant Protection Training Institute. At present 12 divisions/ units are functioning in the Directorate:

i. Locust Warning and Control unit.

ii. Plant Protection Division.
iii. Pest /Disease Surveillance Division.
iv. Biological Control Unit.
v. Plant Quarantine Division,
vi. Implementation of the Insecticide Act. vii. Storage Unit.
viii. Rodent Control Unit.
ix. Pesticide Monitoring Unit.
x. Planning and Coordination Unit.
xi. Documentation Unit.
xii. Plant Protection Training Institute.

20.9 Plant Quarantine system in India

Plant quarantine refers to the legal restriction to prevent the entrance and establishment of a plant disease or insect pest in an area uninfected area. In India, plant quarantine is regulated under DIP Act, 1914. There is various method of plant quarantine like:

i. Inspection at point of destination: Inspection at point of destination of consignment is necessary as many pests can be discovered and accidental introduction of these pests may be eliminated.
ii. Inspection at point of origin: Inspection at point of origin is also necessary to check the insect infestation or plant disease infection, material are allowed to enter the state or the country but they bear a certificate issued by the plant quarantine officer.
iii. Embargoes: Embargoes means the exclusion of the plant material or commodities which can be the hosts of insects. iv. Controlled introduction: The introduction of only a very limited quantity of plant materials are allowed for propagating purpose obtaining, should be free from insect pests and diseases (Rajak *et al.,* 1999). If any infestation or infection is found in the material then it will not allowed for introduction and proceed for rejection. Several countries have strict quarantine laws for importation and spread of pests. African countries import pome fruit from Asia and America is very rigorously controlled because of the danger of importation of San jose scale, *Quadraspidiotus perniciosus*, which is the most destructive orchard pest in Southern USA (Rajak *et al.,* 1999 and Bhalla *et al.,* 2002).

20.10 National Standards for Phytosanitary measures

In India following eight National Standards has been developed some of which confirm to some of the International Standards but a lot more work is needed in this direction for their development:

i. National Standard for Pest Risk Analysis.
ii. Guidelines for certification of forced hot-air treatment facilities for wood packaging material.
iii. Quarantine treatments and application procedures: I. Methyl bromide fumigation.
iv. Guidelines for assessment, audit and accreditation of fumigation agencies for undertaking methyl bromide fumigation.
v. Requirements for establishment of pest free areas for mango nut (seed) weevil (*Sternochaetus mangiferae*) and pulp weevil (*S. frigidus*).
vi. Requirement for establishment of pest free areas for Tephritid fruit flies.
vii. Guidelines on certification of hot water immersion treatment facilities for mango fruits.
viii. Accreditation treatment for ISPM-15 Compliance.

20.11 Pest Risk Analysis in Plant Bio-security

Pest Risk Analysis is a process of investigation, evaluation of information and decision making with respect to a certain pest that starts once it is known or determined that this is a quarantine pest. Subsequently, an evaluation of the potential of introduction of the pest into country is done along with its economic, social and environment consequences. With identification, determination and evaluation done, the process of culminates with decision making to avoid or reduce the probability of entrance or establishment of the pest into the country. Initiation points for PRA:

i. The identification of a pathway, usually an imported commodity that may allow the introduction and spread of quarantine pest.
ii. The identification of a pest that may qualify as a quarantine pest.
iii. The review or revision of phytosanitary policies and priorities. The PRA process as described in the International Standards for Phytosanitary measures (ISPM) is divided into four phases- Pest risk initiation, Pest Risk assessment, Pest risk management and Pest risk communication/documentation (Gupta and Khetarpal, 2004).

20.12 Survey and Surveillance in plant bio-security

To examine the situation of a region or incidence of insect pest or disease Survey and Surveillance of that region is essential. Survey is a procedure to determine the characteristics of a pest population or to determine which species occur in an area and in surveillance collection and data on pest occurrence or absence are maintained for the future incidence of pest in that region (Ram and Diwakar, 1999).

20.12.1 Current Status

Recently "The Agricultural Bio-security Bill , 2013" as introduced in Lok Sabha, on March 11, 2013 by the Minister of Agriculture, Mr. Sharad Pawar. It contains different chapters related to agricultural bio- security authority, liabilities, powers and functions to regulate the Bio-security obligations under international agreements etc. The Bill aims to establish an integrated national bio-security system covering plant, animal and marine issues to combat threats of bioterrorism from pests and weeds. Bio-security bill contains the following chapters and clauses:

Chapter I : **Preliminary**

Clauses

1. Short title, extent and commencement.
2. Definitions.
3. Prohibition of import.
4. Prohibition of export.
5. Domestic quarantine.
6. Duty to inform.

Chapter II : **Agricultural bio-security authority of India**

7. Constitution and incorporation of Authority.
8. Composition of Authority.
9. Qualification for appointment of Director General.
10. Powers of Director General.
11. Conditions of Service of the Director General and members.
12. Meetings of Authority.
13. Vacancies, etc., not to invalidate proceedings of Authority.
14. Removal of Director General, Deputy Director General and members.
15. Secretary of Authority and his powers.
16. Officers and employees of Authority.
17. Director, etc., not to undertake any other work.
18. Appointment of experts.
19. Committees of Authority.

Chapter III : **Transfer of assets, liabilities, etc., of authority**

20. Transfer and vesting in Authority undertakings of Plant Animal Quarantine Stations, etc.
21. General effect of vesting of undertaking in Authority.
22. Special Provision for transfer of employees to Authority.

Chapter IV : **Powers and functions of authority**

23. Functions of Authority.
24. Declaration of quarantine pests.

25. Controlled areas.
26. Action by Authority on report of State Government.
27. Control measures and their implementation.
28. Prohibition on removal of plants, animals, plant products or animal products or other objects, or pests under quarantine.
29. Recovery of expenses from State Government.
30. Accredited laboratories and accredited quarantine facilities.
31. Delegation of powers.
32. Authentication of orders and other instruments of Authority.
33. Contracts by Authority.
34. Mode of Executing Contracts on behalf of Authority.
35. Borrowing powers of Authority.
36. Power of Authority to issue directions.

Chapter V : **Obligations under international agreements**

37. Discharge of obligations under International Agreements.

Chapter VI : **Emergency action**

38. Declaration of bio-security emergency.
39. Effect of declaration of bio-security emergency.
40. Bio-security emergency scheme.

Chapter VII : **Appeal and revision**

41. Appeal.
42. Power of revision.
43. Re-analysis.
44. Procedure for sale of goods and application of sale proceeds.

Chapter VIII : **Finance, Accounts and Audit**

45. Grants by Central Government.
46. Constitution of Agricultural Bio-security Fund.
47. Accounts and audit.
48. Annual report and its laying before Parliament.

Chapter IX : **Import of plants and animals, etc.**

49. Application of law relating to customs and powers of Customs officers.
50. Release of imported products.
51. Revocation or suspension of permit or certificate.
52. Removal of imports.

Chapter X : **Enforcement**

53. Appointment of Designated Officers.
54. Powers of Designated Officers.
55. Duties of Designated Officers.

56. Procedures for sampling and analysis.
57. All agencies to assist.
58. Compensation not to be paid in certain cases.

Chapter XI : **Offences and punishment**

59. Penalty for contraventions for which no specific penalty is provided.
60. Punishment for interfering with seized items.
61. Punishment for false information.
62. Punishment for obstructing or impersonating a Designated Officer.
63. Penalty for contravention in relation to imports and exports.
64. Punishment for subsequent offences.
65. Offences by companies.
66. Cognizance and trial of offences.
67. Recovery of penalty, etc.

Chapter XII : **Control by Central Government**

68. Directions by Central Government.
69. Power of Central Government to prohibit import of pests or plants or animals and its products in public interest.
70. Power to supersede Authority.

Chapter XIII : **Miscellaneous**

71. Act to have effect in addition to other laws.
72. Power of Central Government to obtain reports and returns.
73. Officers and employees of the Authority to be public servants.
74. Protection of action taken in good faith.
75. Bar of jurisdiction of civil courts.
76. Power of Central Government to make rules.
77. Power to make regulations.
78. Laying of rules and regulations before Parliament.
79. Power to remove difficulties.
80. Repeal and savings.

20.13 Agricultural Bio-security Authority of India, Faridabad

Central government specified the direction and policies of Agriculture Bio-security Authority of India, regulating the import and export of plants, animals and related products, preventing the introduction of quarantine pests from another country and implementing post-entry quarantine measures. The Authority may prohibit import material whose use it considers risky. No person shall import any plant, animal, and plant or animal products in contravention of guidelines issued by the bill. In this case phytosanitary certificate will be required, issued by the Authority officers. Authority may notify any pest to be a quarantine pest and shall communicate the quarantine measures that the state government. If any bio-security emergency declares by the central government in an area in case of an outbreak, distribution, or spreading of a pest or organism, which has the potential to cause a significant loss to bio security then centre may give direction to control the situation. The Authority shall act as the national organisation under various international conventions such as the International Plant Protection Convention and may award penalty to any person for loss or damage to non-infested plants, animals or related products incurred by him as a result of any sanitary or phytosanitary measures.

20.14 National Institute of Plant Health Management (NIPHM)

NIPHM, Hyderabad (An autonomous Institution under DAC, Ministry of Agriculture) is a premier capacity building institution in the field of Plant Health Management, Bio-security and Incursion Management and Pesticide Management. The Institution also acts as a policy support centre to the Department of Agriculture and Cooperation. The Institute is the designated nodal centre at the National Level for capacity building in:

i. Plant Health Management,
ii. Bio-security and Incursion anagement,
iii. Rodent Pest Management,
iv. Pesticide Management. NIPHM organise different trainings related to Plant Bio-security and Incursion Management every year in to signify the Pest.

Risk Analysis, Pest Surveillance and diagnostics in bio-security management and threats of invasive alien species in the context of globalization are highlighted in the programme, besides exposure to SPS Measures and the relevant International standards.

20.15 Need for Bio-security in Trade and Exchange Programmes

The introduction of pests into a new locality is brought about in various ways namely:

a. the host may be the carrier;
b. inert materials such as packing material may carry resting stages of the organism;
c. insect vectors and birds may transport it;
d. air currents may carry the pest over long distances; or
e. there may be deliberate, illegal introductions to use them as bioweapons.

The first two modes of distribution lend themselves to curtailment by quarantine measures. The next two are by and large beyond human control and are a major limitation in the control of pest by exclusion. A different degree of alertness and preparedness is required for the last category. The global trade in agricultural commodities and transboundary movement of genetic resources of plants and animals has led to situations in the past which warranted legislative measures to regulate such trade/ exchange. Plant quarantine is the government endeavour enforced through legislative measures to regulate the introduction of planting materials, plant products, soil, living organisms, etc. in order to prevent inadvertent introduction of pests (including fungi, bacteria, viruses, nematodes, insects and weeds) harmful to the agriculture of a country/ state/ region, and if introduced, prevent their establishment and further spread. The devastating effects resulting from diseases and pests introduced along with international movement of planting material, agricultural produce and products are well documented.

1. The historical Irish famine of 1845, caused by late blight of potato introduced from Central America; coffee rust introduced in Sri Lanka in 1875 and its subsequent introduction in India in 1876; fluted scale on citrus introduced from Sri Lanka in 1928; San Jose scale in apple introduced into India in 1930s; bunchy top of banana introduced from Sri Lanka in 1943; the dreaded Golden nematode infesting potatoes introduced in 1960s from the UK and the noxious weed *Lantana camara* introduced in 1809 from Central America are glaring examples that clearly demonstrate that introduction and establishment of quarantine pests including weeds into new areas can severely damage the crop production and economy of a region/country. Infectious diseases in poultry and livestock cause an annual loss of about Rs 50 billion, besides the losses due to export restrictions and the impending danger of public health hazards. In the case of epidemic outbreaks the losses are much more colossal.
2. Notable examples include Swine fever in 1962 causing a mortality of 3500 pigs, African horse sickness in 1960 which resulted in the death of over 22,000 equines between 1960-63 and the introduction of hydro-pericardium (Lichi disease) of poultry through cross-border transmission. The recent case of H5N1 avian influenza virus first reported from Vietnam and China in December, 2003 has since killed 166 people and has been reported from 11 countries, namely Azerbaijan, China, Cambodia, Djibouti, Egypt, Iraq, Nigeria, Indonesia, Thailand, Turkey and Vietnam.
3. Instances of crop pests being used or the credible threat of use for agro-warfare include: the dropping of the dreaded Colorado potato

beetles (*Leptinotarsa decimlineata*) by Germany in the potato fields of the UK during World War II; Cuba accused US of attacking Cuban crops on several occasions through pests like tobacco blue mold, sugarcane rust and *Thrips palmi*; the threat by opponents of the Idi Amin regime to poison Uganda's coffee and tea plantations to deny the government foreign exchange, and recently, a Florida University professor informed the CIA that an outbreak of citrus canker in Florida was a result of the Cuban biological weapons programme.4 Moreover, the trend towards planting monocultures that have a high degree of genetic homogeneity, the concentration of a single crop in one region and the intensive rearing of animals all aid the disease spread. As recently revealed by the US, Iraq had a wheat smut bomb and stockpiles of wheat rust spores while aflatoxin and anthrax spores were also used in powdered forms during the US-Afganistan war. *Clostridium tetani, Clostridium botilinum, Salmonella typhimurium* organisms are being used as weapons in agro-warfare, in case of which cases the targets are vulnerable and the security low.

After the Second World War, FAO convened an International Plant Protection Convention (IPPC) in 1951, to which India became a party in 1956 along with Australia, Sri Lanka, UK, the Netherlands, Indonesia, Portugal and Vietnam. At present, there are more than 160 signatory members of the IPPC. The IPPC aims to develop international cooperation among various countries to prevent the introduction and spread of regulated pests that may accompany international movement of plants and planting material (http://www.ippc.org). The IPPC requires that each country establish a national plant protection organization to discharge the functions specified by it.

20.16 Risk Associated with Imports

Based on the quantity, plant/ planting material are either imported as bulk material for sowing/ planting or for consumption while small samples of trial and germplasm are required for research work. Of these, bulk imports for sowing/ planting carry maximum risk as thorough examination and treatment becomes difficult and the planting area required is also too large. Quarantine processing is often restricted to smaller samples derived from them and based on the results of these samples the whole consignment is rejected/ detained or released after giving fumigation treatment or fungicidal/ insecticidal treatments. The bulk consignments meant for consumption pose lesser hazards. However, certain small samples meant for research purposes are of immense quarantine importance. These samples usually comprise of germplasm material or wild relatives or landraces of a crop, and are thus more likely to carry diverse biotypes/ races/ strains of the pest. Besides, in case of true seed, generally, risks are more due to deep-seated infections than with surface-borne contamination. However, the import of vegetative propagules present a much higher order of risk than true seeds. In case of import of animals too, there is always a risk of the introduction of new diseases/ pathogenic organisms into the country causing serious animal health problems. Once such pathogens are introduced into a country, they can cause epidemics due to the absence of diagnostics, vaccine or effective drugs, lack of resistance in the host animals. Hence, there is a need to take extra precaution in the import of livestock and livestock productions infected with such pathogens. While discussing bio-security, a threat from the exotic destructive pests is of foremost importance in this era of liberalized import under the WTO. However, with the changing conditions, the indigenous pests already existing but having lower damage levels in India are changing their habit and gaining more importance over the years. There are new diseases emerging at an unprecedented rate over the last decade without any clear-cut explanation of the reasons for their emergence or recurrence. These diseases now continue to spread over large areas. These include foliar blight in wheat, necrosis in sunflower, bract mosaic in banana, sheath blight in maize and paddy, *pyrilla* in sugarcane, etc. and the number of such pests/ diseases continues to grow while inflicting increased crop losses. Likewise in the livestock and poultry sector also various diseases have emerged and were responsible for heavy economic losses. These include *peste des petites ruminants* (PPR), Blue Tongue, *Babesiosis, Cysticercosis, Anaplasmosis*, infectious bovine *rhinotracheitis/* infectious pustular *vulvo-vaginitis* (IBR/IPV), *Theileriosis, Contagious agalactiae (Mycoplasmosis),* Enzootic abortion (ovine *chlamydiosis*), Caprine arthritis/ encephalitis (CAE), Maedi-visna, Contagious caprine pluropneumonia (CCPP), Equine influenza, Equine piroplasmosis, Equine viral rhino-pneumonitis, Bovine viral diarrhoea (BVD), Bovine immunodeficiency virus infection (BIV), Infectious hydropericardium (Hydropericardium syndrome), Bovine leucosis, etc. With the change in cropping system there is a resurgence of pest problems such as necrosis in sunflower and groundnut, leaf curl in potato, etc. which are warning signals that cannot be neglected in the attempt to sustain the crop production and food security. Also, with the excessive use of chemical pesticides, a number of resistant strains of pests have evolved which constantly need improved mitigative measures. Apart from the threat posed by resurgence, a large number of diseases are endemic in nature and continue to cause losses in a given area, e.g. Karnal bunt in wheat, brown plant hopper in paddy, yellow mosaics in pulses, etc. in agriculture while in livestock and poultry these include: Foot and mouth disease (FMD), anthrax, clostridiosis, leptospirosis, etc. There is a need to understand the genetics of resistance to major diseases, identify the genes for resistance/ durability of resistance in important crops like rice, maize and grain legumes, by pyramiding both conventional and molecular genes for resistance. Efforts also need to be made to unfold the etiology of certain diseases that remain of unknown etiology and to develop diagnostic protocols for seed certification for use by the certification agencies or seed testing laboratories as they are immensely important in trade.

20.17 Threat of Biowarfare

The potential for dangerous pests not reported from India being misused or mishandled and thereby threatening the ecosystems on a large scale is also an issue of national concern. The agricultural economy is vulnerable to threat from exotic pests/ diseases that have the potential to be used as bioweapons (Table 20.2). Besides, a large number of destructive agricultural pests have strains/ isolates/ biotypes reported which also have a potential for use as bioweapons. These include viruses such as *Rice tungro bacilliform virus* with four variables isolated from South Asia; *Rice tungro spherical virus* whose Indian isolate is different from the South-East Asian isolates, *Cotton leaf curl virus* which causes severe damage in Pakistan but has a limited distribution in India, *Groundnut bud necrosis virus* having a wide host range, *Banana bunchy top virus* with five identified strains and *Tobacco streak virus, Citrus tristeza virus* and *Mungbean yellow mosaic virus* with several strains reported. The pathogens causing serious diseases where variability has been reported are cereal rusts caused by *Puccinia triticina* (whose spores are air-borne of which a number of virulent pathotypes are known), rice blast (*Pyricularia oryzae,* where a high degree of variability has been reported), *Bulkholderia solanacearum* (race 2 of which is not known in India) and *Xanthomonas campestris* pv *malvacearum* (of which the most virulent pathovar in Africa – XcmN is not known in India). The insects where biotypes have been reported include *Bemisia tabaci,* (a highly polyphagous pest which attacks >600 host plant species has 16 known biotypes); brown plant hopper (*Nilaparvata lugens,* where biotypes from India differ from those in other Asian countries); rice gall midge (*Orseolia oryzae,* has six biotypes known from India), red flour beetle (*Tribolium castaneum,* whose strains show variability in level of pesticide resistance). Several races are also reported for nematodes like *Meloidogyne incognita, M. javanica/ M. arenaria* and *Heterodera avenae.*

Table 20.2. Major Threats from Devastating Exotic Plant Pests/ Diseases

Bacterial and Fungal Pathogens	
Bacterial wilt and ring rot of potato	*Clavibacter michiganensis* sub sp. *sepedonicus*
Fire blight of apple and pear	*Erwinia amylovora*
Black pod of cocoa	*Phytophthora megakarya*

Powdery rust of coffee	*Hemileia coffeicola*
Sudden death of oak	*Phytophthora ramorum*
South American leaf blight of rubber	*Microcyclus ulei*
Vascular wilt of oil palm	*Fusarium oxysporum* f sp. *elaedis*
Soybean downy mildew	*Peronospora manshurica*
Blue mold of tobacco	*P. hyocyami* sub sp. *tabacina*
Tropical rust of maize	*Physopella* zeae
Virus, Viroid and Phytoplasm	
Barley stripe mosaic virus	
Coconut cadang cadang (Viroid)	
Palm lethal yellowing (Phytoplasma)	
Nematodes	
Pine wood nematode *(Bursaphelenchus xylophilus)*	
Red ring nematode of coconut *(Rhadinaphelenchus cocophilus)*	
Insects	
Mediterranean fruit fly	*Ceratitis capitata*
Cotton boll weevil	*Anthonomus grandis*
Russian wheat aphid	*Diuraphis noxia*

Ethically, the potential for harm must be weighed against the scientific or commercial freedom for research and to develop microbes for useful reasons. Given the invasions of ecological systems by alien species, as well as the potential for bioterrorism, we need to foster policies and technological capabilities to prevent, detect, and respond to incidents involving such acts. Lack of genetic diversity in certain cases limits natural defenses to disease and to biological agents that are intentionally introduced into an environment. Introduction of naturally-occurring or manmade genetically modified (e.g., recombinant DNA) viruses and experimental biotechnology into weaponry, livestock, crops and medicine is controversial because, if not adequately controlled, these may threaten the well-being of entire populations and ecosystems. The new biological strains of plant pathogens can easily cause significant harm, as often observed in the case of livestock, because they rapidly infect elements of ecosystems that have not yet developed immunities.

20.18 Ensuring Bio-security during Trade and Exchange: National Scenario

Legislation

The Government of India legislated the Destructive Insects and Pests (DIP) Act in 1914. This Act has been amended through various notifications issued from time to time and also has provision for domestic quarantine to restrict the movement of certain planting material from one state to another state. In 1984, a notification was issued under this Act, namely Plants, Fruits and Seeds (Regulation of Import into India) Order popularly known as the PFS Order which was revised in 1989 after the announcement of the New Policy on Seed Development by the Government of India in 1988, proposing major modifications for smooth quarantine functioning. This order has nowbeen superceded by the Plant Quarantine (Regulation for Import into India) Order 2003 as there was an urgent need to fill the gaps in the existing PFS order regarding import of germplasm/ genetically modified organisms (GMO's)/ transgenic plant material/ biocontrol agents, etc. to fulfill India's legal obligations under the international Agreements. Under this order, the need for incorporation of additional/ special declarations for freedom of import commodities from quarantine and invasive alien species (IAS), on the basis of standardized pest risk analysis (PRA), particularly for seed/ planting materials is also taken care of. The other salient features of the order include, prohibition on import of commodities with weed/ alien species contamination as per Schedule VIII; and restriction on import of packaging material of plant origin unless treated. Besides, the agricultural imports have been classified as:

a. prohibited plant species (Schedule IV);
b. restricted species where import is permitted only by authorized institutions (Schedule V);
c. restricted species permitted only with additional declarations of freedom from quarantine/ regulated pests and subject to specified treatment certifications (Schedule VI); and
d. plant material imported for consumption/ industrial processing permitted with normal Phytosanitary Certificate (Schedule VII). So far, i.e., till February 2007, ten amendments of the Plant Quarantine (PQ) Order 2003 have been notified to the WTO, with revised requirements for crops under the Schedules V, VI and VII, of which Schedule VI and VII now include 411 and 284 crops/ commodities, respectively. Issues related to IAS and GMOs are covered under the Environment Protection Act (EPA) 1986 but it does not state in clear terms the modality for restriction and prohibition of their potential threats to the environment.

Infrastructure

The Directorate of Plant Protection Quarantine and Storage (DPPQS) of the Ministry of Agriculture is the apex body for implementation of plant quarantine regulations. It has a national network of 29 plant quarantine stations at different sites, for example: airports (10), seaports (10) and land frontiers (9). In all, two categories of materials are being imported under the PQ Order, 2003:

a. bulk consignments for consumption and sowing/ planting, and
b. samples of germplasm in small quantities for research purposes. The Plant Quarantine Stations under the DPPQS undertake quarantine processing and clearance of consignments of the first category (http:// www.plantquarantineindia.org). The five major Plant Quarantine Stations have been modernized recently under an FAO-UNDP funded project. The National Bureau of Plant Genetic Resources (NBPGR) undertakes the quarantine processing of all plant germplasm and transgenic planting material under exchange for which it has well equipped laboratories and green house complexes and recently, a containment facility has also been established for processing transgenics. The NBPGR also has a well-equipped quarantine station at Hyderabad, which mainly deals with export samples of the International Crop Research Institute for Semi-arid Tropics. Besides the above mentioned plant quarantine set-up, the regulation of import and export of livestock and livestock products, control of exotic disease and certification as per Office International des Epizooties (OIE) regulations is done through the Animal Quarantine and Certification Services (AQ&CS) under the control of the Department of Animal Husbandry & Dairying (DAHD) at New Delhi, Kolkatta, Chennai and Mumbai.

Detection of Pests

During quarantine processing, seeds and plant material are examined for the presence of unwanted weed seeds, plant debris, soil clods, insect and mite pests, plant parasitic nematodes and pathogens including fungi, bacteria, viruses, phytoplasmas, etc. The external feeders and other incidental insect pests infesting the planting material are easily detected visually either by the naked eye or with the help of a magnifying glass or stereoscopic bioinocular microscope. The presence of nematodes is indicated during visual examination by the observation of galls or swellings on roots, tubers and rhizomes; white, yellow or brown pinhead sized round bodies adhering to roots; swollen or malformed leaf, stem or other tissues or root lesions or unusual root proliferation. Fungal infection is indicated by the presence of *sclerotia*, smut balls, malformed seeds and fungal fructifications on seed surface. Presence of yellow discolouration around the hilum is suggestive of bacterial infection. Specialized tests used for detection of different groups of pests are given below:

Insects and mites

1. X-ray radiography is used to detect seeds infested with phytophagous chalcidoids, bruchids and certain other insect groups

that do not exhibit any external symptoms on the seed surface. Based on literature survey and past experience, a list of >340 plant genera has been drawn up (Bhalla *et al.,* 2002), that are compulsorily subjected to X-ray radiography. On developing the X-ray plates, insects if present, are hand-picked and healthy seeds released to the indentor.

2. Transparency method is used for detecting infestation in small seeds and seeds of family Graminae. The seeds are boiled in lactophenol solution (phenol, lactic acid, distilled water and glycerine in the ratio of 2:2:2:1, respectively) for 1-2 hours depending on the hardness of the seeds. This renders them transparent to reveal insect infestation.

Plant Parasitic Nematodes

1. Soaking of seeds known/ suspected to carry seed-borne nematodes in water overnight softens the seeds which are teased/ crushed enabling the nematodes, if present, to come out in water.
2. Soaking of some plant material in water and when sieved through nematological sieves (the finest sieve is of 400 mesh per linear inch) reveals nematodes that are retained on the sieve. These are recovered and examined under the compound microscope for identification.
3. The staining technique is used for quick detection of nematodes in vegetative propagules where a part of the plant tissue (especially roots) is boiled in acid fuchsin lacto-phenol solution for a few minutes and de-stained in clear lacto-phenol. The nematodes, if present, retain the red stain more deeply than the plant tissue and can easily be detected under a stereo microscope
4. Examination of the accompanying soil shows the presence of viable nematodes especially ectoparasites and cysts of cyst forming nematodes.

Fungal and Bacterial Pathogens

1. Examination of suspension after seed washing is used for detecting surface borne pathogens. The seeds are shaken in water and the resultant suspension is examined for spores of smuts, bunts, downy mildew, powdery mildew and some of the fungi like *Protomyces macrosporus* under the compound microscope.
2. The blotter method is used for detection of many fungal and bacterial pathogens capable of producing mycelial growth and fruiting structures under the incubation. Seeds when placed on moist filter paper in plastic petriplates and incubated at 20 + 1°C under fluorescent tubes in alternating cycles of 12 hours light/ darkness for seven days show growth of fungi and bacteria.

Seed-borne viruses

1. Seeds known or suspected to carry seed-borne viruses are grown in insect proof post-entry quarantine net house/ environment controlled screen houses. Seedlings showing viral symptoms are uprooted and burned. Produce from only healthy plants is released to the indentors.
2. Infectivity test is done to assay the presence of virus by inoculating leaf extracts of seedlings showing symptoms on indicator hosts. This method reveals the symptomless or latent infections of plants not observed in grow-out tests.
3. Observation of the sample from a leaf showing viral symptoms under the transmission electron microscope also reveals the size and shape of the virus particles, if present.
4. For serological diagnosis of plant viruses enzyme-linked immunosorbent assay (ELISA), a relatively simple, rapid and sensitive technique is used for simultaneous testing of a large number of samples.

Interceptions in Imported Material

Over the years, during quarantine processing, a large number of pests have been intercepted in imported bulk consignments (http:// www.plantquarantineindia.org/ docfiles/appendix-8) and in germplasm and other research material. These pests intercepted include many that are not known to occur in India, have different races/ biotypes/ strains not known to occur in India; are present on a new host, are from a country from where they were never reported before or are an entirely new pest species hitherto unreported in science or are reported to be present widely in India (Table 20.3). These interceptions, especially of pests and their variability not yet reported from India (Category 1 and 2) signify the importance of the quarantine in preventing the introduction of destructive exotic pests. The third and fourth category of pests are not expected in the sample as per the risk analysis which is literature based and since no records are available on the pest/ host their presence is unexpected and important from quarantine view point. The same is the case with the fifth category of pest as the pest is a new species hitherto unknown to science. The last category-pests with a wide host range are critical and could become invasive in case they find a suitable biotic and abiotic environment. Besides, in the recent past, diseases like Bovine Viral Diarrhea (BVD), Malignant Catarrh Fever (MCF), Rabbit Hemorrhagic Disease (RHD), Avian Influenza (AI) have been diagnosed in the imported livestock and poultry at the entry point of quarantine. If not intercepted and controlled, these diseases would have played havoc with our livestock and poultry.

Table 20.3. Examples of Different Categories of Pests Intercepted in Quarantine

Category of pest intercepted	Pest/ host/ source country
Not known to occur in India	*Peronospora manschurica*/ Soybean/ USA *Uromyces betae*/ Sugarbeet/ USA and Italy *Fusarium nivale*/ Wheat/ U K *Cowpea mottle virus*/ Cowpea/ Philippines *Tomato black ring virus*/ French bean/ CIAT (Colombia) *Heterodera schachtii*/ Sugarbeet/ Denmark *Anthonomus grandis*/ Cotton/ USA *Quadrastichodella eucalypti*/ Eucalyptus/ Australia
Known to occur but the race/biotype/strain intercepted is not	*Helminthosporium maydis*/ race T/ Sorghum/ USA Pea *seed-borne mosaic virus*/ Broadbean *Burkholderia solanacearum biovar 3*/ Groundnut/ Australia
Intercepted on a host on which it was never reported before	*Alternaria zinniae*/ Tobacoo/Japan *Pseudomonas syringae pv* syringae/Hibiscus cannabinus/ Bangladesh *Aphelenchoides besseyi*/ Stylosanthes hamata/ Australia *Merobruchus columbinus*/ Samanea saman/ UK *Bruchus ervi*/ Acacia brachustachva/Australia *Pachymerus lacerdae*/ Orbygnya phalerata/ Italy
Intercepted from a country from where it was never reported before	*Peronospora manschurica*/ Soybean/ Malaysia *Heterodera zeae*/ *Vetiveria zizanioides*/ Tanzania *Bruchus ervi*/ *Acacia brachustachva*/ Australia
A new species hitherto un re ported	*Drechslera pluriseptatal*/ *Eleusine coracana*/ Zambia *Tylenchorhynchus neoclavicaudatus potato*/ *tubers*/ USA *Polenchus minutus*/ Palm plants/ UK
Known to occur in India but possess a wide host range	*Collectotrichum graminicola, Drechslera turcica* and *Gloeocercospora sorghi*/ Sorghum/ Nigeria *Drechslera siccans*/ Soybean/ USA *Claviceps purpurea*/*Avena sativa*/ USA *Drechslera sorokiniana*/*Carthamus tinctorius*/ Italy, USA

Source: Khetarpal and Gupta, 2006a.

20.19 Major Issues in Import of Transgenics

The rules for safe use and release of GMOs have been notified under the EPA. The Department of Biotechnology (DBT) of the Ministry of Science and Technology reviews, permits, monitors the experiments utilizing GMOs and recombinant DNA products, while the MoEF implements their large-scale commercial use through its Genetic Engineering Approval Committee (GEAC). The biosafety guidelines prepared by DBT in 1990 which have subsequently been revised requires that every organization involved in work on GM plants set up its Institutional Biosafety Committee (IBSC) for interaction with the government, various committees have been formed by the DBT with well-defined objectives for implementing the biosafety guidelines. The Review Committee for Genetic Manipulation (RCGM) a national committee under the DBT reviews the ongoing R&D projects on GMOs, undertakes field visits of experimental sites and issues clearance for import/ export of GMOs for research and training. The RCGM has also constituted a Monitoring cum Evaluation Committee to monitor the limited field trials for which permission for multi-locational field-testing has been accorded. In addition, the concerned committees of each state and district are also involved in the monitoring of the experiments at their respective field sites. As per the PQ Order 2003, the NBPGR is responsible for ensuring that imported transgenic material is free from pests and terminator gene technology. More than 7000 samples of transgenic germplasm have been imported till date. Several interceptions have also been made in these imported transgenic samples including downy mildew caused by *Peronospora manschurica* in soybean from USA and *Cryptolestis ferrugenius* in paddy from Singapore, which are yet not reported from India. The interceptions, especially of pests not yet reported from India signify the role of quarantine in preventing the introduction of such exotic pests.

The National Plant Protection Organization needs to be upgraded in terms of manpower, infrastructure and capabilities to raise it to international standards as the increase in imports and the stipulation of WTO has brought about additional challenges to be faced by the plant protection personnel. Strengthening should not only ensure prevention of exotic pests but would also check the interstate spread of indigenous pests and diseases of both plants and animals by effective implementation of domestic quarantine regulations/ certification services against certain important pests and diseases which have been introduced/ detected in the country in the recent years and which are likely to spread fast. The new PQ Order is as an attempt to comply with the various provisions of the Agreement on Application of Sanitary and Phytosanitary (SPS) Measures of the WTO (of which India is a signatory member) and to promote trade and not to use plant quarantine measures as a technical barrier to trade. In order to prevent the ingress of exotic diseases and ensure an effective check against the spread of infection/ diseases in animals, the amendment of existing Livestock Importation Act needs to be undertaken as a priority. This may be modified as per the regulations of the OIE following the standards laid down for Australia and New Zealand which pay more attention to animal husbandry and safe export of livestock and livestock products. Pest risk analysis is a procedure by which quarantine services can technically justify new regulations, phytosanitary guidelines or safeguards, operational procedures, entry status of imported articles and resource allocations. The DPPQS has taken an initiative in undertaking PRA as per the WTO/ SPS requirements and norms but needs to be technically supported on the matter as far as expertise is concerned, as this would be the most ideal way of warding off exotic diseases by rejecting certain imports with prior knowledge and/ or permitting the imports with proper prophylactic treatments. DAHD through its Technical Committee on Animal Health is responsible for making the risk analysis and for preparing the health protocols required as per OIE norms for the import of livestock and livestock products. It should also recognize that risk analysis procedures should provide a basis for bio-security and not create barriers to trade.

There is a need to reprioritize our research projects based on the critical gaps identified to meet the requirement for export and import. A few that need immediate emphasis are as follows:

- Survey and surveillance of disease/ pests of national and international importance to have endemic pest data.
- Generation of comprehensive epidemiological data on important pests so that *tolerance limits* can be defined, to develop PRA and to identify PFAs as per the WTO norms.
- Development of diagnostic protocols using molecular techniques for detection of exotic pests and their variants not easily detectable otherwise but are nevertheless of grave economic concern.
- Development of digitized keys for identification of exotic pests.
- Evaluation of the biosafety risks in import of transgenics and beneficial biocontrol agents.
- Development of methods to detect the presence of transgenic/ GM contamination (deliberate or inadvertent mixing of GM seedwith non-GM seed) in unknown samples during quarantine processing. There is also a dire need to develop an accessible platform for getting information on bio-security for the policy makers, administrators and the industry groups. A website http://www.plantquarantineindia.org consisting of a national database on legislation, quarantine procedures, methodologies, etc. designed by DPPQS is available. Compilations on potential quarantine pests of various crop groups are being taken up at the NBPGR. However, an internet-based portal mechanism for exchange of official information on food safety, and vanimal and plant health and the environment (like the International Portal for Food Safety and Animal and Plant Health), to facilitate communication among countries in these sectors needs

to be developed for easy accessibility of information. To address the issue of emerging pests in a proper perspective recently, there was an endeavour by the government to establish a National Institute for Emerging and Exotic Plant Pests with a focused mandate for the development of an early warning system for pest outbreaks, and models for pest/ disease forecasting and risk analysis; development of novel diagnostic approaches for detecting pests, imparting multiple resistance to enhance the durability of resistance by pyramiding genes or developing management strategies by developing transgenics resistant to diseases. This institute could possibly work in a network mode to facilitate research by those laboratories that are not well equipped and may boost research in the case of many regional and chronic plant diseases. Finally, awareness about bio-security among the scientific fraternity and general public is required to generate respect for the regulations in the interest of national security.

20.20 Developing a Comprehensive Strategy

In the light of various issues as highlighted above, a holistic concept to address the issue of bio-security needs to be evolved. Internationally, the Agreement on the Application of Sanitary and Phytosanitary Measures of the WTO, governs SPS measures in relation to international trade. The Codex Alimentarius Commission (Codex), the IPPC and the Office International des Epizooties (OIE) provide international standards for food safety, plant health, and animal health, respectively. Further, the Cartagena Protocol of the Convention on Biological Diversity (CBD) applies to the transboundary movement, transit, handling and use of Living Modified Organisms (LMOs). Guidelines on the management of invasive alien species have been developed under the SBSTTA (Subsidiary Body on Scientific, Technical and Technological Advice) of CBD. This group of international agreements, organizations and programmes are part of a loose international framework for bio-security, and reflect the sectorial approach to regulate this area. It is also clear that under the present international scenario, the plant protection specialists have a major role to play not only in promoting and facilitating export and import in the interest of their respective nations but also in protecting the environment from the onslaughts of invasive alien pests and unforeseen ill-effects of the introduction and trading in GMOs. Besides, the threat to national bio-security from the use of such instruments as bioweapons to create agro-terrorism is a possibility that requires preparedness. It is because of these reasons that the FAO has recognized the growing importance of bio-security, and has included it as one of its sixteen Priority Areas for Inter-disciplinary Action. Bio-security was also included in the Medium Term Plan which aims at "promoting, developing and reinforcing policy and regulatory frameworks for food, agriculture, fisheries and forestry" (http://www.fao.org.COAG/2003/9.htm). At the national level efforts are being made for the development of a coherent bio-security strategy for the country. A number of meetings have recently been held to discuss the issue and during one such meeting held recently at the National Institute of Advanced Studies, Bangalore, it was highlighted that there was a need for "convergence in the efforts of all departments and ministries" to develop a coherent bio-security strategy. It was also emphasized that education, regulation and social mobilization were the three pillars necessary to formulate a bio-security strategy. The holistic approach to ensure bio-security seeks to use the synergies of various existing sectors at the national level, without necessarily creating new structures. It further recognizes the need for integration of various aspects of bio-security and the institutions involved. There is a need to establish a National Plant Bio-security System comprising a National Plant Bio-security Centre (NPBC) to deal with plants, animals, living aquatic resources, and agriculturally important microorganisms as its four major Divisions with the Ministry of Home Affairs (MoHA) at the helm of affairs. MoHA has the National Disaster Management Authority that can properly monitor and regulate the bio-security issues. Besides, MoHA is also the nodal point of the National Crisis Management Plan 2003. Therefore, if we have to put in place an effective bio-security system in future, it has to be under the MoHA so that matters related to both bio-security and bioterrorism (agroterrosim) can be dealt with more effectively and the funds can be invested in the right direction. The emergency action plan and the rapid response team that are critical in the outbreak of an epidemic also needs to be with MoHA. The MoA in its Indian Council of Agricultural Research (ICAR) has excellent researchers and laboratories in the field of plant protection, which would support and work in a networking mode for diagnostics and control of diseases/ pests in the case of emergent situations. Hence, a network of various ministries like MoA, MoEF, Ministry of Health and Family Welfare (MoHFW), Ministry of Defence (MoD), Ministry of Food (MoF) and Ministry of Science and Technology (MoST) could work in close collaboration with the MoHA. A holistic model depicting the components of bio-security including biowarfare, production, trade, environment (biodiversity and biosafety) as proposed recently is given in Figure 24.1. It also shows the regulatory mechanism including various committees in place at the national level [Export Import Policy (EXIM), EPA, Biological Diversity Act (BDA), Protection of Plant Varieties and Farmers Rights Act (PPVFRA), RCGM and GEAC] for implementation of the international agreements [Biological Weapons Convention, CBD, Cartagena Protocol (CP) and SPS/ WTO]. The various stakeholder ministries involved are also depicted in the Figure 20.1. A coordinated effort and synergy is all that is required to institute an effective bio-security system in the country. Models to rationalize regulatory functions among sectors in a quest for improved effectiveness and efficiency have appeared in a number of countries. For example, New Zealand has a Bio-security Act since 1993 and a Bio-security Minister and Council since 1999. In the US, the Department of Homeland Security was created in 2002 with as many as 22 agencies including the US Department of Agriculture's Animal and Plant Health Inspection Service. Likewise, the Australian Government has established Bio-security Australia. In Belize, food safety, and animal and plant quarantine and environmental issues, are dealt with by a single authority, the Belize Agricultural and Health Authority. For a biosecure India, what we need is a synergy of expertise from various organizations under the Ministries of agriculture, environment and forests, food, science and technology, home affairs, commerce and industry and defence. More so, we need the political will.

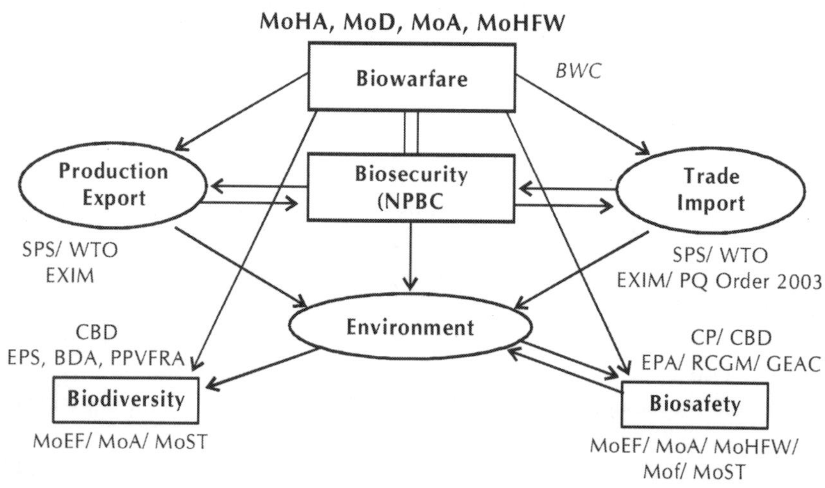

Figure 20.1 A Holistic Model for National Plant Bio-security System

20.21 Recent bio-security threats

Recent years have seen a range of bio-security problems which are notable for their high costs and public profile. With respect to livestock, an outbreak of classic swine fever in 1997 cost The Netherlands approximately £2.4bn (Whiting, 2003). The foot and mouth disease (FMD) outbreak in 2001 cost the UK approximately £7bn (Thompson *et al.,* 2002), while the appearance of bovine spongiform encephalopathy (BSE) in Canada and the USA in 2003 is estimated to have cost each of those countries $3–4bn in lost trade revenue (Anon, 2004). Until 2003, avian influenza had been considered a relatively rare animal disease, but in that year a devastating outbreak occurred in The Netherlands and a series of outbreaks began in Asia which have continued and spread to other regions (Harder & Werner, 2006). The explosive growth of aquaculture and mariculture worldwide has led to the spread of many serious diseases and parasites of fishes and shrimp (Hedrick, 1996; Hill, 2000). In Europe, the recent spread of Gyrodactylus salaris, a small, leech-like parasite of salmonids, threatens salmon fishing (Peeler *et al.,* 2003). Insect pests of animals have also been moving worldwide. The introduction of the New World screwworm fly, Cochliomyia hominivorax, into Africa in 1988 led to a successful emergency eradication programme coordinated by the UN (Lindquist *et al.,* 1992). While accidental, human-assisted movement of pests and pathogens appears to be the major cause of animal bio-security problems, other mechanisms of introduction are emerging. For instance, the spread of bluetongue disease of sheep in Europe appears to be related to range extension of its culicoid fly vectors (Defra, 2002), probably as a result of climate change. The African bont tick, *Amblyomma variegatum,* a potential vector of important cattle diseases, was accidentally introduced into the Caribbean in the 1800s, but its rapid spread into new countries in recent decades has been stimulated by another introduction, that of the cattle egret, which spreads the tick to new island countries (Pegram *et al.,* 2004). The great diversity of crops and their rich and often cryptic insect and pathogen complexes guarantee a continuing and high level of new pest and disease introductions. In recent years, particularly worrying introductions have occurred on four crops: wheat, rice, maize and potatoes that constitute 50% of the world's food supply. Karnal bunt, *Tilletia indica,* an Asian fungal disease of wheat, was introduced into the US in 1996, where it is under containment in southwestern states (USDA, 2004). Potato late blight, Phytophthora infestans, responsible for the Irish potato famine that led to the death or emigration of millions of rural poor in the mid-1800s, continues to evolve and spread new virulent forms (Goodwin *et al.,* 1994). In East Africa, wheat stem rust, *Puccinia graminis,* has recently re-emerged after decades of suppression with resistant varieties, creating a global bio-security risk if it now spreads (CIMMYT, 2005). In the UK, a booming trade in horticultural plants has led to many new introductions (Independent, 2003) one of which, the fungus *Phytophthora ramorum,* may threaten indigenous forest trees (Brasier *et al.,* 2004; Defra, 2005c). Forestry in general has seen a dramaticpattern of new disease and pest introduction, particularly through the recent opening of trade between East Asia and other regions (Cock, 2003). This brief snapshot of agricultural bio-security threats illustrates their diversity and considerable potential impact on agriculture. It also illustrates a policy-relevant phenomenon. Since bio-security problems are usually one-off, distinctive, time-bound events, they tend to be presented as clusters of cases as above (see also Bright, 1998; Baskin, 2002). Taken all together, we may get the impression of an immense and urgent crisis. Further, discovery of one new threat can lead to greater surveillance which in turn increases the likelihood of finding another, creating a rising spiral of new problems. In fact, at a national level, truly new introductions may be relatively few and far between, while chronic, indigenous or long-established pest or disease problems may be of greater economic significance than those posed by new bio-security threats. One of the challenges of bio-security research is to move beyond anecdote into rigorous analysis of the nature and impact of these diverse threats in the proper context.

20.22 Conclusion

Developing countries, focusing on to maximum food production to meet the requirement of large number of population and to gain maximum profit, in this context if the bio-security program continues to receive inadequate attention across people then unaffordable loss to global environment, agriculture and biodiversity would take place. Investment of resources and approaches in a systematic manner to establish bio-security program is the most effective option to achieve food security goals.

Models to rationalize regulatory functions among sectors in a quest for improved effectiveness and efficiency have appeared in a number of countries. For example, New Zealand has a Bio-security Act since 1993 and a Bio-security Minister and Council since 1999. In the US, the Department of Homeland Security was created in 2002 with as many as 22 agencies including the US Department of Agriculture's Animal and Plant Health Inspection Service. Likewise, the Australian Government has established Bio-security Australia. In Belize, food safety, and animal and plant quarantine and environmental issues, are dealt with by a single authority, the Belize Agricultural and Health Authority. For a biosecure India, what we need is a synergy of expertise from various organizations under the Ministries of agriculture, environment and forests, food, science and technology, home affairs, commerce and industry and defence. More so, we need the political will.

The PQ measures acts as an important tool in excluding pests from the crop. Effective implementation of quarantine is highly emphasized for management of pests, which in turn helps in maintaining the productivity of crops. In order to be food secured, we must be bio-secured. As a matter of fact, unsafe food is no food. Therefore, whatever we produce and eat, we must ensure that it is free from bio-risks. Notwithstanding the usefulness of regional and global partnerships, bio-security and food security must essentially be priority programmes of individual countries and locally owned research, information, strategies and management systems must be in place in each country. Local policy makers and service providers should be committed to the movement and bio-security should be mainstreamed in national policy and political agenda. It is hoped that the Government of India will take immediate action in setting up a National Agricultural Bio-security Council, National Centre for Agricultural Bio-security and a National Agricultural Bio-security Network. The recommendations made above relating to strategic interventions and strategic partnerships also need immediate attention. Above all, a National Agricultural Bio-security Fund will help to strengthen our infrastructure, introduce new molecular techniques of identification and verification, derive benefits from our animal genetic resources, and provide needed and timely help to the affected families.

References

Anonymous. 2004. Cow with BSE likely to be one infected animal in the herd. *Journal of American Veterinary Medicine Association.*

Baskin, Y. 2002. A plague of rats and rubbervines: the growing threat of species invasions. Washington, DC: Island Press.

Bhalla, S., Kapur, M. L., Lal, B., Verma, R. and Singh, C. 2002. "Quarantine Risk Associated with Exchange of Plant Genera carrying Hidden Infestation", *Indian Journal of Plant Genetic Resources*, 15(2): 160-163.

Brasier, C. M., Denman, S., Brown, A. and Webber, J. 2004. Sudden oak death (*Phytophthora ramorum*) discovered on trees in Europe. *Mycology Research,* 108: 1108–1110.

Bright, C. 1998. Life out of bounds: bioinvasion in a borderless world. New York, NY: W.W. Norton.

CIMMYT. 2005. Sounding the alarm on global stem rust: an assessment of race Ug99 in Kenya and Ethiopia and the potential for impact in neighboring regions and beyond. Expert Panel on the Stem Rust Outbreak in Eastern Africa. Mexico City: International Maize and Wheat Improvement Center (CIMMYT) (September 8).

Cock, M. J. W. 2003. Bio-security and forests: an introduction-with particular emphasis on forest pests. FAO forest health and bio-security working paper FBS/2E, 2003.

Defra. 2005a. Bio-security guidance to prevent the spread of animal diseases. Defra. 2005c. *Phytophthora ramorum.* Plant health. Information sheet.

FAO. 2003. Bio-security in food and agriculture. Report on the 17th Session of the committee on agriculture, Rome 31 March–4 April 2003.

Goodwin, S. B., Cohen, B. A., Deahl, K. L. and Fry, W. 1994. Migration from Northern Mexico as the probable cause of recent genetic changes in populations of *Phytophthora infestans* in the United States and Canada. *Phytopathology,* 84: 553–558.

Gupta, K. and Khetarpal, R. K. 2004. Concept of Regulated Pests, their Risk Analysis and the Indian Scenario, *Annual Review of Plant Pathology*, 3: 409-441.

Harder, T. C. and Werner, O. 2006. Avian influenza. In Influenza report (eds B. S. Kamps, C. Hoffmann & W. Preiser). Flying Publisher.

Hawkes C. and Ruel, M. 2006. The links between agriculture and health: an intersectoral opportunity to improve the health and livelihoods of the poor, *Bulletin of the World Health Organization*, 84(12).

Hedrick, R. P. 1996. Movement of pathogens with the international trade of live fish: problems and solutions. *Revue Scientifique et Technique,* 15: 523–531.

Hill, B. J. 2000. International trade in farmed fish and shellfish: the impact of disease spread. In Safeguarding animal health in global trade (moderators: Thierman, A., Moenig, M.). Online Conf. on Sustainable Agricultural Production. Braunschweig, Germany: Federal Agricultural Research Centre, FAL. Independent 2003 Gardeners sound alarm over unprecedented invasion by foreign pests.

Josling, T., Roberts, D. and Orden, D. 2003. Food regulation and trade. Towards a safe and open global system. Washington, DC: Institute for International Economics. Khetarpal, R. K. and Gupta, K. 2005. Status of Plant Protection in India in the wake of International Agreements, *Indian Journal of Plant Protection*, 33: 153-163.

Khetarpal, R. K. and Gupta, K. 2006. Dev Usha and Joshi Nidhi., "Plant Bio-security in India- Status and Strategy". Discussion meeting on *Setting up a National Agenda for Bio-security* during November 23-24, by Ministry of Agriculture, at NIAS, Bangalore.

Koblentz, G. D. 2010. "Bio-security Reconsidered: Calibrating Biological Threats and Responses". *International Security,* 34 (4): 96–132.

Lindquist, D. A., Abusowa, M. and Hall, M. J. 1992. The New World screwworm fly in Libya: a review of its introduction and eradication. *Medical and Veterinary Entomology*, 6: 2–8.

Mumford, J. D. 2002. Economic issues related to quarantine in international trade. *European Review of Agricultural Economics,* 29: 329–348.

Nugent, R., Benwell, G., Geering, W. K., McLellan, J., Mumford, J. D., Quinlan, M. and Zelasny, B. 2001. Economic impacts of transboundary plant pests and animal diseases. *In The state of food and agriculture 2001*. Part III. Rome, Italy: FAO Information Division.

O'Toole, T. and Inglesby, T. V. 2003. Toward bio-security. *Biosec. Bioterror. Biodef. Strat. Prac. Sci.,* 1: 1–3.

Peeler, E., Thrush, M. and Gardiner, R. 2003. A preliminary risk analysis for the transmission of the exotic fish parasite Gyrodactylus salaris between river catchments in England and Wales. Dorset, UK: The Centre for Environment, Fisheries and Aquaculture Science.

Pegram, R., Indar, L., Eddi, C. and George, J. 2004. The Caribbean Amblyomma program: some ecologic factors affecting its success. *Ann. NY Acad. Sci.,* 1026: 302–311.

Raghubanshi, A. S., Raj, L. C., Gaur, J. P. and Singh, J. S. 2005. Invasive alien species and biodiversity in India, *Current Science*, 88: 539-540.

Rajak. R. L., Radhey, S., Kumar, U. and Chattargee. G. 1999. PRA- amilstone in plant quarantine functioning of India, *Plant Protection Bulletin,* 51: 1-2.

Ram, A. and Diwakar M. C. 1999. Role of pest surveillance in plant quarantine, *Plant Protection Bulletin,* 51: 32-34.

Rana, R. S., Dhillon, B. S. and Khetarpal, R. K. 2004. "Invasive Alien Species: The Indian Scene". *Indian Journal of Plant Genetic Resources*, 16(3): 190-213.

Singh R. B. (2008). Towards a Food Secure India: making Hunger History. In souvenir –Science-based Agricultural transformation towards Alleviation of Hunger and Poverty in SAARC countries, 1-46, IFFCO Foundation.

Thompson, D., Muriel, P., Russell, D., Osborne, P., Bromley, A., Rowland, M., Creigh-Tyte, S. and Brown, C. 2002. Economic costs of the foot and mouth disease outbreak in the United Kingdom in 2001. *Revue Scientifique et Tecnique,* Office International des Epizooties 21: 675–687.

USDA. 2004. Karnal Bunt: a fungal disease of wheat. USDA APHIS plant protection and quarantine factsheet March 2004.

Whiting, T. L. 2003. Foreign animal disease outbreaks, the animal welfare implications for Canada: risks apparent from international experience. Special report. *Canadian Journal of Veterinary,* 44: 805–815.

21 RISK TO BIODIVERSITY BY EXOTIC SPECIES INTRODUCED FOR BIOLOGICAL CONTROL

D. P. Abrol

21.1 INTRODUCTION

Biological control has been practiced since 120 years resulting in import of hundreds of species of exotic natural enemies, their mass rearing and release for successful control of number of species of pests More than 2000 species of natural enemies have been released resulting in successful management of 165 species of pests worldwide (Greathead 1995; van Lenteren 2000, 2003; van Lenteren *et al.* 2003; Wratten and Gurr 2000). In recent past, negative effects of these releases were relatively unknown (Lynch and Thomas 2000; Lockwood *et al.* 2001; Lynch *et al.* 2001). Undesirable direct or indirect non-target impacts of biological control agents are the greatest current concern. This has grown out of an ecological literature that has discussed many potential negative non-target impacts associated with deliberate biological control releases. This has been a popular topic of discussion and speculation as the global society itself becomes increasingly risk averse, particularly with regard to the remaining natural environment and biodiversity. Studies conducted by Howarth (1985, 1991) on non-target effects of these releases were seriously criticized by Follett *et al.* (2000). However, of late many studies have confirmed that the release of exotics is not without adverse effects at times. Unfortunately potential risks of biological control agents released for pest control have not been properly studied in pre-release evaluations (van Lenteren and Woets 1988, van Lenteren *et al.* 2003, van Lenteren and Woets, 1988; Waage, 1997; Barratt *et al.*, 1999). However, there are several cases where nontarget species testing has been taken very carefully such as control of *Sitonia* beetles in lucerne in New Zealand (Barratt *et al.* 1999) and cassava mealybugs in Africa (Neuenschwander and Markham 2001). The alternative approach to reduce the risk of introduction of exotic natural enemies is to limit the number of releases and give more attention to explore native enemies. Unfortunately during the last four decades many exotic biocontrol agents have been imported and released without proper evaluation (van Lenteren 2000, 2003).

It is only recently that potential risks of releases of exotic natural enemies have been realized and many countries have applied risk evaluation procedures before importing or releasing a new natural enemy. More than 25 countries have already formed their regulation regarding import of exotic biocontrol agents. Some countries like Australia, New Zealand and Hawaii (Lockwood *et al.* 2001) have such strict regulations that import and release of exotic natural enemies is extremely difficult. However, in some other countries there are no regulations at all resulting in import or release of any species (van Lenteren 2000, 2003). Such countries pose a cross border risk to neighbouring countries. But the awareness is increasing and is expected to significantly increases in coming years as a result of the agreements reached during the sixth UN meeting of the Convention on Biological Diversity (CBD 2001) to prevent the spread of invasive alien species.

The aim of this discussion is to publicize both strengths and weaknesses and to help encourage existing assessments to be fair to all without blunting the value of biological control as an effective tool against pests. Improving the analytical approaches used by countries runs a high risk of grinding biological control releases to a halt in a world where the precautionary approach has been adopted with respect to threats from exotic organisms on biodiversity. The benefits of biological control remain poorly understood by the public, allowing the risks to attain disproportionate attention.

21.2 Risks from non-indigenous control agents

Introduction of non-indigenous control agents was suggested by the National Academy of Sciences which advocated that biological control method should become the primary pest control method in United States (National Academy of Sciences 1987). However, the concerns were expressed that non-indigenous species would wipe out the native species from native ecosystems (Howarth 1983, 1985, 1991). Other investigators also expressed similar concerns in which potential problems from non-indigenous control agents had been insufficiently considered (e.g. Ehler 1991; Nafus 1992; Simberloff 1992; Simberloff and Stiling 1996, Lockwood 1993a,b). Still a group of practioners believed that these fears are unfounded and the current procedures are sufficient enough to minimize the unforeseen disasters (Funasaki *et al.* 1988; Lai 1988; Gonzalez and Gilstrap 1992; Messing 1992; Carruthers and Onsager 1993), however, they admit that early projects especially with generalized vertebrate predators had negative consequences but were due to primitive procedures compared to present day standards. DeLoach (1991) reported that current practices in United States are sufficient to prevent the damage from nontarget species.

Simberloff (1992) stated that earlier biological control projects used generalized predators that attacked nontarget species. For instance, the fish *Gambusia affinis* and *G. holbrooki,* introduced into many regions to control mosquitoes, proved counterproductive resulting in decreased populations of numerous native fish species through predation (Lloyd 1990; Simberloff 1992). Likewise, polyphagous herbivores introduced to control weeds affected nontarget native species. Some of the most damaging of such introductions are those of freshwater fishes to attack introduced aquatic weeds which have greatly reduced vegetation that the native fish community structure was highly modified, but other possible effects of these fishes still remain to be known (Moyle *et al.* 1986; Courtenay and Williams 1992; Courtenay 1993). Groves (1989) stated that insects introduced for control of terrestrial weeds have never eliminated nontarget native species, but there are some examples where their impact was negative. For example, the weevil *Rhinocyllus conicus* was introduced to Canada in 1968 and Virginia in 1969 to control weedy, non-indigenous thistles (Ehler 1991). It was then released in many other regions of North America (Schroder 1980). In the 1980s, it was discovered feeding on several native thistle species *(Cirsium* spp.) (Turner *et al.* 1987), of which three species are classified as category 2 species under the Endangered Species Act. Similarly, in North Carolina, a curculionid was introduced in 1990 for control of non-indigenous pasture thistles which threatened the host-specific little metal mark butterfly *Calephelis virginiensis* by reducing its food plant *Cirsium horridulum* and also two rare native thistles. Howarth (1991) in a similar study in Hawaii found that the native plant *Tribulus cistoides* has become rare owing to herbivory by seed weevils *Microlarinus* spp. introduced to control puncture vine *T. terrestris.* The cactus moth *Cactoblastis cactorum* introduced in 1957 to the Lesser Antilles to control *Opuntia* spp., arrived in the Florida Keys by 1989 and has severely threatened a category 2 species in the wild, the semaphore cactus *Opuntia spinosissima* (Habeck and Bennett 1990; Simberloff 1992). In each of these cases, the presence of a common alternative host maintains high numbers of the introduced phytophage, so that density dependence does not operate between the insect and the native nontarget host as populations of the latter decline. Funasaki *et al.* (1988) did not consider that that parasitoids brought to

Hawaii for biological control have eliminated native lepidopterans. They contend that virtually all non-indigenous parasitoids recorded from native moth species reached the Hawaiian islands on their own. they believed that current procedures are so greatly improved over those of early practitioners that environmental concern is unwarranted.

Resource competition is even more difficult to demonstrate than predation and parasitism, so one cannot discount the possibility that introduced biological control agents compete with native species simply because very few cases have been documented. Elliott *et al.* (1996) reported that the Palearctic seven-spot lady beetle *Coccinella septempunctata* is strongly suspected of replacing native coccinellids in South Dakota and Washington, based on precipitous declines in populations of some native species exactly when the seven-spot invaded. How the decline in native coccinellids will affect other species, such as their prey, is still unexplored.

21.3 Effects on community and ecosystem

The precise phonologies of host and native natural enemies are often not well-studied, resulting in adverse effects in community and exosystem as a whole. For example, coconut leaf-mining beetle *Promecotheca reichei* in Fiji was originally controlled by two native parasitoids (DeBach 1974). An accidentally introduced mite (*Pediculoides ventricosus*) was found to attack beetle larvae and pupae, but not eggs or adults. This mite species in the dry season, destroyed all larvae and pupae, then laid eggs and died thereby allowing the synchronous build up of overlapping beetle generations. Now that there were no beetle larvae and pupae, the mite population crashed. Another negative consequence of continuous beetle population was elimination of native parasitoids, adapted to feed on all stages of the beetle at all times. Thereafter neither the mite nor the native parasitoid could prevent a subsequent beetle population explosion. Evidently, the beetle population could only be controlled by the introduction of a parasitoid that was not so specific about the developmental stage of the host when it oviposits (Taylor 1937). This could imply that phonologies of host and native natural enemies must be known. One of the best examples of ecological ripples includes the the introduction of a scelionid wasp (*Scelio parvicornis*) and fungus (*Entomphaga praxibuli*) from Australia to control native grasshoppers in western rangeland (Carruthers and Onsager 1993) in USA because the target constitute a keystone taxon (Lockwood 1993a,b). As there are no no-indigenous grasshoppers in that habitat and the natives ones are not harmful. Although no complete studies are available to define their myriad ecological roles, yet some species are known to benefit the native communities. Change in numerical taxa of such a dominant species is likely to generate many other effects in the community.

21.4 Neoclassical biological control

Lockwood (1993a) described to the control of grasshoppers as 'neoclassical biological control' because a non-indigenous enemy is introduced to attack a native species, while classical biological control refers to introducing enemies of non-indigenous pests (Nechols and Kauffman 1992). Neoclassical control predicts that new associations will probably lead to more effective control of the target pest because the pest has not coevolved with its enemies (Hokkanen and Pimentel 1984, 1989). Waage and Greathead (1988), however did not consider that most effective biological control programmes involve new associations between host and enemy (Hokkanen and Pimentel 1984, 1989), and that often they involve host-range expansion by a species that had been considered monophagous or oligophagous. The rejection was not on the grounds that that new associations and host-range expansions do not occur, but rather on the basis of the definition of 'effective' and the precise fractions of projects that constitute new associations. The species that are especially likely to constitute successful neoclassical biological control introductions are those that are pre-adapted to attack, survive, and reproduce on new hosts that would be most likely to affect nontarget species (Roderick 1992).

21.5 Cost-benefit and risk analysis

Simberloff (1992) stated that conservation issues do not lend themselves very well to cost-benefit analyses because the value of a population or species is extremely difficult to assess, in economic or other terms. The economic cost of losses to a plant pest can be measured, and so can the cost of pesticides and manpower. For biological control, the cost of collecting trips and testing for efficacy and specificity is also straightforward. On the other hand, even if we could predict that events such as extinction of a population or species or change in community composition or perturbation of an ecosystem process would occur, it is not obvious how to assign costs to them. Beyond the fact that we usually cannot predict very well the subsequent ripple effects of these events on other populations, species, communities, and ecosystems, how much is the very existence of these entities worth? Economists and philosophers who have grappled with this matter have yet to reach consensus (Norton 1987; Rolston 1991). This problem is particularly acute for biological control because an established population can almost certainly not be eradicated.

The ecological costs to nature, of particular invasions are so devastating that it is unimaginable that unintended consequences of a biological control introduction could be worse (Center 1995). For example, the catastrophic effects of purple loosestrife *Lythrum salicaria* on native plant communities of the Midwestern and northeastern US and adjacent Canada and the insufficiency of mechanical and chemical control are seen as justifying a biological control programme (Malecki *et al.* 1993); in this instance, host-specificity testing is mandated, but the implication is that entire communities will be lost anyway if something is not done. For the imported fire ant *Solensopsis invicta* in the south-eastern US, Buren (1983) suggested importing 20–30 ant species from South America and constructing an 'artificial' ant fauna on the grounds that the fire ant is destroying most of the native ant fauna anyway and no other control is working.

When an invasion affects an entire community, is spreading rapidly, and appears to be producing irreversible effects, even the most cautious observers would suggest that biological control be considered (Howarth 1991; Simberloff 1992). However, it is still important to attempt to list the suspected ecological costs of the invasion and the potential problems that a control agent might cause. Furthermore, the fact that living organisms need not stay in the habitat or region to which they are introduced must be considered. Introductions of the New World carnivorous snail *Euglandina rosea* to control the giant African snail *Achatina fulica* were made in agricultural communities because *Achatina* was seen as an agricultural pest; however, the snail has wiped out populations and entire species of endemic snails of forested habitats in the Tahitian and Hawaiian islands (references in Civeyrel and Simberloff 1996). Similarly, the cactus moth discussed above dispersed from its initial site of introduction, where it is not known to have caused ecological damage, to other islands where it threatens at least one endemic species with extinction. So a risk assessment for a biological control introduction must consider dispersal.

21.6 Inconclusive evidence and risk

The lack of evidence that problematic evolution of biological control agents has occurred is perhaps unsurprising when one considers the depth of research that had to be conducted to determine that, for example, a host-shift was genetically determined (Bush 1975; Prokopy *et al.* 1988; Feder and Bush 1989). It is no mean undertaking to demonstrate that many traits, especially behavioural ones, are genetically determined, much less than changes in such traits are caused by genetic changes. Even the ranges and population densities of most insect species are very inadequately monitored (Center *et al.* 1994). Just to have detected that a trait changed would be remarkable in this context. To show that the change is genetic would be astounding. In other words, it is possible that such changes are commonplace even though we have no ironclad examples. Many investigators (Lai 1988, Funasaki *et al.* 1988 Center 1995) view the paucity of known ecological damage from biological control introductions in comparison to the numerous projects as proof that biological control is generally safe. However, according to Simberloff (1992) this argument is unconvincing if one considers the remarkable circumstances surrounding discovery of some of the known problems.

The introduction of the predatory snail *Euglandina rosea* for biological control resulted in the extinction of seven endemic land snails of the genus *Partula* on the Tahitian island of Moorea (Clarke *et al.* 1984; Murray *et al.* 1988; Murray 1993). The Moorean *Partula* have long been recognized as a superb system for studying aspects of evolution, particularly divergence. Thus a team of researchers were closely studying the genetics and evolution of *Partula* on Moorea just as *Euglandina* was spreading. If these *Partula* had not been the subject of intense long-term biological research, we might not have known the existence or extinction of all these species. Furthermore, the *Partula* on Moorea and Tahiti are culturally and economically important to the indigenous people who may be interested to know changes in their status. Unfortunately the studies are very limited and it is not known for how many invertebrates and plants species, the long-term studies and cultural interest is under way. Only a small fraction of all taxa are observed more than casually even by scientists. Snails, butterflies, and birds are beautiful, and large groups of amateurs collect them or their sightings. Few species receive this attention. Evidently, the argument of Funasaki *et al.* (1988) that few harmful effects are known for biological control introductions should not comfort us (Simberloff 1992). The post-release effort to study nontarget hosts and community and ecosystem properties are extremely limited which prevents us from understanding the decline and disappearance of a species (Simberloff 1994). Groombridge (1992) stated that many known extinctions of animals since 1600 are based on guess and not even assigned a 'possible cause. This is especially true for invertebrates. Thus, the fact that, apart from a few cases such as that of the *Partula* snails, few extinctions are known to have been caused by biological control agents, should not make us complacent. Disruption of community or ecosystem processes is even less likely to be observed than is extinction.

21.7 Whether biocontrol is safe?

The harmful effects of biocontrol agents as discussed above, some of which are indirect and unpredictable may probably be the tip of the iceberg (Simberloff and Stiling 1996). In view of the inherent unpredictability and usual irreversibility of biological introductions

(Simberloff 1991), it is a matter of concern whether biological control is safe. And whether any procedure could ensure against ecologically harmful impacts? But this does not mean that no biological control should be practiced. However, the unpredictability and irreversibility do suggest that bioflogical control must be viewed as risky and that specific projects should not be assumed to be innocuous until substantial effort has been expended to support this assumption.

There is difference of opinion and scepticism over the merit of more rigorous testing and research on biological control agents which will cost so much and negate the entire approach. Because testing of pharmaceuticals and is very expensive and some of the pharmaceuticals that would otherwise have been developed and useful were never produced because of the regulatory costs. In a similar analogy, one can also argue that some technologies that might be of great use are so risky that they should not be developed. For example, some critics would enlist nuclear power in this category but the positive side needs to be exploited.

Unfortunately problematic projects would continue so long as professional biological control advocates and downplay its risks, no matter how carefully programmes are conducted and officially regulated. Most of these projects are mounted by a public pressure unconcerned with risks, eager to solve perceived pest problems cheaply. This is not a new problem, the examples can be found dating back to 19th century when the small Indian mongoose *Herpestes auropunctatus* was privately introduced for rat control by sugarcane growers in both the West Indies and the Hawaiian islands (Tomich 1986), causing much subsequent harm to native species (Simberloff 1992). Similar private actions brought stoats *Mustela erminea* and weasels *Mustela nivalis* from Britain to New Zealand in the 19th century for rabbit control (King 1984). In the 1930s, a private citizen introduced feral hogs *Sus domesticus* onto Santa Catalina Island, CA, solely to attempt to control rattlesnakes (Mayer and Brisbin 1991).

Warren (1994) has reported that some of the projects in recent past include the release of the goldenhorn marisa or wheeled apple snail *Marisa cornuarietus* in Florida in attempts to control hydrilla *Hydrilla verticillata* and water hyacinth *Eichhornia crassipes*in spite of the fact that it feeds indiscriminately on many desirable native plant species and causes severe damage to aquatic animals whose habitat is native macrophytes. Similarly, there are other innumerous instances where these releases have brought more (Courtenay and Robins 1973 Courtenay *et al.* 1974, Emerson and Jacobson 1976 Courtenay 1994). Most state laws in the United States encourage the use of biological control and constrain it minimally (Abrams 1990; Miller and Aplet 1993). For example,

California has no provision for damage to ecosystems while Florida limits biological control agents to those that do not have 'an unreasonable effect on the environment. However, the Wisconsin law appears strongest because it requires a permit for any biological control agent, even for intrastate movement, and expresses concern for safety and injury to 'useful' plants and animals, but natural ecosystems or biodiversity is not mentioned.

Perhaps, Australia is the only nation with a specific law for classical biological control, the Australian Biological Control Act of 1984. Its primary purpose is to bring concerns into public view, and it approves an introduction only upon a finding of no significant harm to any person or to the environment, although 'significant' is undefined and variously interpreted. This law arose because of a conflict between beekeepers and ranchers (Delfosse 1985) that could not be resolved by a mechanism, similar to the current United States procedures, in force at the time. The proposal to introduce insects to control Paterson's curse *Echium plantagineum* a range pest, elicited opposition from apiarists who prized the same plant for its high pollen/nectar ratio. Thus, concern for nontarget organisms or ecosystems as a whole was secondary, although the Act is certainly a vehicle for airing such concerns.The problems as outlined above will continue, no law can allow biological control and yet completely eliminate the possibility of collateral damage. However, some of the investigators (Campbell 1993, Miller and Aplet 1993) believe that, at least United States could enact laws that would be far superior to the current patchwork of regulations and would allow biological control while greatly lowering the risk of unintended harmful consequences of biological introductions.

21.8 Risks of releasing of exotic biological control agents

Van Lenteren *et al.* (2006b) has very nicely reviewed the risks to human health, the economy, and the environment arising from the introduction of exotic biological control agents (Table 21.1). According to Ehlers (2003) health risks to production and or application personnel and consumers are not unique to exotic invertebrates, and none are known for any macroorganisms, apart from some cases of allergy in the mass production of predatory mites or nematodes. Goeden and Louda (1976) reported that the economic losses may also occur sometimes if introduced natural enemies attack previously introduced biological control agents. The most important risks from biological control introductions relate to possible changes in the distribution and abundance of native organisms. It also depends on the feeding pattern, if occasional feeding on a nontarget does not result in changes in the distribution or abundance of the nontarget, it may not be regarded a problem. On the other hand, if a biological control agent affects exotics other than those introduced on purpose, the impact should not be considered negative. The true environmental risks include local or global extinction of a native species (target or nontarget); large reductions in either abundance or distribution of native organisms; interference in the efficacy of native natural enemies of pests via intraguild interactions or competitive displacement; vectoring of pathogens harmful to native organisms; loss of biodiversity and identity of native species via hybridization between close relatives; and in general, any major shifts in the balance of native species via direct or indirect mechanisms. The most serious threat is that of global extinction of a species – all other possible impacts are relative and may be reversible (Louda *et al.* 2003).

21.9 Nontarget effects of exotic biological control agents

It is essential to have an overview of the abundance and nontarget impacts of all nonindigenous organisms before considering those caused by exotic natural enemies. A brief overview of these nontarget impacts is given in (Table 21.1 and 21.2). In general, accidentally introduced organisms have always outnumbered those that have been intentionally introduced everywhere, often by a factor of 10 or more. Pimentel *et al* (2002) reported that of the approximately 50,000 accidentally introduced organisms in the United States, 10% to 50% have caused ecological problems. If we look specifically at biological control introductions of insects, it becomes clear that these introductions have been relatively safe, particularly in light of the general safety patterns of exotic organisms: Less than 1% appears to have caused population-level effects in nontargets, and only 3% to 5% may have caused some smaller effects. However, only a minority of introductions have included a careful evaluation of nontarget effects. On the other hand, if strong nontarget effects had appeared, they would have been observed and reported. Many inundative releases of arthropods appear to have some transient effects on no targets particularly polyphagous agents but none have caused population level effects, probably because of their transient nature (Louda *et al.* 2003). Table 21.1 gives examples of known cases of suspected nontarget impacts on native insects resulting from biological control introductions. Many of these are subject to controversy because there is no strong evidence for the prime reasons for the observed population declines. All the possible global extinctions that have been linked to biological control introductions are disputed by some experts: Extinctions have not been proven, and other factors may have contributed to them much more than the exotic agents did. However, in several cases, successful biological control agents have caused population declines not only in their target hosts, but also in some nontargets, including those that have depended on the target host as their major source of food. Lynch *et al.* (2001) presented exhaustive data on more than 5000 recorded biological control cases and have underlined our ignorance of the degree to which nontarget effects may occur. Therefore, environmental risk analyses need to become standard procedures in each biological control project (Coombs 2003). This does not necessarily result in fewer introductions of exotic agents, as has been shown by an evaluation of the IPPC Code of Conduct (Kairo *et al.* 2003). However, it does lead to higher costs and delays introductions. Examination of the commonly cited 'rogue' or unwanted biological control agents presented in Table 21.1 clearly demonstrates that these biological control projects were ill-conceived, not necessarily because the pests were unsuitable targets, but primarily because the natural enemies selected had very broad host ranges and substantial nontarget impacts should have been predictable. In some instances, agricultural interest groups (e.g., sugarcane growers and farmers) carried out the projects listed in Table 21.1 with little or no scientific grounding, and government oversight was lax either because of non-involvement, or there was no regulatory infrastructure (i.e., governing legislation) by which to identify suitable targets and to assess the safety of imported natural enemies before release or their subsequent redistribution following establishment.

Table 21.1. Cases with Clearest Evidence for Nontarget Effects on Native Insects of Released Exotic Biological Control Agents

Suspected effect	**References**	**Introduced natural enemy**	**Affected native species**	**Location, time**
Possible global extinction	Howarth 1991; Kuris 2003; Sands 1997; Tothill *et al.* 1930	*Bessa remota* (Aldrich)	*Levuana iridescens* Bethune-Baker (target pest)	Fiji, 1925
Possible global extinction	Sands 1997; Tothill *et al.* 1930	*Bessa remota* (Aldrich)	*Heteropan dolens* Druce (nontarget)	Fiji, 1925
Reduction in distribution range	Louda *et al.* 2003; Sands and van Driesche 2000	*Cotesia glomerata* (L.)	*Pieris napi oleraceae* Harris	Massachusetts, late 1800s
Reduction in population levels	Aubert and Quilici 1983	*Tetrastichus dryi* Waterston	*Trioza eastopi* Orian	La Reunion, 1978
Reduction in population levels	Louda *et al.* 2003	*Compsilura concinnata* (Meigen)	Several species of native Lepidoptera (Saturniidae)	North America, 1906 onward
Partial displacement of native natural enemies	Elliott et al. 1996; Wheeler and Hoebeke 1995	*Coccinella septempunctata* (L.)	*Coccinella novemnotata* Herbst; *Coccinella transversoguttata* Fald.; Adalia bipunctata (L.)	North America, 1973 onward
Partial displacement of native natural enemies	Viggiani 1994	*Cales noacki* Howard	Native parasitoids of *Aleurotuba jelinekii* (Frauenfeld)	Southern Italy, 1980–1994
Interbreeding with native natural enemies	Hopper *et al.* 2006	*Aphelinus varipes* Foerster, species complex	*Aphelinus varipes* Foerster, species complex	North America, 1987 onward
Lowering effectiveness of an introduced natural enemy for weed control	Goeden and Louda 1976	*Cryptolaemus montrouzieri* Mulsant	*Dactylopius opuntiae* (Cockerell)	Mauritius, late 1940s

Source : van Lenteren *et al.* 2006b.

Table 21.2. Generalist Biological Control Agents, Target Pests, Country and Year of First Introduction and Unintended Impacts on Non-Target Wildlife

Biological control agent	Target pest	Country and year of first introduction	Non-target impacts	References
European red fox, *Vulpes vulpes* L. (Native range: Palaearctic regions)	Rabbits, *Oryctolagus cuniculus* L.	Australia, 1871	Foxes eat native marsupials, birds, and lambs	Saunders *et al.* 1995
Stoat, *Mustela erminea* L. (Native range: Eurasia and North America)	Rabbits	New Zealand, 1884	Stoats eat native birds, insects, and lizards	King 1990a
Ferret, *Mustela furo* L. (Native range: Central Europe and the Mediterranean)	Rabbits	New Zealand, 1879	Ferrets eat native birds	Lavers and Clapperton,1990
Weasel, *Mustela nivalis vulgaris* Erxleben (Native range: Eurasia and North America)	Rabbits	New Zealand, 1884	Weasels eat native birds, insects, and lizards	King 1990b
Small Indian mongoose, *Herpestes javanicus* (Saint-Hilaire), (=auropunctatus [Hodgson]) (Native range: Iraq to the Malay Peninsula)	Rats, *Rattus* spp.	Trinidad 1870; Jamaica, 1872; Cuba, 1886; Puerto Rico, 1877; Barbados, 1877; Hispaniola, 1895; St Croix, 1884; Surinam, 1900; Hawaii, 1883	Mongoose eats native birds and reptiles	Hinton and Dunn 1967; Loope *et al.* 1988

Biological control agent	Target pest	Country and year of first introduction	Non-target impacts	References
Cane toad, *Bufo marinus* L. (Native range: Northwestern Mexico through southern Brazil)	White grubs, *Phyllophaga* sp.; sweet potato hawk moth, *Agrius convolvuli* L.; Grey backed cane beetle, *Dermolepida albohirtum* (Waterhouse)	Jamaica, 1844; Bermuda 1855; Puerto Rico 1920; Hawaii, 1932; Australia, 1935; Fiji, 1936; Guam, 1937; New Guinea, 1937; Philippines; 1934	Cane toads eat native insects, amphibians, and reptiles. Toads are toxic to native wildlife that consume it, and *B. marinus* outcompetes native amphibians for shelter and breeding sites	Easteal 1981; Freeland 1985; Hoddle 1999.
Mosquito fish, *Gambusia affinis* (Baird and Girard) (Native range: eastern USA and Mexico)	World-wide dissemination of *G. affinis* for control of mosquito larvae promoted by the World Health Organization until 1982	Intensive releases began worldwide around 1900 and approximately 70 countries now have permanent *G. affinis* populations including: Afghanistan, Australia, Canada, China, Ethiopia, Grand Cayman Island, Greece, Hawaii, Iran, Korea, New Zealand, Somalia, Turkey, Ukraine.	Substantial nontarget attacks on native aquatic invertebrates and vertebrates outside its native range	Diamond 1996; Gamradt and Kats 1996; Legner 1996; Meisch 1985; Rupp 1996; Walton and Mulla 1991

Biological control agent	**Target pest**	**Country and year of first introduction**	**Non-target impacts**	**References**
Predatory snail, *Euglandina rosea* (Ferrusac) (Native range: southeastern USA)	Giant African snail, *Achatina fulica* Bowdich	Hawaii, 1955; Tahiti, 1974; Moorea, 1977; New Caledonia, 1978; Guam, 1957; Vanuatu, 1973; Papua New Guinea, 1952; Japan, 1958; Taiwan, 1960; Madagascar, 1962; Seychelles, 1960; Mauritius, 1961; Reunion, 1966; India, 1968; Bermuda, 1958	Predation of native snails (e.g., native *Achatinella* spp. and *Partula* spp.), probably leading to extinction of some native species	Clarke *et al.* 1984; Davis and Butler 1964; Griffiths *et al.* 1993; Kinzie 1992; Murray *et al.* 1988; Simmonds and Hughes 1963
Tachinid fly, *Compsilura concinata* Meigen (Native range: Europe)	Gypsy moth, *Lymantria dispar* (L.) and other lepidopteran pests	USA, 1906	Regional declines of native saturniid moths because of heavy parasitism of larvae	Boettner *et al.* 2000
Tachinid fly, *Bessa* (= Ptychomyia) *remota* (Aldrich) (Native range Indo-Malay archipelago)	Coconut moth, *Levuana iridescens* Bethune-Baker	Fiji, 1925	Possible extirpation of the native Fijian zygaenid *Heteropan dolens* and reduction in abundance of other native zygaenids	Howarth 1991; Robinson 1975; Sands 1997; Tothill *et al.* 1930

Biological control agent	Target pest	Country and year of first introduction	Non-target impacts	References
Seven spotted lady beetle, *Coccinella septempunctata* L. (Native range: Palearctic regions)	Species of pest aphids	USA, 1957	Competitive displacement of native aphidophagous coccinellids in agricultural crops, and nontarget predation of native lepidopterans	Elliott *et al.* 1996; Obrycki *et al.* 2000
Cactus moth, *Cactoblastis cactorum* (Bergroth) (Native range: Northern Argentina, Uruguay, Paraguay, and southern Brazil)	*Opuntia* spp. of cacti	Australia, 1925; Caribbean, 1957; Hawaii, 1950; Mauritius, 1950; South Africa, 1932; accidental introduction into the USA in 1989	Invaded mainland USA from the Caribbean in 1989 and attacks native *Opuntia* spp. thereby threatening the survival of endangered native species	Bennett and Habeck 1992; Holloway 1964; Pemberton 1995
Flowerhead weevil, *Rhinocyllus conicus* Fröelick (native to Eurasia)	*Carduus* spp. thistles	USA, 1968	Attacks on seed heads of native *Cirsium* spp. thistles potentially limiting regeneration of plants, and displacing native thistle head feeders	Louda *et al.* 1997; Strong 1997; Louda 2000; Gassmann and Louda 2001

Source: Hoddle (1999).

21.10 Ecological and economic advantages of biological control

The reaction of the biological control community to evidence of nontarget impacts by biological control agents and subsequent criticism of the use of this technology by reputable biologists have been largely defensive. Two recent books *Nontarget Effects of Biological Control* (Follet and Duan 2000) and *Evaluating Indirect Ecological Effects of Biological Control* (Wajnberg *et al.* 2001) advocate improved host specificity testing procedures, more rigorous field assessments of potentially rogue agents with food web analyses, greater legislative guidance, more use of theoretical community assemblage studies, and ownership of past mistakes with a vision to strive towards improved safety and greater efficacy. Retrospective analyses of data sets that have well quantified nontarget impacts may be invaluable in guiding studies and legislation that aim to improve the safety of biological control programmes (Louda *et al.* 2003). A major publication by respected biological control scientists that compiles and documents the positive ecological and economic benefits derived from classical biological control programmes is needed to temper some of the current criticism directed at the use of natural enemies for pest suppression. These case studies need not be limited to terrestrial arthropods and weeds and the scope should be expanded to include new projects focusing on the biological control of 'non-traditional' targets. New targets being evaluated for biological control include marine algae, green crabs, zebra mussels, brown tree snakes, and vertebrates with sterilizing viruses.

Biological control projects should be carefully selected and conducted by trained professionals. Host specific natural enemies should be used, neo-classical biological control programmes should be regarded warily and subjected to close scrutiny, and the legal or illegal movement of generalist natural enemies by individuals seeking quick fixes for environmental and agricultural problems must not be condoned. Furthermore, the 'shotgun approach' which releases large numbers of different species of natural enemies increases the risks of generalists establishing as 'environmental winnowing' will most likely select polyphagous agents that do not exhibit high host and habitat fidelity. These kinds of projects are seen as reckless and bring biological control into disrepute with ecologists and conservationists who should be considered supporters of biological control and proponents of the careful use of natural enemies. Consequently, greater regulatory guidelines will probably be developed to mitigate the adverse effects of biological control agents of arthropods and to provide criteria for selecting 'safe' natural enemies for release. Biological control is unreservedly an ally of agriculture and conservation in its attempts to reduce pesticide use, as a habitat management tool, and biological control presents itself as the only sustainable and cost effective technology for pest management when the risks from the 'do nothing' approach are unacceptably high.

21.11 Legal framework of environmental risk assessment

A number of international legal frameworks which regulate the introduction of exotic species into new environments are available in addition to several countries that have national regulations (CBD 1992; CBD 2001; Fasham and Trumper 2001; Genovesi and Shine 2003; Sheppard *et al.* 2003; Shine *et al.* 2000). Countries with extensive experience in CBC (Australia, Canada, New Zealand, South Africa, UK, and the United States) had legislation and procedures in place relatively early to regulate imports and to analyze risks of exotic biological control agents (Sheppard *et al.* 2003). For those countries that had not, the FAO Code of Conduct served as a guideline (Kairo *et al.* 2003). The new version of this code of conduct (ISPM3) (IPPC 2005), which will become the international standard, has extended its range from CBC to ABC, native natural enemies, microorganisms, and other beneficial organisms, and also includes evaluation of environmental impacts. Several other organizations have recently developed guidelines for import, evaluation, and release of biological control agents, and evaluation of environmental effects forms a central element of these guidelines (EPPO 1999, 2000; IPPC 2005; NAPPO 2001; Sheppard *et al.* 2003). Also, a growing number of countries apply ERA procedures before the import and/or release of new natural enemies (Baker *et al.* 2005; Bigler *et al.* 2005; Sheppard *et al.* 2003). To facilitate decision making on introductions, some organizations recommend working towards a regional species listing system based on higher biogeographic units, consistent with international law (Genovesi and Shine 2003). For natural enemies, such lists are already used in some regions/countries (ANBP 2004; EPPO 2002), but these are seldom the result of a thorough ERA procedure. Guidelines mentioned above aim to facilitate procedures for a proper risk assessment, but they do not yet provide working instructions for the risk assessment itself. For many other purposes, risk assessments are widely accepted as a tool for decision making and for evaluating economic and environmental costs and benefits (Genovesi and Shine 2003; IPPC 2005). Aspects of risk assessments have been developed and applied during the past two decades for exotic natural enemies, though often in a preliminary way (van Lenteren *et al.* 2003). Methods have generally been derived from existing pest risk analysis protocols developed by regional organizations (EPPO 1997; NAPPO 2001). In others, methods are based on domestic regulative measures, largely as amendments of regulation relating to plant health, pesticide use, and/or biodiversity (DEFRA 2000). Most procedures are not yet tailored for the intentional release of biological control agents. Some assessment procedures, such as those of Australia, New Zealand, and Hawaii, are so strict that import and release of exotic natural enemies are extremely difficult. In contrast, other countries have no regulations.

Currently, there is a trend towards more stringent regulatory requirements (Barratt *et al.* 2000). Novel strategies are necessary (Baker *et al.* 2005), more ecological information should be used to increase the precision of risk assessment (Louda *et al.* 2003), and both the risks and the benefits of biological control applications should be considered in the evaluation (Sheppard *et al.* 2003). The Organisation for Economic Cooperation and Development (OECD) developed a harmonized ERA protocol (OECD 2004), and the IOBC-WPRS (International Organization for Biological Control of Noxious Animals and Plants West Palearctic Regional Section) has drafted a detailed guideline on information requirements (Bigler *et al.* 2005). Both documents provide guidelines to the preparation of dossiers in support of applications and assist reviewers in a balanced risk/benefit evaluation of future biological control releases. Today, the major challenge in developing risk assessment methodologies is to produce protocols that will help prevent serious mistakes through release of harmful exotics while promoting safe forms of biological control. The most critical ecological issues are to estimate the probabilities of attack on nontarget organisms and the dispersal and establishment capacities of the biological control agent. Few natural enemies are strictly monophagous (Zwolfer 1971), but many are oligophagous and thus have a restricted host/prey range. Others are polyphagous and have a wide host/prey choice. It is expected that future ERAs will integrate information on the potential of an agent to establish in an environment, its abilities to disperse, its host range, and its direct and indirect effects on nontargets. Contrary to existing risk assessments and to be of practical use, future risk assessments should be (a) quantifiable, so that the environmental effects of different biological control agents can be compared and informed decisions made, and (b) consist of a stepwise procedure so that demonstrably safe agents will be identified early in the process, with lowest possible costs involved. The ERAs discussed below form an element of a general risk assessment framework for the regulation of import and release (van Lenteren *et al.* 2003) and consist of the following basic elements: (a) characterization and identification of the biological control agent, and determination of (b) health risks for humans, (c) environmental risks, and (d) efficacy.

21.12 Methodology for risk assessment

Van Lenteren *et al.* (2003) developed a methodology for risk assessment as a basis for regulation of import and release of exotic natural enemies used in inundative forms of biological control. They found that the parameter 'host range' forms a central element in the whole risk evaluation process, as the lack of host specificity might lead to unacceptable risk if the agent establishes and disperses widely, whereas, in contrast, a monophagous biological control agent is not expected to create serious risk even when it establishes and disperses well. When dealing with mono- or oligophagous natural enemies, establishment and dispersal to nontarget habitats do not lead to increased risk. However, high values for establishment and dispersal are considered negative when the biological control agent has a wide host range. Extreme care should be taken with exotic agents that establish across the area of introduction, that spread actively by dispersal or migration, that have a broad host range and that have nontarget habitats in common with related indigenous species, such as for instance, some coccinellid predators like *Harmonia* spp. Records from Palaearctic species introduced into North America have shown range expansions over the past 30 years (Hoebeke and Wheeler Jr 1996; McCorquodale 1998), resulting in nontarget effects (Lynch and Thomas 2000). Some currently commercially available biological control agents obtain high-risk indices because of their wide host range and, thus, supposedly many nontarget effects. Recent studies indicate that, apart from its high effectiveness on target species, *Lysiphlebus testaceipes*, for instance has become the predominant parasitoid of a number of indigenous aphid species, either pests or indifferent species, in all environments from agro-ecosystems to climax ecosystems such as forests (Stáry *et al.* 1988a,b) across the Mediterranean area (Nicoli and Burgio 1997). The successful introduction of *Cales noacki* as a classical biological control agent against the woolly whitefly *Aleurothrixus floccosus* (Homoptera: Aleyrodidae) may have led to the complete displacement of all competing parasitoids in populations of the nontarget whitefly species *Aleurotuba jelinekii* in some parts of the Mediterranean area (Viggiani, 1994). Although both are excellent control agents against their respective target pests, new releases of these agents should be reconsidered and for the time being not be disseminated commercially into areas where they do not yet occur.

The risk assessment methodology allows for ranking of natural enemies that are evaluated for control of the same pest. For example, for control of whitefly, one would conclude that it is safer to use the exotic parasitoid *Encarsia formosa*, than *Encarsia pergandiella* or *Chrysoperla carnea*. In present day biological control, governments that rigidly regulate the introduction of biological control agents, such as for example Australia and New Zealand, usually require that candidate agents undergo host-range testing to ensure that they will not become pests or threaten desirable species (Charles 2001). This results in a general preference for highly specific natural enemies, as was already common sense among most biological control workers. The commercial biological control industry, however, often prefers generalist, less host specific natural enemy species.

Implementation of a registration procedure for natural enemies is currently a topic of hot debate among the biological control industry and regulators. The biological control industry foresees lengthy, cumbersome procedures leading to high costs, and thus, in some cases, the possibility that potentially interesting natural enemies will not enter the market because costs are too high. Regulators within ministries of environment and agriculture want to prevent unnecessary and risky releases of exotic organisms. The history of arthropod biological control shows that very few mistakes have been made until now (Lynch *et al.* 2001). This is a point in favour for the biological control industry, and is in strong contrast with the problems that have been created by accidental importation of pests and diseases on infested plant material by others. The risk assessment procedure proposed by Van Lantern *et al.* (2003) is a step towards a registration procedure. With this risk assessment procedure, biological control experts will be able to put the natural enemies under evaluation for release into different risk categories, which makes it possible (a) to choose the safest control agent when more candidates are available, (b) to decide if more, and what kind, of information is needed for a full risk assessment, and (c) to conclude that certain natural enemies are not suitable for release.

21.13 Risk factors determining the impact of an introduced agent

Van Lenteren *et al.* (2003) identified the significance of five risk factors of natural enemies which include host range assessment, establishment, dispersal, and direct and indirect effects on nontargets and also provided the guidelines for their quantification as given below:

21.13.1 Host range assessment

According to several investigators (Kuhlmann *et al.* 2006; van Driesche and Reardon 2004; Van Lenteren *et al.* 2006b) the first and most important and critical step in host range assessments is the selection of appropriate nontarget species. The centrifugal phylogenetic method based on phylogenetic/taxonomic affinities of nontarget species, which has been used for more than 30 years in weed biological control programmes (Wapshere 1974), is usually the starting point when evaluating natural enemies of arthropods. However, this phylogenetic approach is insufficient in arthropod biological control for a number of reasons, including poor knowledge of arthropod phylogeny, unreliable host lists, the greater number of taxa that can be attacked by natural enemies, and other parameters such as the same feeding niche or the common habitat of target and nontarget species that may be more meaningful as nontarget indicators (Kuhlmann *et al.* 2006; van Driesche and Reardon 2004). In addition, species of conservation concern may need to be included in nontarget testing. As a result, the initial nontarget test lists are often composed of large numbers of species, and testing of all these species would be impossible. Host range testing in weed biological control may include 40 to more than 100 plant species and is feasible because of the ease of storing plant seeds and plant rearing. However, arthropod hosts and natural enemies may not be amenable to laboratory culture, or their rearing is much more complicated. Thus, it has been proposed that nontarget species be selected from the following three categories to design initial test species lists (Kuhlmann *et al.* 2006): (a) species with phylogenetic/taxonomic affinities, (b) species with ecological similarities (habitat and/or niche overlap), and (c) species of conservation concern (beneficial, or rare/endangered). If this first selection results in an impractically long list, species can be filtered out on the basis of attributes such as nonoverlapping geographical distribution, different climate requirements, phenological asynchronization, or host size outside the range accepted by the candidate biological control agent. Other species might need to be removed from the test list because they are not available in large enough numbers. Furthermore, rare and endangered species might be replaced by congenors as surrogates for testing. The filtering process results in a provisional test list of 10 to 20 species from which species can be removed or added during a reiterative process that evaluates the outcome from the ongoing host range testing programme (Kuhlmann *et al.* 2006).

21.13.2 Establishment

Depending on the control programme and the environment into which an exotic agent is released, establishment may be the intended outcome or an undesirable event. Establishment is determined by abiotic and biotic factors. In CBC, preliminary studies are normally conducted to assess the similarity of climatic conditions between the countries of collection and release. Similarly, if an exotic species escapes after release into a greenhouse, establishment will depend on the ability of the organism to survive in a climate that is usually colder than that into which it was released and from where it was collected, and on sources of native hosts.

21.13.2.1 Abiotic factors

The main abiotic factor affecting the establishment of exotic species in new environments is temperature. Although humidity may play a role, it alone does not determine establishment. Many successes in CBC have involved introductions into similar climates, and an inability to survive through winter is a common cause of failure (DeBach, 1964). Although 'climate matching' is a useful guide to the similarities and differences between native and intended release areas, this approach is limited and is not a reliable basis for predicting survival and establishment (Hart *et al.* 2002a,b). Temperature has a major influence on the key processes that determine establishment and subsequent spread of introduced natural enemies. For all arthropods, there is a temperature below which there is no development, i.e., the 'developmental threshold.' A number of methods can be used to estimate the threshold temperature, which may lead to variation in the threshold (Hart *et al.* 1997, 2002 a, b).

The thermal budget is the number of day degrees required to complete one generation and, in combination with the threshold temperature, can provide valuable information on the likelihood of establishment at any intended release site. The threshold temperature can be related to temperature records to indicate periods in the year when there will be no development (such as winter), and the organism may have to enter a dormant state (diapause) to survive. Also, the thermal budget can be compared with estimates of the annual number of available day degrees above the developmental threshold to determine the likely voltinism at the intended release site.

21.13.2.2 Biotic factors

The availability of target host/prey in space and time, occurrence of alternative host/prey, presence of competitors, and access to food are important biotic factors that influence establishment. Biotic factors are more difficult to assess than abiotic factors and information from both areas of origin and introduction are often lacking. The ability of predators and parasitoids to synchronize their life cycle with their host/prey (target or alternative) is a fundamental prerequisite for establishment.

This is critical for specialist natural enemies because they cannot use alternative hosts/prey to sustain populations when the target is not present or not in the right stage. The degree of synchrony often reflects the likelihood of permanence of host parasitoid relationships and thus the probability of establishment (Doutt *et al.* 1976). In temperate climates, permanent establishment of parasitoids is possible only if overwintering hosts are available. Native hosts may not be suitable for overwintering of the introduced parasitoid either because they are attacked at a physiological stage that does not allow development of the overwintering natural enemy or because the type of dormancy is unsuitable for the parasitoid (Boivin 1994). The probability of establishment in temperate climates is affected by mortality due to climatic extremes and to the difficulty of adjusting and synchronizing life cycles, especially when hosts enter quiescent or diapause stages during these unfavourable periods (Bale 1991a,b). Availability of adequate food increases parasitoid efficacy, but it is difficult to demonstrate the influence of food sources on establishment (Jervis *et al.* 1993). The level of dependence on food is species-specific but also a function of availability and quality, influencing life cycle parameters, behaviour, competitive interactions, and niche partitioning of parasitoids and predators (Jervis *et al.* 1996). Parasitoids and predators are carnivorous, but most species also use plant-derived foods, such as pollen, nectar, plant sap, and honeydew. Plant foods can have a strong effect on survival and population dynamics, particularly for obligatory plant-derived food consumers (Wäckers 2003, 2005). Facultative consumers can use plant foods to bridge periods of low host/prey availability (Limburg and Rosenheim 2001). Plant-derived food is important, but we lack a full understanding of the effects of such factors on establishment.

21.13.2 Dispersal

The probability that a biological control agent will find a potential nontarget host depends strongly on the dispersal capacity of the agent. Below the dispersal characteristics and methods to measure dispersal are summarized.

21.13.2.1 Modes of dispersal and associated risks

Good searching and dispersal abilities are considered valuable indicators of successful biological control agents (van Lenteren and Woets 1988). Knowledge of dispersal characteristics is also essential for researchers to assess to what extent the agent could disperse into nontarget habitats. Dispersal is affected by trivial movements, host foraging, or migratory behaviours and can be separated into short-distance and long-distance movements. The scale (numbers and distance) of dispersal may vary considerably between taxa, populations, and regions (Nathan *et al.* 2003). Natural enemies disperse actively by flight or locomotion, and distance is positively correlated with body size (Tscharntke and Brandl 2004). Dispersal by small-bodied agents such as entomopathogens (Hokkanen *et al.* 2003; Hokkanen and Lynch 1995), parasitoids (Bellamy and Byrne 2001; Kuske *et al.* 2003), and predatory mites (Jung and Croft 2001) is limited in distance and time. However, such agents can become part of the aerial plankton and thus become widely dispersed. More rarely, adult parasitoids are actively vectored by their hosts (Fatouros *et al.* 2005). Migration, on the other hand, typically involves a mass movement to a distant location and is unidirectional, persistent, and undistracted (Dingle 1996).

The encounter probability between the agent and its host or prey depends on the mechanism of dispersal, its life span, the local climate, and habitat conditions in the area of release (Kuske *et al.* 2003; Louda *et al.* 2003; Tscharntke and Brandl 2004). In ABC, in which natural enemies are released in large numbers for immediate control of the target pest, direct dispersal (overflow, drift) from the release area into the surrounding environment is of main concern for direct nontarget effects whether or not the natural enemy species is exotic. Moderate dispersal ability may greatly reduce the risks of nontarget effects (Kimberling 2004). Habitat connectivity, in combination with the use of nontarget hosts, however, may enable an agent to spread rapidly through a landscape (Pons *et al.* 2004). Case histories of CBC, in which establishment is intended, indicate that introduced agents are able to disperse and reproduce on nontarget hosts at large distances from their target habitats (Henneman and Memmott 2001; Louda *et al.* 2003).

21.13.3.2 Methods determining dispersal

Knowledge of the dispersal characteristics of a natural enemy is crucial for understanding the probability of temporal and spatial encounters between the agent and nontarget species (van Lenteren *et al.* 2003). The radius of potential nontarget impacts is determined by the dispersal ability and the potential population-level impacts by the density of the dispersers (Mills *et al.* 2006). A wide variety of materials and techniques exist for estimating short- and long-distance dispersal (Nathan *et al.* 2003), but mark-release- recapture experiments are most suitable to study local dispersal of mass released agents (Mills *et al.* 2006). To distinguish dispersing agents from the wild population, molecular markers are increasingly used (Kuske *et al.* 2003). Abiotic factors (weather, wind) and biotic features of the agent (longevity, body size, wing forms, sex, mating status, egg-load) and the environment (availability of resources, density of native hosts, landscape elements) largely modify the density distance curve obtained. For mass-released agents, empirical or diffusion-based models give the best estimates of mean and quantile dispersal distance, direction, and density (Kuske *et al.* 2003; Mills *et al.* 2006). More importantly, analysis of dispersal data can provide qualitative and quantitative scales for environmental impact assessment (van Lenteren *et al.* 2003).

23.13.3 Impact of released organisms on other organisms in ecosystems

21.13.3.1 Effects on nontarget herbivores

Sometimes the established exotic biological control agent may affect the abundance of native nontarget species. These species are not usually the preferred hosts of the agent, and therefore their abundance is seldom affected. When an agent does affect the abundance of a nontarget, it is highly unlikely that it will, on its own, lead to extinction. Natural enemies generally leave a host patch before they have found all hosts. Hosts also escape their enemies in space and time, reducing the chances of extinction. Furthermore, asynchrony between local dynamics allows for large-scale persistence even when local extinctions occur (Hanski and Singer 2001). As a result, pests have seldom if ever been exterminated in the more than 100 years of insect biological control.

However, effective lowering of population densities in small habitat patches or tropical islands will increase the likelihood that other factors, including chance events, will destroy a population completely, in contrast to patterns observed on large islands or continents (MacArthur and Wilson, 1967). Therefore, it can be expected that some species utilized by biological control agents will become part of the natural extinction/colonization dynamics. It is unlikely that the extinction rates of these species differ significantly from that of other species in the same ecosystems.

21.13.3.2 Effects on other trophic levels

When the exotic agent is able to also attack natural enemies of the target herbivores [intraguild predation (Rosenheim *et al.* 1995) or facultative hyperparasitism (Sullivan, 1987)], it might interfere with the regulation of these herbivores. An example is a study on the trophic and guild interactions of five species of indigenous and introduced bethylid wasps used to control the coffee berry borer and lepidopteran pests of coconuts and almonds (Perez-Lachaud *et al.* 2004). One of the parasitoids not only killed conspecific and allospecific eggs and larvae, but could also develop as a facultative hyperparasitoid, threatening the establishment and survival of other parasitoid species. Intraguild predators or hyperparasitoids may lead to relaxed predation (Brodeur and Rosenheim 2000; Rosenheim *et al.* 1995), possibly resulting in temporal outbreaks of the prey. Ultimately, it may increase populations of natural enemies that have been released from their competitors (Polis and Holt 1992).

An exotic control agent itself may serve as food for other organisms (enrichment), increasing these populations, but possibly influencing indirectly populations of the target. Introduction of such an exotic may temporarily release or increase the predation pressure on the carnivore's prey, depending on the behavioural response (Holt and Kotler, 1987). On a longer timescale, the natural enemy population is expected to increase. Ultimately, this may result in a decrease in its prey (apparent competition between the agent and the other prey) (Holt, 1977; Messing *et al.* 2006). Apparent competition mediated by a generalist hyperparasitoid has been investigated by studying how the relationship between a specialist braconid wasp, *Cotesia melitaearum*, and its generalist hyperparasitoid, *Gelis agilis*, was influenced by the introduction of a second braconid host of *G. agilis*, *Cotesia glomerata*, into the field (van Nouhuys and Hanski 2000). The addition of the new primary parasitoid increased populations of the hyperparasitoid, which then reduced populations of *C. melitaearum*, to the extent that some of the original populations went extinct (van Nouhuys and Hanski 2000). Sometimes the natural enemy vectors pathogens or hyperparasitoids (Bjørnson and Schütte 2002; Goettel *et al.* 2006). For example, insects always carry various eubacteria. Some of these are opportunistic, whereas others are highly evolved pathogens such as *Wolbachia*, an intracellular parasite that can have profound negative effects on the reproductive fitness of biological control agents (Goettel *et al.* 2006).

21.13.3.3 Competition

When an introduced agent attacks and reduces a herbivore population, it may negatively affect other natural enemies that exploit the same resource. Ultimately, only one of them may survive. For example, successive introductions of braconid parasitoids of fruit flies into Hawaii led to the complete displacement and local extinction of the African species *Psyttalia humilis* by the Australian wasp *Diachasmimorpha tryoni* and, subsequently, the substantial displacement of *D. tryoni* in turn by the later arriving Southeast Asian species *Fopius arisanus* (Messing *et al.* 2006).

21.13.3.4 Other indirect effects and hybridization

Any of the described effects may result in further changes in the ecosystem, for example, by affecting non-food requirements (e.g., protection, pollination, and seed dispersal) of other species. Besides ecological changes, an exotic biological control agent may also cause genetic changes in other populations in the ecosystem. One possibility is hybridization between the agent and indigenous biotypes of the same or closely related species. An example is the *Aphelinus varipes* species complex, in which biotypes from various parts of the world hybridize, resulting in changed ecological preferences (Hopper *et al.* 2006).

21.14 Risk assessment future strategies

That the broader society considers biological control to be risky, however, also results from poor public and political perceptions of the benefits biological control has provided and can provide. For biological control to achieve its full potential, agencies responsible for its practice should consider redressing this to ensure policy discussions relating to improved risk analysis lead to fair and open assessment processes balancing risks to benefits in the future. Some controversy had developed over the last two decades regarding these potential non-target impacts (Follett and Duan, 2000; Bigler *et al.*, 2006; Van Lenteren *et al.*, 2006). Simberloff and Stiling (1996) summarized the controversy and highlighted potential risks such as predation or parasitism of non-target species, competition with native species, community and ecosystem effects, and unexpected effects such as loss of species dependent on the target of biological control efforts. The significance and practical impacts of these potential non-target impacts has been thoroughly debated in the literature (Simberloff and Stiling, 1996, 1998; Frank, 1998; Follett and Duan, 2000; Bigler *et al.*, 2006), and conclusions vary depending on individual perspective. Many biological control scientists view these impacts as a real concern, but primarily as a problem of the past currently considered and dealt with by existing rules and regulations. They also feel that the benefits provided by importation biological control far outweigh the few negatives resulting from occasional cases of non- target impacts. Simberloff and Stiling (1996) argued that the few documented cases of non-target impacts, compared with the number of natural enemy introductions, may have been more the result of a lack of monitoring and documentation than a lack of actual impacts. This suggestion may be supported by the database on non-target effects of importation and augmentation compiled by Lynch and Thomas (2000). These authors found that from the relatively few cases where data had been collected in biocontrol projects, there appeared to be a number of non-target effects, although these were primarily from very early importation efforts and were mostly relatively minor.

The most important risks from biological control introductions are threats to the environment, including global extinction of species. Biological control introductions of arthropods have been remarkably safe. Exotic natural enemies seem to have rarely caused permanent and strong population declines in nontarget organisms. Risks of nontarget effects caused by exotic natural enemies are of recent but growing concern. Ecological factors determining the environmental impact of an exotic agent are increasingly included in current evaluations. Host range data are now used to reject or accept introductions (Van Lenteren *et al.* 2006b), and determination of the potential establishment of exotics based on their thermal biology could provide a relatively simple methodology of determining whether exotics can establish in the target area (Boivin *et al.* 2006). Methods for assessment of dispersal and of direct and indirect effects on nontargets have not been used to a great extent in arthropod biological control but are currently being developed (Bigler *et al.* 2006; Mills *et al.* 2006).

Regulation of exotic natural enemies is a subject keenly debated by the biological control industry, scientists, and regulators (Blum *et al.* 2003; van Lenteren *et al.* 2003). Industry fears lengthy, cumbersome procedures leading to high costs and thus, in some cases, the inability to market a safe natural enemy. On the other hand, regulators within ministries of environment and agriculture have a responsibility to prevent unnecessary and risky releases of exotic organisms. When ecological data are used in quantitative, stepwise risk assessment procedures (van Lenteren and Loomans 2006), decisions on whether to release can be taken in a tiered system approach and lengthy procedures can be avoided. To allow continued use of safe biological control agents, a quick-scan method can be applied to agents already in use. Such scans may result in white lists of supposedly safe species for specified (eco)regions. Availability of such lists might stimulate the wider application of biological control. The present activities concerning ERA will hopefully result in a light and harmonized regulation procedure that is not prohibitive to the biological control industry and in the selection of safe natural enemies.

References

Abrams, J. C. 1990. Biological control agents in integrated pest management: Are they regulated – A research guide. *Pace Environ. Law Rev.,* 8, 89–115.

ANBP. 2004. *USDA-APHIS permitted beneficials imported into the United States from other countries as of 27 Feb. 2004.* http://www.anbp.org/beneficial%20list.htm

Baker R, Cannon R, Bartlett P, Barker I. 2005. Novel strategies for assessing and managing the risks posed by invasive alien species to global crop production and biodiversity. *Ann. Appl. Biol.* 146: 177–91.

Bale JS. 1991a. Implications of cold hardiness for pest management. *In: Insects at Low Temperature* (RE Lee, DL Denlinger Eds), pp. 461–498. New York: Chapman and Hall.

Bale JS. 1991b. Insects at low temperature: a predictable relationship? *Funct. Ecol.* 5: 291–298.

Barratt BIP, Ferguson CM, Goldson SL, Phillips CB, Hannah DJ. 2000. Predicting the risk from biological control agent introductions: a New Zealand approach. *In: Biological Control for a Small Planet: Nontarget Effects of Biological Control Introductions* (PA Follett, JJ Duan Eds), pp. 59–75. Dordrecht: Kluwer.

Barratt BIP, Ferguson CM, McNeil MR, Goldson SL. 1999. Parasitoid host specificity testing to predict field host range. *In: Recommendations for Host Specificity Testing in*

Australia: Towards Improved Assays for Biological Control (TM Whithers, L Barton Browne, JN Stanley Eds), pp. 70–83. Brisbane: CRC for Tropical Pest Management.

Bellamy DE, Byrne DN. 2001. Effects of gender and mating status on self-directed dispersal by the whitefly parasitoid *Eretmocerus eremicus*. *Ecol. Entomol.* 26: 571–577.

Bennett, F. D. and D. H. Habeck. 1992. *Cactoblastis cactorum*: a successful weed control agent in the Caribbean, now a pest in Florida? In: Delfosse, E.S. and R. R. Scott (Eds), *Proceedings of the Eighth International Symposium on Biological Control of Weeds.* 2–7 February 1992, Lincoln University, Canterbury, New Zealand, pp. 21–26.

Bigler F, Babendreier D, Kuhlmann U, Eds 2006. *Environmental Impact of Invertebrates in Biological Control of Arthropods: Methods and Risk Assessment.* Wallingford, UK: CAB Int., 300 pp.

Bigler F, Bale J, Cock MJW, Dryer H, GreatRex R, *et al.* 2005. Guideline on information requirements for import and release of invertebrate biological control agents (IBCAs) in European countries. *IOBC/WPRS Work. Group Harmon. Regulation of Invertebr. Biol. Control Agents.*

Bjørnson S, and Schütte C. 2002. Pathogens of mass-produced natural enemies and pollinators. In: van Lenteren JC, (Ed) 2003. *Quality Control and Production of Biological Control Agents: Theory and Testing Procedures.* Wallingford, UK: CAB Int, pp. 133–165.

Blum B, Ehlers R, Haukeland-Salinas S, Hokkanen H, Jung K, *et al.* 2003. Letter to the editor. Biological control: safety and regulatory policy. *BioControl* 48: 477–484.

Boettner, G. H., J.S. Elkinton, and C.J. Boettner 2000. Effects of biological control introduction on three nontarget native species of saturniid moths. *Conserv. Biol.* 14: 1798–1806.

Boivin G. 1994. Overwintering strategies of egg parasitoids. In: *Biological Control with Egg Parasitoids* (E. Wajnberg, SA Hassan Eds), pp. 219–244. Wallingford, UK: CAB Int.

Boivin G, Kölliker U, Bigler F, and Bale J. 2006. Factors that determine establishment of natural enemies and evaluation methods. *In:* Bigler F, Babendreier D, Kuhlmann U, eds. 2006. *Environmental Impact of Invertebrates in Biological Control of Arthropods: Methods and Risk Assessment.* Wallingford, UK: CAB Int pp. 98–113.

Brodeur J, Rosenheim JA. 2000. Intraguild interactions in aphid parasitoids. *Entomol. Exp. Appl.* 97: 93–108.

Buren, W. F. 1983. Artificial faunal replacement for imported fire ant control. *Fla. Entomol., 66,* 93–100.

Bush, G. L. 1975. Sympatric speciation in phytophagous parasitic insects. *In: Evolutionary strategies of parasitic insects and mites* (P. W. Price Ed.). Plenum, New York, pp. 187–206.

Campbell, F. C. 1993. Legal avenues for controlling exotics. In: *Biological Pollution: The Control and Impact of Invasive Exotic Species* (B. N. McKnight Ed.). Indiana Academy of Sciences, Indianapolis, pp. 243–250.

Carruthers, R. I. and Onsager, J. A. 1993. Perspective on the use of exotic natural enemies for biological control of pest grasshoppers (Orthoptera: Acrididae). *Environ. Entomol.,* 22, 885–903.

CBD 1992. Convention text UNEP. http://www.biodiv.org/doc/legal/cbd-en.pdf

CBD 2001. *Review of the efficiency and efficacy of existing legal instruments applicable to invasive alien species.* CBD Tech. Ser. No. 2, Montreal, SCBD. http://www.biodiv.org/doc/publications/ cbd-ts-01.pdf

Center, T. D. 1995. Selection criteria and ecological consequences of importing natural enemies (book review). *Biodivers. Conserv.,* 4, 524–526.

Center, T. D., Frank, J. H. and Dray, F. A., Jr. 1994. Scientific methods for managing invasive non-indigenous species – biological control. In: *An assessment of invasive non-indigenous species in Florida's 'public lands'* (D. C. Schmitz and T. C. Brown Eds). Florida Dept. of Environmental Protection, Tallahassee, pp. 209–247.

Charles, J., 2001. Introduction of a parasitoid for mealybug biocontrol: a case study under new environmental legislation. *New Zealand Plant Prot.* 54: 37–41.

Civeyrel, L. and Simberloff, D. 1996. A tale of two snails: Is the cure worse than the disease? *Biodivers. Conserv.* 5, 1231–1252.

Clarke, B., Murray, J. and Johnson, M. S. 1984. The extinction of endemic species by a program of biological control. *Pacif. Sci.,* 38, 97–104.

Coombs M. 2003. Post-release evaluation of *Trichopoda giacomellii* (Diptera: Tachinidae) for efficacy and non-target effects. In *Proc. Int. Symp. Biol. Control Arthropods, Honolulu, Hawaii, 14– 18 Jan. 2002*, ed. RG van Driesche, pp. 399–406. Morgantown, WV:USDAFor. Serv.

Courtenay, W. R., Jr 1993. Biological pollution through fish introductions. In: *Biological Pollution: The Control and Impact of Invasive Exotic Species* (B. N. McKnight Ed.). Indiana Academy of Science, Indianapolis, pp. 35–61.

Courtenay, W. R., Jr 1994. Non-indigenous fishes in Florida. In: *An Assessment of Invasive Non-indigenous Species in Florida's Public Lands* (D. C. Schmitz and T. C. Brown Eds). Florida Dept. of Environmental Protection, Tallahassee, pp. 57–63.

Courtenay, W. R., Jr and Robins, C. R. 1973. Exotic aquatic organisms in Florida with emphasis on fishes: a review and recommendations. *Trans. Am. Fish. Soc.,* 102, 1-12.

Courtenay, W. R., Jr and Williams, J. D. 1992. Dispersal of exotic species from aquaculture sources, with emphasis on freshwater fishes. In: *Dispersal of Living Organisms into Aquatic Systems* (A. Rosenfield and R. Mann Eds), MD Sea Grant College, College Park, pp. 49–81.

Courtenay, W. R., Jr, Sahlman, H. H., Miley, W. W. II and Herrema D. J. 1974. Exotic fishes in fresh and brackish waters of Florida. *Biol. Conserv.,* 6, 292–302.

Davis, C.J. and G.D. Butler 1964. Introduced enemies of the giant African snail, *Achatina fulica* Bowdich, in Hawaii (Pulmonta: Achatinidae). *Proc. Hawaiian Entomol. Soc.* 18: 377–389.

DeBach P. 1964. *Biological Control of Insect Pests and Weeds.* London: Chapman and Hall. DeBach, P. 1974. *Biological Control by Natural Enemies.* Cambridge University Press, London.

DEFRA. 2000. The Regulation and Control of the Release of Non-Native Animals and Plants into the Wild in Great Britain. A guide to sections 14 and 16 of the Wildlife and Countryside Act 1981. http://www.defra.gov.uk/environment/gm/ nonnav/ index.htm

Delfosse, E. S. 1985. *Echiurn plantagineum* in Australia: effects of a major conflict of interest. *In: Proceedings of the VI International Symposium on Biological Control of Weeds* (E. S. Delfosse Ed.). Agriculture Canada, Ottawa, pp. 293–298.

DeLoach, C. J. 1991. Past successes and current prospects in biological control of weeds in the United States and Canada. *Nat. Areas J.* 11, 129–142.

Diamond, J. M. 1996. A-bombs against amphibians. *Nature* 383: 386–387.

Dingle H. 1996. *Migration: The Biology of Life on the Move.* Oxford, UK: Oxford Univ. Press

Doutt RL, Annecke DP, Tremblay E. 1976. Biology and host relationships of parasitoids. *In: Theory and Practice of Biological Control* (CB Huffaker, PS Messenger Eds), pp. 143–168. New York: Academic Press.

Easteal, S. 1981. The history of introductions of *Bufo marinus* (Amphibia: Anura); a natural experiment in evolution. *Biol. J. Linn. Soc.* 16: 93–113.

Ehler, L. E. 1991. Planned introductions in biological control. *In: Assessing Ecological Risks of Biotechnology* (L. R. Ginzburg Ed.). Butterworth-Heinemann, Boston, pp. 21–39.

Ehlers R-U. 2003. Biocontrol nematodes. *In: Environmental Impacts of Microbial Insecticides,* ed. HMT Hokkanen, AE Hajek, pp. 177–220. Dordrecht: Kluwer

Elliott, N., R. Kieckhefer, and W. Kauffman. 1996. Effects of an invading coccinellid on native cocinellids in an agricultural landscape. *Oecologia* 105: 537–544.

Emerson, W. K. and Jacobson, M. K. 1976. *Guide to Shells.* Knopf, New York. EPPO. 1997. Pest risk assessment scheme. *Bull. OEPP/EPPO Bull.* 27: 281–305.

EPPO. 1999. Safe use of biological control: first import of exotic biological control agents for research under contained conditions. EPPO *Standard* PM6/1(1). http://www.eppo.org/

EPPO. 2000. Safe use of biological control: import and release of exotic biological control agents. *EPPO Standard* PM6/2(1). http://www.eppo.org/

EPPO. 2002. List of biological control agents widely used in the EPPO region: PM6/3(2). *Bull. OEPP/EPPO Bull.* 32: 447–461.

Fasham M, and Trumper K. 2001. Review on Non-Native Species Legislation and Guidance. St. Ives, UK: Ecoscope, 145 pp.

Fatouros NE, Huigens ME, van Loon JJA, Dicke M, Hilker M. 2005. Chemical communication: butterfly antiaphrodisiac lures parasitic wasps. *Nature* 433: 704.

Feder, J. L. and Bush, G. L. 1989. Gene frequency clines for host races of *Rhagoletis pomonella* (Diptera: Tephritidae) in the midwestern United States. *Heredity,* 63, 245–266.

Follett, P.A., and J.J. Duan (Eds), 2000. *Nontarget Effects of Biological Control.* Kluwer Academic Publishers, Boston, Massachusetts, 316 pp

Follett, P.A., J. Duan, R.H. Messing and V.P. Jones, 2000. Parasitoid drift after biological control introductions: re-examining Pandora's box. *Am. Entomol.* 46: 82–94.

Frank, J.H. 1998. How risky is biological control? Comment. *Ecology* 79:1829–1834. Freeland, W. J. 1985. The need to control cane toads. *Search* 16: 211–215.

Funasaki, G. Y., Lai, P. Y, Nakahara, L. M., Beardsley, J. W. and Ota, A. K. (1988). A review of biological control introductions in Hawaii: 1890 to 1985. *Proc. Hawaiian Entomol. Soc.,* 28, 105–160.

Gamradt, S.C. and L.B. Kats 1996. Effect of introduced crayfish and mosquitofish on California newts. *Conserv. Biol.* 10: 1155–1162.

Gassmann, A. and S.V. Louda 2001. *Rhinocyllus conicus*: Initial evaluation and subsequent ecological impacts in North America. *In:* Wajnberg, E., J. K. Scott, and P. C. Quimby (Eds). *Evaluating Indirect Ecological Effects of Biological Control.* CABI Publishing, Wallingford, UK, pp. 147–183.

Genovesi P, Shine C, Eds 2003. European strategy on invasive alien species. Convention on the conservation of European wildlife and natural habitats. Counc. Eur., 23rd Meet. Stand. Comm., Strasbourg, 1–5 Dec. http://www.coe.int/t/e/Cultural Co-operation/Environment/Nature and biological diversity/Nature protection/sc23 tpvs07e.pdf?L=E

Goeden RD and Louda SM. 1976. Biotic interference with insects imported for weed control. *Annu. Rev. Entomol.* 21:325–42

Goettel M, Ehlers R-U, Inglis D. 2006. Risks posed by contaminants and methods for determination. In: Bigler F, Babendreier D, Kuhlmann U, (Eds). 2006. *Environmental Impact of Invertebrates in Biological Control of Arthropods: Methods and Risk Assessment.* Wallingford, UK: CAB Int pp. 145–165.

Gonzalez, D. and Gilstrap, F. E. 1992. Foreign exploration: Assessing and prioritizing natural enemies and consequences of pre introduction studies. In: *Selection Criteria and Ecological Consequences of Importing Natural Enemies* (W. C. Kauffman and J. E. Nechols Eds). Entomological Society of America, Lanham, MD, pp. 51–70.

Greathead, D. J., 1995. Benefits and risks of classical biological control. *In:* H.M.T. Hokkanen and J.M. Lynch (Eds), *Biological Control: Benefits and Risks.* Cambridge University Press, Cambridge, pp. 53–63.

Griffiths, O., A. Cook, and S.M. Wells 1993. The diet of the introduced carnivorous snail *Euglandina rosea* in Mauritius and its implications for threatened island gastropod faunas. *J. Zool.* 229: 79–89.

Groombridge, B. (Ed.). 1992. *Global Biodiversity.* Chapman and Hall, London.

Groves, R. H. (1989). Ecological control of invasive terrestrial plants. In: *Biological Invasions: A Global Perspective* (J. A. Drake, H. A. Mooney, F. diCastri, R. H. Groves, F. J. Kruger, M. Rejmanek and M. Williamson Eds). Wiley, Chichester, pp. 437–461.

Gurr G, Wratten S, Eds 2000. *Measures of Success in Biological Control.* Dordrecht: Kluwer, 448 pp.

Habeck, D. H. and Bennett, F. D. 1990. *Cactoblastis cactorum* Berg. (Lepidoptera: Pyralidae), a phycitine new to Florida, Florida Dept. of Agric. and Consumer Services, Div. of Plant Industry, Entomol. Circ. No. 333.

Hanski I, Singer MC. 2001. Extinction colonization dynamics and host-plant choice in butterfly metapopulations. *Am. Nat.* 158: 341–353.

Hart AJ, Bale JS, Tullett AG, Worland MR, Walters KFA. 2002a. Effects of temperature on the establishment potential of the predatory mite *Amblyseius californicus* McGregor (Acari: Phytoseiidae) in the UK. *J. Insect Physiol.* 48: 593–599.

Hart AJ, Tullett AG, Bale JS, Walters KFA. 2002b. Effects of temperature on the establishment potential in the UK of the non-native glasshouse biocontrol agent *Macrolophus caliginosus. Physiol. Entomol.* 27: 112–123.

Henneman ML, Memmott J. 2001. Infiltration of a Hawaiian community by introduced biological control agents. *Science* 293: 1314–1316.

Hinton, H. E., and A. M. S. Dunn 1967 *Mongooses: Their Natural History and Behaviour.* University of California Press, Berkeley, CA, USA, 144 pp.

Hoddle, M. S. 1999. Biological control of vertebrate pests. *In:* Bellows, T.S., Jr and T. W. Fisher (Eds). *Handbook of Biological Control.* Academic Press, San Diego, CA, USA, pp. 955–974.

Hoebeke, E.R. and A.G. Wheeler, Jr, 1996. Adventive lady beetles (Coleoptera: Coccinellidae) in the Canadian maritime provinces, with new eastern U.S. records of *Harmonia quadripunctata. Entomol. News* 107: 281–290.

Hokkanen HMT and Lynch JM (Eds) 1995. *Biological Control: Benefits and Risks.* Cambridge, UK: Cambridge University Press.

Hokkanen, H. and Pimentel, D. 1984. New approach for selecting biological control agents. *Can. Entomol.,* 116, 1109–1121.

Hokkanen, H. and Pimentel, D. 1989. New associations in biological control: Theory and practice. *Can. Entomol.* 121: 829–840.

Hokkanen HMT, Babendreier D, Bigler F, Burgio G, Kuske S, *et al.* 2003. Evaluating environmental risks of biological control introductions into Europe. Final Report EU-FAIR5 CT97-3489 (ERBIC), Comm. Eur. Commun., Brussels.

Holloway, J. K. 1964. Projects in biological control of weeds. *In:* DeBach, P. (Ed.). *Biological Control of Insect Pests and Weeds.* Reinhold Publishing Corporation, New York, pp. 650–670.

Holt RD. 1977. Predation, apparent competition, and the structure of prey communities. *Theor. Popul. Biol.* 12: 197–229.

Holt RD and Kotler BP. 1987. Short-term apparent competition. *Am. Nat.* 130: 412–430.

Hopper KR, Britch SC, Wajnberg E. 2006. Risks of interbreeding between species introduced for biocontrol and native species and methods for evaluating its occurrence and impact.In: Bigler F, Babendreier D, Kuhlmann U, eds. 2006. *Environmental Impact of Invertebrates in Biological Control of Arthropods: Methods and Risk Assessment.* Wallingford, UK: CAB Int. pp. 78–97.300

Howarth, F. G. 1983. Classical biological control: panacea or Pandora's box? *Proc. Hawaiian Entomol. Soc.* 24: 239–244.

Howarth, F. G. 1985. Impacts of alien land arthropods and mollusks on native plants and animals in Hawaii. In: *Hawaii's Terrestrial Ecosystems: Preservation and Management* (C. P. Stone and J. M. Scott Eds). University of Hawaii Cooperative National Park Resources Study Unit, Honolulu, pp. 149–178.

Howarth, F. G. 1991. Environmental impacts of classical biological control. *Annu. Rev. Entomol.,* 36, 485–509.

Jervis MA, Kidd NAC, Fitton MG, Huddleston T, Dawah HA. 1993. Flower visiting by hymenopteran parasitoids. *J. Nat. Hist.* 27: 67–105.

Jervis MA, Kidd NAC, Heimpel GE. 1996. Parasitoid adult feeding behaviour and biocontrol: a review. *Biocontrol News Inform.* 17(1): 11N–26N.

Jung C, Croft BA. 2001. Ambulatory and aerial dispersal among specialist and generalist predatory mites (Acari: Phytoseiidae). *Environ. Entomol.* 30: 1112–1118.

Kairo MTK, Cock JW, Quinlan MM. 2003. An assessment of the use of the Code of Conduct for the Import and Release of Exotic Biological Control Agents (ISPM No. 3) since its endorsement as an international standard. *Biocontrol News Inform.* 24: 15N–27N.

Kimberling DN. 2004. Lessons from history: predicting successes and risks of intentional introductions for arthropod biological control. *Biol. Inv.* 6: 301–318.

King, C. 1984. *Immigrant Killers.* Oxford University Press, Auckland.

King, C. M. 1990a. Stoat. *In:* King, C.M (Ed.). *The Handbook of New Zealand Mammals.* Oxford University Press, Auckland, New Zealand, pp. 288–312.

King, C. M. 1990b. Weasel. *In:* King, C.M (Ed.). *The Handbook of New Zealand Mammals.* Oxford University Press, Auckland, New Zealand, pp. 313–320.

Kinzie, R.A. III. 1992. Predation by the introduced carnivorous snail *Euglandia rosea* (Ferussac) on endemic aquatic lymnaeid snails in Hawaii. *Biol. Conserv.* 60: 149–155.

Kuhlmann U, Schaffner U, Mason PG. 2006. Selection of non-target species for host specificity testing. In: Bigler F, Babendreier D, Kuhlmann U, (Eds). 2006. *Environmental Impact of Invertebrates in Biological Control of Arthropods: Methods and Risk Assessment.* Wallingford, UK: CAB Int. pp. 15–37.

Kuris AM. 2003. Did biological control cause extinction of the coconut moth, *Levuana ridescens,* in Fiji? *Biol. Inv.* 5:133–41

Kuske S, Widmer F, Edwards PJ, Turlings T, Babendreier D, Bigler F. 2003. Dispersal and persistence of mass released *Trichogramma brassicae* (Hym., Trichogrammatidae) in nontarget habitats. *Biol. Control* 27: 181–193.

Lai, P.-Y. 1988. Biological control: A positive point of view. *Proc. Hawaiian Entomol. Soc.,* 28, 179–191.

Lavers, R. B. and B. K. Clapperton. 1990. Ferret. *In:* King, C.M (Ed.). *The Handbook of New Zealand Mammals.* Oxford University Press, Auckland, New Zealand, pp. 320–330.

Legner, E. F. 1996. Comments on adverse assessments of *Gambusia affinis. J. Am. Mosquito Control Assoc.* 12: 161.

van Lenteren, J.C. (ed), 2003. *Quality Control and Production Control Agents: Theory an Testing Procedures.* CABI Publishing, Wallingford, UK (in press).

van Lenteren, J.C. and J. Woets, 1988. Biological and integrated pest control in greenhouses. *Annu. Rev. Entomol.* 33: 239–269.

van Lenteren, J.C., 2000. Measures of Success in Biological Control Of Arthropods By Augmentation Of Natural Enemies. In: S.Wratten and G. Gurr (eds), *Measures of Success in Biological Control.* Kluwer Academic Publishers, Dordrecht. pp. 77–103.

Limburg DD, Rosenheim JA. 2001. Extrafloral nectar consumption and its influence on survival and development of an omnivorous predator, larval *Chrysoperla plorabunda* (Neuroptera: Chrysopidae). *Environ. Entomol.* 30: 595–604.

Lloyd, L. 1990. Native fishes as alternatives to the exotic fish, *Gambusia,* for insect control. *In: Introduced and Translocated Fishes and Their Ecological Effects* (D. A. Pollard Ed.). Department of Primary Industries and Energy, Bureau of Rural Resources, Canberra, pp. 115–122.

Lockwood, J. A. 1993a. Environmental issues involved in biological control of rangeland grasshoppers (Orthoptera: Acrididae) with exotic agents. *Environ. Entomol.,* 22, 503–518.

Lockwood, J. A. 1993b. Benefits and costs of controlling rangeland grasshoppers (Orthoptera: Acrididae) with exotic organisms: Search for a null hypothesis and regulatory compromise. *Environ. Entomol.,* 22, 904–914.

Lockwood, J.A., F.G. Howarth and M.F. Purcell (Eds), 2001. *Balancing Nature: Assessing the Impact of Importing Non-Native Biological Control Agents.* Thomas Say Publications in Entomology: Proceedings, Entomological Society of America, Lanham, MD.

Loope, L., O. Hamann, and C.P. Stone 1988. Comparative conservation biology of oceanic archipelagos. *BioScience* 38: 272–282.

Louda, S.M. 2000. Negative ecological effects of the musk thistle biological control agent, *Rhinocyllus conicus*, pp. 215–243. *In:* Follet, P.A. and J. J. Duan (Eds). *Nontarget Effects of Biological Control.* Kluwer Academic Publishers, Boston, MA, USA.

Louda, S.M., D. Kendall, J. Connor, and D. Simberloff 1997. Ecological effects of an insect introduced for the biological control of weeds. *Science* 277: 1088–1090.

Louda, S.M., R.W. Pemberton, M.T. Johnson, and P.A. Follett 2003. Nontarget effects – the Achilles' heel of biological control? Retrospective analyses to reduce risk associated with biocontrol introductions. *Annu. Rev. Entomol.* 48: 365–396.

Lynch, L.D. and M.B. Thomas, 2000. Nontarget effects in the biocontrol of insects with insects, nematodes and microbial agents: the evidence. *Biocontrol News Inform.* 21: 117–130.

Lynch LD, Hokkanen HMT, Babendreier D, Bigler F, Burgio G, *et al.* 2001. Insect biological control and non-target effects: a European perspective. *In:* E. Wajnberg, J.C. Scott and P.C. Quimby (Eds), *Evaluating Indirect Ecological Effects of Biological Control.* Wallingford, UK: CAB Int., pp. 99–125.

MacArthur RH, Wilson EO. 1967. *The Theory of Island Biogeography.* Princeton, NJ: Princeton Univ. Press, 203 pp.

Malecki, R. A., B. Blossey, B., Hight, S. D., Schroeder, D., Kok, L. T. and Coulson, J. R. 1993. Biological control of purple loosestrife. *BioScience,* 43, 680–686.

Mayer, J. J. and Brisbin, I. L. Jr 1991. *Wildpigs in the United States.* University of Georgia

Press, Athens, GA.

McCorquodale, D.B., 1998. Adventive lady beetles (Coleoptera: Coccinellidae) in the eastern Nova Scotia, Canada. *Entomol. News* 109: 15–20.

Meisch, M. V. 1985. *Gambusia affinis affinis. In:* H.C. Chapman (Ed.). Biological Control of *Mosquitoes.* American Mosquito Control Association Bulletin No. 6, pp. 3–17.

Messing, R. H. 1992. Biological control in island ecosystems: Cornerstone of sustainable agriculture or threat to biological diversity? *Pacif. Sci.,* 46, 387–388.

Messing R, Roitberg B, Brodeur J. 2006. Effects of competition, displacement and intraguild predation in biocontrol and evaluation methods. *In:* Bigler F, Babendreier D, Kuhlmann U, eds. 2006. *Environmental Impact of Invertebrates in Biological Control of Arthropods: Methods and Risk Assessment.* Wallingford, UK: CAB Int pp. 64–77.

Miller, M. and Aplet, G. 1993. Biological control: a little knowledge is a dangerous thing. *Rutgers Law Rev.,* 45, 285–334.

Mills NJ, Babendreier D, Loomans AJM. 2006. Significance and assessment of dispersal. *In:* Bigler F, Babendreier D, Kuhlmann U, eds. 2006. *Environmental Impact of Invertebrates in Biological Control of Arthropods: Methods and Risk Assessment.* Wallingford, UK: CAB Int pp. 114–131.

Moyle, P. B., Li, H. W. and Barton, B. A. 1986. The Frankenstein effect: impact of introduced fishes on native fishes in North America. *In: Fish Culture in Fisheries Management* (R. H. Stround Ed.). American Fisheries Society, Bethesda, MD, pp. 415–426.

Murray, E. (1993). The sinister snail. *Endeavour* (New Series), 17(2), 78–83.

Murray, J., E. Murray, M. S. Johnson, and B. Clarke. 1988. The extinction of *Partula* on Moorea. *Pacif. Sci.* 42: 150–153.

Nafus, D. M. 1992. Impact of intentionally and accidentally introduced biological control agents on unintended hosts, *Hypolimnas anomala* and *H. bolina* (Lepidoptera: Nymphalidae) *on Guam. Pacif. Sci.,* 46, 394–395.

NAPPO 2001. Guidelines for petition for release of exotic entomophagous agents for the biological control of pests. RSPM No. 12. http://www.nappo.org/Standards/OLDSTDS/RSPM12-e.pdf

Nathan R, Perry G, Cronin JT, Strand AE, Cain ML. 2003. Methods for estimating long-distance dispersal. *Oikos* 103: 261–273.

National Academy of Sciences 1987. Research Briefings 1987. *Report of the Research Briefing Panel on Biological Control in Managed Ecosystems.* National Academy Press, Washington, DC.

Nechols, J. R. and Kauffman, W. C. 1992. Introduction and overview. *In: Selection Criteria and Ecological Consequences of Importing Natural Enemies* (W. C. Kauffman and J. E. Nechols Eds). Entomological Society of America, Lanham, MD, pp. 1–5.

Neuenschwander, P. and Markham, 2001. Biological control in Africa and its possible effects on biodiversity. In: E. Wajnberg, J.C. Scott and P.C. Quimby (Eds), *Evaluating Indirect Ecological Effects of Biological Control.* CABI Publishing, Wallingford, pp. 127–146.

Nicoli, G. and G. Burgio, 1997. Mediterranean biodiversity as source of new entomophagous species for biological control in protected crop. *Bulletin IOBC/WPRS* 20: 27–38.

Norton, B. G. 1987. *Why Preserve Natural Variety?* Princeton University Press, Princeton, NJ.

Obrycki, J.J., N.C. Elliott, and K.L. Giles 2000. Coccinellid introductions: potential for and evaluation of nontarget effects. *In:* Follet, P. A. and J. J. Duan (Eds). *Nontarget Effects of Biological Control.* Kluwer Academic Publishers, Boston, MA, USA, pp. 127–145.

OECD. 2004. Guidance for information requirements for regulation of invertebrates as biological control agents. *OECD Ser Pestic.* 21. http://www.oecd. org/dataoecd/6/20/28725175.pdf

Pemberton, R. W. 1995. *Cactoblastis cactorum* (Lepidoptera: Pyralidae) in the United States: an immigrant biological control agent or an introduction of the nursery trade? *Am. Entomol.* 41: 230–232.

Perez-Lachaud G, Batchelor TP, Hardy ICW. 2004. Wasp eat wasp: facultative hyperparasitism and intra-guild predation by bethylid wasps. *Biol. Control* 30: 149–155.

Pimentel D, Lach L, Zuniga R, Morrison D. 2002. Environmental and economic costs associated with non-indigenous species in the United States. *In: Biological Invasions,* ed. D Pimentel, pp. 285– 303. Boca Raton, FL: CRC Press

Polis GA, Holt RD. 1992. Intraguild predation: the dynamics of complex trophic interactions. *Trends Ecol. Evol.* 7: 151–154.

Pons X, Lumbierres B, Stáry P. 2004. Expansíon de *Lysiphlebus testaceipes* (Cresson) (Hym., Braconidae, Aphidiinae) en el Noreste de la Península Ibérica. *Bol. Sanid. Vegétal. Plagas* 30: 547–552.

Prokopy, R. J., Diehl, S. R. and Cooley, S. S. 1988. Behavioral evidence for host races in *Rhagoletis pomonella* flies. *Oecologia, Berl.,* 76, 138–147.

Robinson G. S. 1975. *Macrolepidoptera of Fiji and Rotuma*. Classey, Oxon, UK, 362 pp.

Roderick, G. K. 1992. Postcolonization evolution of natural enemies. *In: Selection Criteria and Ecological Consequences of Importing Natural Enemies* (W. C. Kauffman and J. E. Nechols Eds). Entomological Society of America, Lanham, MD, pp. 71–86.

Rolston, H., III 1991. Environmental ethics: values in and duties to the natural world. *In: Ecology, Economics, Ethics: The Broken Circle* (F. H. Bormann and S. R. Kellert Eds). Yale University Press, New Haven, CT, pp. 73–96.

Rosenheim JA, Kaya HK, Ehler LE, Marois JJ, Jafee BA. 1995. Intraguild predation among biological control agents: theory and evidence. *Biol. Control* 5: 303–335.

Rupp, H. R. 1996. Adverse assessments of *Gambusia affinis*: an alternate view for mosquito control practitioners. *J. Am. Mosquito Control Assoc.* 12: 155–166.

Sands D. P. A. 1997. The safety of biological control agents: assessing their impact on beneficial and other non-target hosts. *Mem. Mus. Victoria* 56: 611–616.

Sands DPA, van Driesche RG. 2000. Evaluating host specificity of agents for biological control of arthropods: rationale, methodology and interpretation. *In: Proc. Session: Host Specificity Testing of Exotic Arthropod Biological Control Agents: The Biological Basis for Improvement in Safety. 10th Int. Symp. Biol. Control Weeds,* 4–14 July 1999, Bozeman, MT (RG Van Driesche, TA Heard, AS McClay, R Reardon Eds), pp. 69–83. Morgantown, WV: USDA For. Serv.

Saunders, G., B. Coman, J. Kinnear, and M. Braysher. 1995. *Managing Vertebrate Pests: Foxes.* Australian Government Printing Service, Canberra, 141 pp.

Schroder, D. 1980. The biological control of thistles. *Biocontrol News Inform.* 1, 9–26.

Sheppard AW, Hill R, DeClerck-Floate RA, McClay A, Olckers T, *et al.* 2003. A global review of risk-benefit-cost analysis for the introduction of classical biological control against weeds: a crisis in the making? *Biocontrol News Inform.* 24: 91–108N.

Shine C, Williams N, Gündling L. 2000. *A Guide to Designing Legal and Institutional Frameworks on Alien Invasive Species.* Gland, Switzerland: IUCN, 138 pp.

Simberloff, D. 1991. Keystone species and community effects of biological introductions. *In: Assessing Ecological Risks of Biotechnology* (L. R. Ginzburg Ed.). Butterworth- Heinemann, Boston, pp. 1–19.

Simberloff, D. 1992. Conservation of pristine habitats and unintended effects of biological control. In: *Selection Criteria and Ecological Consequences of Importing Natural Enemies* (W. C. Kauffman and J. E. Nechols Eds). Entomological Society of America, Lanham, MD, pp. 103–117.

Simberloff, D. 1994. The ecology of extinction. *Acta Palaeontol. Pol.,* 38, 159–174.

Simberloff, D. and Stiling, P. 1998. How risky is biological control? Reply. *Ecology* 79:1834–1836

Simberloff, D., and P. Stiling. 1996. Risks of species introduced for biological control. *Biol. Conserv.* 78: 185–192.

Simmonds, F.J. and I.W. Hughes 1963. Biological control of snails exerted by *Euglandia rosea* (Ferrussac) in Bermuda. *Entomophaga* 8: 219–222.

Sullivan DJ. 1987. Insect hyperparasitism. *Annu. Rev. Entomol.* 32: 49–70.

Stáry, P., J.P. Lyon and F. Leclant, 1988a. Post-colonisation host range of *Lysiphlebus testaceipes* in the Mediterranean area (Hymenoptera, Aphidiidae). *Acta Entomol. Bohem.* 85: 1–11.

Stáry, P., J.P. Lyon and F. Leclant, 1988b. Biocontrol of aphids by the introduced *Lysiphlebus testaceipes* (Cress.) (Hymenoptera, Aphidiidae) in Mediterranean France. *J. Appl. Entomol.*, 105: 74–87.

Strong, D. R. 1997. Fear no weevil? Science 277: 1058–1059.

Taylor, T. H. C. 1937. *The biological control of an insect in Fiji. An account of the coconut leaf-mining beetle and its parasite complex.* Imperial Institute of Entomology, London.

Tomich, P. Q. 1986. *Mammals in Hawaii,* 2nd edn. Bishop Museum Press, Honolulu.

Tothill J. D., T. H. C. Taylor, and R. W. Paine. 1930. *The Coconut Moth in Fiji: A History of its Control by Means of Parasites.* Imperial Bureau of Entomology, London, 269 pp.

Tscharntke T, Brandl R. 2004. Plant insect interactions in fragmented landscapes. *Annu. Rev. Entomol.* 49: 405–430.

Turner, C. E., Pemberton, R. W. and Rosenthal, S. S. 1987. Host utilization of native *Cirsium* thistles (Asteraceae) by the introduced weevil *Rhinocyllus conicus* (Coleoptera: Curculionidae) in California. *Environ. Entomol.,* 16, 111–115.

van Driesche RG, Reardon R. 2004. *Assessing Host Ranges for Parasitoids and Predators Used for Classical Biological Control: A Guide to Best Practice.* Morgantown, WV: USDA For. Serv.

van Lenteren JC, and Loomans AJM. 2006. Environmental risk assessment: methods for comprehensive evaluation and quick scan. *In:* Bigler F, Babendreier D, Kuhlmann U, eds. 2006. *Environmental Impact of Invertebrates in Biological Control of Arthropods: Methods and Risk Assessment.* Wallingford, UK: CAB Int. pp. 254–272.

van Lenteren JC, and Woets J. 1988. Biological and integrated pest control in greenhouses. *Annu. Rev. Entomol.* 33: 239–269.

van Lenteren JC, Babendreier D, Bigler F, Burgio G, Hokkanen HMT, *et al.* 2003. Environmental risk assessment of exotic natural enemies used in inundative biological control. *BioControl* 48: 3–38.

van Lenteren JC, Cock MJW, Hoffmeister TS, Sands DPA. 2006. Host specificity in arthropod biological control, methods for testing and interpretation of the data. In: Bigler F, Babendreier D, Kuhlmann U(Eds) 2006. *Environmental Impact of Invertebrates in Biological Control of Arthropods: Methods and Risk Assessment.* Wallingford, UK: CAB Int. pp. 38–63.

van Lenteren JC, ed. 2003. *Quality Control and Production of Biological Control Agents: Theory and Testing Procedures.* Wallingford, UK: CAB Int., 327 pp.

van Lenteren, J.C. 2000. Measures of success in biological control of arthropods by augmentation of natural enemies. *In:* S. Wratten and G. Gurr (Eds), *Measures of Success in Biological Control.* Kluwer Academic Publishers, Dordrecht, pp. 77–103.

van Lenteren, J.C., Bale, J., Bigler, F., Hokkanen, H.M.T., Loomans, A.J.M., 2006b. Assessing risks of releasing exotic biological control agents of arthropod pests. Annu. Rev. Entomol. 51, 609–634

van Lenteren, J.C., Cock, M.J.W., Hoffmeister, T.S., Sands, D.P.A., 2006a. Host specificity in arthropod biological control, methods for testing and interpretation of the data. *In:* Bigler, F., Babendreier, D., Kuhlmann, U. (Eds.), *Environmental Impact of Invertebrates in Biological Control of Arthropods:* Methods and Risk Assessment CAB Int., Wallingford, UK, pp. 38–63.

van Nouhuys S, and Hanski I. 2000. Apparent competition between parasitoids mediated by a shared hyperparasitoid. *Ecol. Lett.* 3: 82–84.

Viggiani, G., 1994. Recent cases of interspecific competition between parasitoids of the family Aphelinidae (Hymenoptera: Chalcidoidea). *Norw. J. Agric. Sci., Supplement* 16: 353–359.

Wäckers FL. 2003. The parasitoids' need for sweets: sugars in mass rearing and biological control. In: van Lenteren JC, ed. 2003. *Quality Control and Production of Biological Control Agents: Theory and Testing Procedures.* Wallingford, UK: pp. 59–72.

Wäckers FL. 2005. Suitability of (extra-) floral nectar, pollen and honeydew as insect food sources. *In: Plant-Provided Food for Carnivorous Insects* (FL Wäckers, PCJ van Rijn, J Bruin Eds), pp. 17–74. Cambridge, UK: Cambridge Univ. Press.

Waage, J. K. and Greathead, D. J. (1988). Biological control: Challenge and opportunities. *Phil. Trans. R. Soc. Lond., B* 318, 111–128.

Waage, J.K., 1997. Global developments in biological control and the implications for Europe. *In:* I.M. Smith (Ed.), *EPPO/CABI Workshop on Safety and Efficacy of Biological Control in Europe.* Blackwell Science Ltd., Oxford, pp. 5–13.

Wajnberg, G., Scott, J.K. and Quimby, P.C. (Eds) 2001. *Evaluating Indirect Ecological Effects of Biological Control.* CABI Publishing, Wallingford, UK, 288p

Walton, W.E. and M.S. Mulla 1991. Integrated control of *Culex tarsalis* larvae using *Bacillus sphaericus* and *Gambusia affinis*: effects on mosquitoes and nontarget organisms in field mesocosms. *Bull. Soc. Vector Ecol.* 16: 203–221.

Wapshere AJ. 1974. A strategy for evaluating the safety of organisms for biological weed control. *Ann. Appl. Biol.* 77: 201–211.

Warren, G. L. 1994. Ecologically disruptive non-indigenous freshwater invertebrates in Florida. In: *An Assessment of Invasive Non-Indigenous Species in Florida's Public Lands* (D. C. Schmitz and T. C. Brown Eds). Florida Dept. of Environmental Protection, Tallahassee, pp. 104–109.

Wratten, S. and G. Gurr (Eds), 2000. *Measures of Success in Biological Control.* Kluwer Academic Publishers, Dordrecht, 448 pp.

Zwolfer H. 1971. The structure and effect of parasite complexes attacking phytophagous host insects. In: *Dynamics of Numbers in Populations* (PJ den Boer, GR Gradwell Eds). Wageningen: Pudoc, pp. 405–416.

EXTENSION METHODOLOGIES FOR DISSEMINATION OF IPM TECHNOLOGIES

22

P. S. Slathia, Rakesh Kumar and S. K. Kher

22.0 INTRODUCTION

Different extension methodologies play significant role in the dissemination of various IPM technologies among farming community as well as in the adoption of these technologies in different crops as per their suitability and mode of application. IPM offers a very compatible and eco-friendly way of controlling the insect-pests of various crop plants and knowledge about different IPM technologies is must from farmers point of view because they are the end users of any new agricultural technologies and information about these technologies should be disseminated in such a way that farmers may fully understand it from theoretical and practical point of view.

22.1 Extension is concerned with two basic assignments:

1. The dissemination of useful and practical information relating to agriculture
2. The practical application of such knowledge to farm.

   The extension methods employed by the extension worker directly influence the effectiveness of his efforts for dissemination of IPM technologies. This is true whether the extension worker is a village extension worker or a subject-matter specialist or whether the learner is a farmer, farm woman, farm youth, or nonfarm person. An understanding of the capabilities and limitations of the available teaching tools is essential to their intelligent selection and efficient use. The extension worker must recurringly exercise judgment in choosing the working tools which he considers appropriate to accomplish the task at hand. The method or combination of methods is sought which is likely to be more effective than other methods in attaining the desired goal. The problem faced by the extension worker in choosing appropriate extension methodologies for dissemination of different IPM technologies is not an easy one. The extension worker is attempting to influence the behavior of large numbers of farmers in life situations which are subject to continual change as the result of economic and social developments. The farmers have diverse interests. They vary greatly in educational training, age levels, and other characteristics that influence their response to educational stimuli.

22.2 Different extension methods suitable for dissemination of IPM technologies are broadly classifieds into following three heads.

A. Individual Contact methods
B. Group Contact Methods
C. Mass contact Methods

Among the above three mentioned methods which method the extension worker needs to use depends upon the nature; number, education etc of the clients i.e farmers. However different extension methods suitable for dissemination of IPM technologies under different situations are explained below.

22.2.1 A. Individual Contact methods

Popular methods suitable for establishing individual contacts for spreading knowledge about IPM technologies are as below:-

22.2.1.1 Farm and home visits

Farm and home visit is a direct, face to face contact by the extension agent with the farmers at their farm for extension work. Under farm and home visit individual contacts are made by the extension worker with the farmer at his/her farm or home for addressing a specific problem e.g how to use IPM technology and offering expert advice to the farmers. The farm or home visit may serve a variety of purposes. It may be in the nature of a service call made upon request to give advice or assistance on a wide range of farm or home problems. The farm or home visit may be merely to extend the agent's acquaintance or it may be part of a planned effort to interest those in the total extension clientele who do not participate in organized extension teaching activities.

Broadly speaking, the farm or home visit is made for the purpose of giving information or obtaining information. From the stand point of the extension agent the visit to the farm or home makes possible the modification of general recommendations about IPM to fit specific situations, thereby increasing the likelihood of their use. Farm and home visits contribute greatly to the effectiveness of the teaching done about Integrated Pest Management.

22.2.1.2 Farmer'call or Office call

Farmer's call is a call made by a farmer at the working place of the extension agent for obtaining information and assistance. Like the farm or home visit the office call involves direct personal contact between the extension worker and the individual desiring information or assistance. The fact that the farmer or homemaker calls on the agent indicates recognition of the problem to be solved and a strong desire to solve it.

22.2.1.3 Personal letters or correspondence

Letter asking for information about IPM should be answered promptly. The person writing the letter has more than passing interest in the matter and will be likely to use information which provides a satisfactory solution to his problem. A good letter should be complete, concise, clear, correct and readable from the farmers point of view about different IPM technologies.

22.3 Group contact methods

The second category of methods involves contact with a substantial number of people assembled in an isolated group or in one of a series of related groups. These include all kinds of meetings for all kinds of purposes. These methods also involve face-to-face contacts.

22.4 Result demonstrations

Result demonstration is a method of motivating the people for adoption of a new practice by showing its distinctly superior results. Results demonstrations play a significant role in the dissemination of IPM technologies among farmers because it presents the results before farmers and they become fully convinced. During the early years of extension work farmers and farm women had to be convinced that extension workers were practical. The result demonstration was unquestionably one of the most convincing arguments employed. Result demonstrations establish proof that the improved practice advocated is applicable locally. Under the direction of the extension agent, the farmer, farm woman, carry out the demonstration. Such demonstrations require a substantial period of time, comparisons are usually necessary, and records are essential. The success of the demonstration depends upon evidence which shows that the new practice is definitely superior to the one it replaces. The more complex the demonstration the greater the difficulty in evaluating the results attributable to each of the practices involved. The simple, clear-cut result demonstrations are, therefore, much more preferable for dissemination of IPM technologies.

The chief purpose of the result demonstration is the establishment of confidence on the part of both the extension worker and the farmer. Having successfully demonstrated that the findings of research apply to a specific farm problem, the extension worker can speak and write of the practice with the conviction of experience. The fact that the practice has been successfully carried out by a neighbor builds confidence of farmers in the soundness of the practice and in the extension worker who directed the demonstration. The result demonstration is one of the most expensive methods of extension teaching due to the large amount of the extension worker's time and travel required to plan the demonstration, select the demonstrator, start the demonstration, keep in touch with its progress, see that it is properly completed, and that sufficient records have been kept to establish proof of the value of the IPM technology demonstrated.

Advantages

1. Gives agent extra assurance that recommendation is practical and furnishes specific local evidence of its advantages.
2. Increases confidence of farmers in Extension and in agent's recommendations regarding IPM.
3. Useful in introducing a new IPM practice.

22.4 Method demonstration

A method demonstration is given before a group of people to show how to carry out an entirely new practice or an old practice in a better way. In contrast to the result demonstration conducted by the farmer or other person under the supervision of the extension worker to prove that the extension recommended practice will work locally, the method demonstration is given by an extension worker or other trained leader for the purpose of teaching a skill to a group. The method demonstration is not concerned with proving the worth of a practice but with "how-to-do" something. That something may be the making of nets for trapping the insects, the adjustment of a tractor plough for deep ploughing. In the role of a skilled technician the extension worker or leader shows the step-by-step procedure in the operation, explaining each succeeding step as he proceeds. The learners watch the process, listen to the oral explanation, and ask questions during or at the close of the demonstration to clear up points about IPM practice in which there is uncertainty, where practicable one or more members of the group repeats the demonstration in the presence of the others. This helps to fix the process in the minds of the audience and increases confidence in their ability to master the technique.

Where the group is small it is frequently possible for all members to participate by performing each step as the demonstration progresses. Each member learns and practices the new skill under the supervision of the teacher. The combining of "seeing" and "hearing" makes a strong impression, further strengthened by practice through participation in the demonstration. The listing of the steps on the blackboard or flannel graph for rapid review and the distribution of outlines, blueprints, or leaflets at the close of the method demonstration are also useful in fixing the steps in the farmer's mind about different IPM practices.

22.5 General meetings

Meetings are held to communicate interesting and useful information about different IPM practices to a large number of farmers at a time. The term "general meetings" includes all kinds of meetings held by extension workers other than method demonstrations and leader training meetings. Geographically, the meeting may be held in a neighborhood, community, or village. The method of presentation may be the lecture or formal talk, informal or formal discussion, or the showing of slides or a motion picture film depending upon the type of farmers. The meeting may be held in a hall, home, or field. Meetings at which method demonstrations are given make it possible for groups of people to learn new or improved skills. General meetings (other than method demonstrations) make it possible for large numbers of people to acquire information about IPM technologies. The group approach makes it possible to share knowledge and experience with others, thereby strengthening learning about IPM. Such meetings serve to attract attention, arouse interest, and strengthen the conviction that something should be done of vital concern to farmers.

22.6 Group discussion

Discussion may be described as the process whereby two or more persons express, clarify, and pool their knowledge, opinions, and feelings. The combined and cooperative thinking of several persons is likely to be superior to that of the individual. Group discussion stands in contrast to the method demonstration used to teach skills and the lecture meeting used to present new technical information about IPM. It is employed to develop desirable attitudes, relate knowledge to experience, deepen understanding, reach agreement, and plan action. Group discussion also facilitates the feedback of the previously disseminated technologies.

22.7 Field Tours

The field tour is in reality a series of field and demonstration meetings arranged in sequence. The scheduling of several such meetings for a single day gets wide publicity and attracts more than usual attention and interest. The tour may be devoted to a single IPM activity. Tour may be planned to include a variety of subjects selected to attract wide participation and acquaint the farmers with important aspects of the IPM program.

22.8 Mass Methods

The third category of methods includes the media employed by extension workers to disseminate information about IPM to large numbers of people. In addition to the personal contact methods and the face-to-face group teaching methods, mass methods enable extension workers to greatly increase their teaching efficiency. Publications, news stories, circular letters, radio, television, ICTs, exhibits, and posters, provide helpful repetition for those contacted personally or through groups. They also make possible the dissemination of information about IPM to a much larger and different category of farmers.

22.8.1 Farm Publications (Bulletins, Pamphlets, Circulars and Leaflets)

Farm publication is a class of publications prepared by the extension agency in printed form, containing information relating to the improvement of farm. Bulletins, pamphlets, circulars, and leaflets dealing with agricultural subject matter continue to play an important role in the present day highly developed system of mass communication available to extension teachers for dissemination of IPM technologies. Publications fit in with and reinforce other methods of reaching and influencing farmers. They are distributed in connection with office calls, farm and home visits, and extension meetings. They are convenient way to answer requests for information received by letter or phone about IPM. They are supplied to local extension leaders and cooperating commercial agencies. They amplify and reinforce the news story, the radio talk, and the telecast; in fact they supply the subject-matter base for many news stories, radio and television programs. The definite and detailed information set forth on the printed page can be studied at leasure, reread at intervals, and kept for later reference.

22.8.2 Radio

Radio is the most widely accessible of all mass media. More than any other medium it has the ability to disseminate information to the largest number of farmers in the shortest time. Extension workers use the radio to inform the public regarding the Integrated Pest Management programmes. Radio programmes are helpful in:-

a. Creating awareness among the people.
b. Help in changing their attitude.
c. Motivating farmers by highlighting the achievements of other farmers.
d. Enhances credibility among the people.
e. Reinforce learning.

22.8.3 Television

Television is an electronic audio-visual medium which provides pictures with synchronized sound. It is more personal than the radio. The viewer meets the speaker in a simulated face-to-face situation even though this relationship is not shared by the speaker. Opportunity is afforded the members of the audience to both see and hear which greatly strengthens the likelihood of learning about any new IPM practice. It combines visual with oral. Over television the extension worker can give a "how-to-do-it" method demonstration about any IPM practice and reach an audience many times larger than the attendance at a meeting. Close up pictures may even make it possible for the viewer to see key operations more clearly than many of those present where the method demonstration is being given. The short talk about IPM can be made more effective with visual aids. Through use of IPM related films and other visual aids, relationships can be clarified and comparisons made between the old and new, the before and after, the good and bad, and the right and wrong.

22.8.4. Exhibits

Exhibits are the visual display of different agricultural technologies. In discussing the exhibit as a distinct method of disseminating the IPM technologies it is necessary to exclude those exhibits which are merely visual aids employed to strengthen the oral presentation at the extension meeting or make the demonstration, the office call, or the farm visit more convincing. The influence of such visual aid exhibits was included in the meeting or other method they supported. Consideration should be given only to those types of exhibits which are primary methods of disseminating IPM related information. The most common such extension exhibit is the one made at the local farmer fairs.

Apart from these, use of ICTs in dissemination of IPM technologies is the most effective and quickest method for reaching to farming community in bulk. ICT help in:-

a. Empowering rural people.
b. Educating rural people.
c. In-depth information.
d. Reducing loss in technology dissemination.
e. Updating the farmers with IPM technology.

Other than radio and television the oldest tool of ICT, new options are now available for dissemination of information about IPM technologies. These include Expert system, video-confercing, tele-confercing, MobileSMS, KissanCallcentres, webportals, websites, telecentres, computer, internet etc.

22.9 Expert system for dissemination of IPM technologies

Expert system

An expert system is a computer programme, which mimics the behaviour of an expert in particular area of knowledge. It is a computer application that solves complicated problems that would otherwise require extensive human expertise.

It has three main components i.e a knowledge base (KB), an inference engine (IE) and a control strategy (CS). For e.g. SOYPEST. The expert system has developed with an objective to provide IPM decision support to the farmer through the internet. (Saini *et al.* 2002)

Advantages of expert system

1. Can be used by extension personnel, researchers and farmers to identify crop diseases.
2. Provide consistent answers to repetitive decisions.
3. Hold and maintain significant level of information.
4. Reduce employee training costs.
5. Centralize the decision making process.
6. Reduce the amount of human errors.
7. Review transactions that human experts may overlook.

Indigenous technological knowledge (ITK) as a tool of farmer to farmer extension for dissemination of IPM technologies.

ITK is an accumulated knowledge, know-how which is unique to a culture, society and community and provide valuable insights to farming system process. It is holistic in nature because it treats entire farming system as an integrated part of environment and bio-diversity. Traditional knowledge lying with the farmers should be exploited for controlling the insect pests of crops in the traditional ways without leaving any adverse impact on the environment. This traditional knowledge should be brought in the fold of IPM because of its numerous merits.

Need for documentation of indigenous technological knowledge/ Practices

1. Sharing the accumulated knowledge of IPM with the farmers help in restoring their pride in their innovative potential.
2. Traditional IPM approaches are environmental healthy and most suitable in area specific farming systems.
3. Dissemination of IPM technologies through farmer to farmer communication would become more feasible if IPM related ITKs of one region is shared in other regions.

22.10 Traditional IPM Practices

1. Inter cropping of marigold (on the bunds) in tomato Marigold is raised on bunds in tomato crop. Farmers believe that attack of the insect-pests lessened in tomato crop and shifted to marigold raised on bunds.
2. Use of *Adhatoda vesica* (Branker) leaves in raising paddy nursery.

Adhatoda leaves are mixed with soil at the time of preparation of field nursery raising. This practice is effective as the leaves of Adhatoda increase soil fertility and act as insecticide and also makes the uprooting of seedlings easier.

22.10.1 Dusting cucurbits with ash to check insect pests.

Cucurbit plants have major problem of insect pest infestation causing damage to plants. Under this measure cucurbits plants are dusted with ash. Ash coming in contact with plant surface forms an outer layering. Dusting of cucurbits with ash has been reported to have beneficial effect in the sense that it effectively checks insect-pests from damaging the plant. It prevents the insects from eating the plant by making it un - palatable to them.

A regular evaluation of IPM is essential to determine the success of pest management strategies, to assess its productivity, effectiveness and efficiencies and to evaluate its achievements.

Extension education stimulates people to make changes that result in better farming. One learns best when there is a strong desire to learn, when the learning is purposeful, and when the effort put forth to learn yields the desired satisfactions.

22.11 Conclusion

- ITK should be used efficiently for promoting IPM.
- Computer literacy should be more strengthened so that farmers can make use of Information Communication Technologies in farm sector as it plays a crucial role in spreading extension reforms in the field of IPM.
- Result and method demonstration should be conducted on large scale for promoting IPM.
- Awareness programme should be conducted on regular basis for the popularization of IPM technologies.
- Exposure visits should be conducted in IPM adopted areas.

References

Dahama O.P and Bhatnagar O.P. 1987. *Education and Communication for Development.* Oxford and IBH Publishing House Co. Ltd., New Delhi.

Kumar and Hansra B. 2000. *Transfer of Technology in Agriculture.* Extension Education for Human Resource Development.

Ray G.L. 1998. *Extension Communication and Management.* Naya Prokosh, 206 Bidhan Sarani, Calcutta.

Saini, H.S., Kamal R. and Sharma A.N. 2002. *Web Based Fuzzy Expert system for Integrated Pest Management in Soyabean.* International Journal of Information Technology. Vol. 8, No. 1 Aug, 2002.

Wilson M.C and Gallup. G. *Extension teaching methods and other factors that influence adoption of agricultural and home economics practices.* Extension Service circular, 495, Dec. 1964 Federal Extension service U.S. Department of Agriculture.

ROLE OF TRANSITION ECONOMY IN INDIAN AGRICULTURE

23

Sudhakar Dwivedi

23.0 INTRODUCTION

India is a global agricultural powerhouse. Agriculture plays an essential role in the process of economic development of less developed countries like India. Besides providing food to nation, agriculture releases labour, provides saving, contributes to market of industrial goods and earns foreign exchange. Agricultural development is an integral part of overall economic development. In India, agriculture was the main source of national income and occupation at the time of Independence. Agriculture and allied activities contributed nearly 50 percent to India's national income. Around 72 percent of total working population was engaged in agriculture. Since independence India has made much progress in agriculture. Indian agriculture, which grew at the rate of about 1 percent per annum during the fifty years before Independence, has grown at the rate of about 2.6 percent per annum in the post-Independence era. Expansion of area was the main source of growth in the period of fifties and sixties after that the contribution of increased land area under agricultural production has declined over time and increase in productivity became the main source of growth in agricultural production (Tripathi, 2009).

Evidence of the presence of wheat and some legumes in the 6th millennium BC have been found in the Indus Valley. Oranges were cultivated in the same millennium. The crops grown in the valley around 4000 BC were typically wheat, peas, sesame seed, barley, dates and mangoes. By 3500 BC cotton growing and cotton textiles were quite advanced in the valley. By 3000 BC farming of rice had started. Other monsoon crops of importance of the time were cane sugar. By 2500 BC, rice was an important component of the staple diet in Mohenjodaro near the Arabian Sea. The Indus Plain had rich alluvial deposits which came down the Indus River in annual floods. This helped sustain farming that formed basis of the Indus Valley Civilization at Harappa. The people built dams and drainage systems for the crops.

By 2000 BC tea, bananas and apples were being cultivated in India. There was coconut trade with East Africa in 200 BC. By 500 AC, egg plants were being cultivated. After independence, considering India's growing population, the government took steps to increase the food production. Yields per unit area of all crops have grown since 1950. The 1970s saw a huge increase in India's wheat production. This is known as the Green Revolution in the country. Reasons for the growth are the special emphasis placed on agriculture and steady improvements in irrigation, technology, application of modern agricultural practices and provision of agricultural credit and subsidies. Operation Flood was the name of a rural development programme started by the National Dairy Development Board (NDDB) in 1970 with the objective of creating a nationwide milk grid. This movement followed the and helped in alleviating poverty and famine levels from dangerous proportions in India during the era. It resulted in India becoming the largest producer of milk and milk products, so it is also called the White Revolution of India. India is the largest producer in the world of milk, cashew nuts, coconuts, tea, ginger, turmeric and black pepper. It also has the world's largest cattle population (193 million). India ranks second worldwide in farm output. It is the second largest producer of wheat, rice, sugar, groundnut and inland fish. It is the third largest producer of tobacco. India accounts for 10 per cent of the world fruit production with first rank in the production of banana and sapota. Despite high growth, international comparisons reveal that the average yield in India is generally 30% to 50% of the highest average yield in the world. The low productivity in India is a result of the following factors:

Illiteracy, general socio-economic backwardness, reforms and inadequate or inefficient finance and marketing services for farm produce. The average size of land holdings is very small (less than 20,000 m^2) and are subject to fragmentation, due to land ceiling acts and in some cases, family disputes. Such small holdings are often over-manned, resulting in disguised unemployment and low productivity of labour. Adoption of modern agricultural practices and use of technology is inadequate, hampered by ignorance of such practices, high costs and impracticality in the case of small land holdings. Irrigation facilities are inadequate, as revealed by the fact that only 53.6% of the land was irrigated in 2000–01, which result in farmers still being dependent on rainfall, specifically the Monsoon season. A good monsoon results in a robust growth for the economy as a whole, while a poor monsoon leads to a sluggish growth. Farm credit is regulated by NABARD, which is the statutory apex agent for rural development in the subcontinent.

In the last few decades several farmers have committed suicide especially in the states of Andhra Pradesh, Maharashtra, Karnataka and Kerala. Combating this has become a major challenge for these governments. Some of the causes for the deaths include indebtedness of small and marginal farmers and repeated crop failures.

At the time of Independence, Indian agriculture was not able to fulfil the demands of the people and food production was low. There was widespread hunger. The situation was worse in years of flood and drought as production fell further and with it the incomes of a large number of rural people. People died of disease and hunger and the British government did not organise sufficient food stocks for the poor in rural areas. Agriculture (about 55% of the land is arable) makes up some 25% of the gross domestic product (GDP) and employs almost 70% of the Indian people. Vast quantities of rice are grown wherever the land is level and water plentiful; other crops are wheat, pulses, sugarcane, jowar (sorghum), bajra (a cereal), and corn. Cotton, tobacco, oilseeds, and jute are the principal non-food crops. There are large tea plantations in Assam, Karnataka, Kerala, and Tamil Nadu. The opium poppy is also grown, both for the legal pharmaceutical market and the illegal drug trade; cannabis is produced as well. Fragmentation of holdings, outmoded methods of crop production, and delays in acceptance of newer, high-yielding grains were characteristic of Indian agriculture in the past, but since the Green Revolution of the 1970s, significant progress has been made in these areas. Improved irrigation, the introduction of chemical fertilizers, and the use of high-yield strains of rice and wheat have led to record harvests, and India became a net exporter of grain in the early 1980s. The subsistence-level existence of village India, ever threatened by drought, flood, famine, and disease, has been somewhat alleviated by government agricultural modernization efforts, but although India's gross food output has been generally sufficient for the needs of its enormous population, government price supports and an inadequate distribution system still threaten many impoverished Indians with hunger and starvation. An estimated 40% of the population is too poor to afford adequate nourishment regularly. India has perhaps more cattle per capita than any other country, but their economic value is severely limited by the Hindu prohibition against their slaughter. Goats and sheep are raised in the arid regions of the west and northwest

India is predominantly an agrarian country with nearly three fourths of the people dependent on agriculture or rural economy. The most outstanding achievement of Indian agriculture since independence is the phenomenal growth of food grains output. During the last three decades, Indian agriculture has experienced a revolutionary breakthrough in food grain production leading the country from deficit and import arena to the positive situation of self-sufficiency and buffer stocks. The food grain production in the country increased from 50.8 million tonnes in 1950-51 to 152.37 million tonnes in 1983-84, but the growth of Indian agriculture still continues to be linked with the vagaries of nature. Some of the states in the country have come across the unprecedented draught of the century for the fourth successive year which has caused tremendous hardship to the people as also loss of production of food grains in those states.

Nearly seventy percent of the total production of food grains in India is retained at farm level where the unscientific and faulty storage conditions enhance the chances of fungal attack and thereby mycotoxin production. The decomposers of food grains i.e. fungi, bacteria etc. are always present on food grains in dormant conditions (usually as spores) and grow under favourable climatic and other conditions. The fungal growth may cause decrease in germinability, discolouration of grain, heating and mustiness, loss in weight, biochemical changes and production of toxins. All these changes may occur before the responsible fungi could be detected on visual examination The fungi produce a large number of mycotoxins in food grains and their products. Mycotoxins are a group of highly toxic secondary metabolises of the fungi produced under certain favourable environmental conditions Because of their potent toxic nature and fairly common occurrence under natural conditions, mycotoxins have attracted world- wide attention in the recent years. The diseases or physiological abnormalities resulting due to ingestion of mycotoxins are known as "mycotoxicosis".

Looking back into the history of Indian agriculture after independence, the first major challenge nation faced was to produce adequate food grains for its growing population. We had to depend on aid and grants of agriculturally developed nations and international organisations for our food requirements. After a long and sustained effort, real independence from the hunger was achieved by the end of sixties when our grain production touched 108 million tonnes in 1970-71 after 'Green revolution', when compared to 50.8 million tonnes of grain production in 1950-51. This was a real scientific breakthrough in plant genetic & breeding. The growth trend of production continued with 130mt, 176mt, 198mt in the year 1980-81, 1990-91 and 1996-97 respectively. India has certainly emerged as a winner on the front of self-sufficiency of the food production. While going through the history of Indian agriculture, the organized extension dates from 1940's when the nation felt problem of feeding the growing population and faced problem of food supply.

i. The first phase dates from year 1948-1960 as Extensive Extension Programmes, which included:
   a. Grow More Food Campaign - 1948,
   b. Community Development Programme - 1952.
ii. The second phase, Intensive Extension Programmes (1960-1974), included:
   a. Intensive Agricultural District Programme (IADP)-1960,
   b. Intensive Agricultural Area Programmes(IAAP)-1966,
   c. High Yielding Variety Programme (HYVP) -1966.
iii. The third phase started with programmes for Research based Extension Methodology viz:
   a. National demonstration programme - 1965,
   b. Operational Research Project (ORP) - 1971,
   c. Lab to Land Programme (LLP) - 1979
iv. and the fourth phase with introduction of World Bank aided Training & Visit (T&V) approach for extension, through following three projects:
   a. State Agricultural Extension Projects (T & V), 1974-75
   b. National Agricultural Research Project (NARP), 1980-88
   c. National Agricultural Extension Project (NAEP), 1985-88 and
   d. the latest National Agricultural Technology Project initiated in year 1998.

In fiscal year 1950, agriculture, forestry, and fishing accounted for 58.9 percent of the gross domestic product (GDP *see* Glossary) and for a much larger proportion of employment. Manufacturing, which was dominated by the jute and cotton textile industries, accounted for only 10.3 percent of GDP at that time.

Ten Measures for developing the Agricultural sector in India Since the dawn of independence, several steps have been taken to develop the agricultural sector of the country. The major breakthrough has been achieved in food grains production.

The production of food grains which was 550 lakh tonnes in 1950 substantially moved to 1991 lakh tonnes in 1995. However, the various measures employed from time to time can be discussed as:

The various technical measures employed to develop agriculture are as under:

1. **Multiple Cropping:** Multiple cropping aims at maximizing production per unit of land and per unit of time by taking three or four crops in a year. By adopting multiple cropping, there are two advantages as of getting increased returns and economy of the farm resources.
2. **Expansion of Irrigation Facilities:** Irrigation facilities have increased manifold over time. Several, minor, medium and major irrigation projects have been launched in the country. At the inception of First Five Year Plan, India had only 18% of total irrigated area which at present increased to about 33.9 percent. Moreover, dry farming has also been introduced in those areas where means of permanent irrigation cannot be installed. In 1994-95 the country witnessed total irrigated area of 876 lakh hectares.
3. **Use of HYV Seeds:** HYV seeds have absolutely revolutionized Indian agriculture by increasing yield per acre. Among these, mention may be made of dwarf varieties of wheat PU-18, Kalyan Sona 227, Sona Lika, Hybrid maize, Vijay, Rice I R-8, Jhona 351, Padma and Jaya etc.
4. **Plant Protection:** Considerable efforts have been made to protect the crops from the insects and pests. For this purpose, 14 Central Plant Protection Centres have been set up by the Govt.
5. **Scientific Methods of Cultivation:** In the planning period, stress has been laid on the scientific methods of cultivation. It has been emphasized to adopt superior agricultural technology in respect of crop rotation, selection of quality seeds, use of proper manure, treatment of soil, selection of crops etc. In this regard, Govt has initiated Intensive Agricultural Area Programme. Moreover, several Agricultural research centers and universities have also been established. In this regard, Haryana Agricultural University Hissar, Punjab Agricultural University Ludhiana, Himachal Agricultural University Palampur, ICAR, Delhi is playing a pioneer role to develop agriculture.
6. **Use of Mechanization:** Mechanization is another noteworthy step employed to develop agriculture. Small farmers are assisted with cheap credit facilities through co-operative societies, community development blocks to purchase machinery and other modern equipments.
7. **More Use of Chemical Fertilizers:** Use of chemical fertilizers has also contributed significantly to the growth of agricultural output. Several steps have been taken to encourage the use of cow- dung as manure rather than as fuel. In 1950-51, 0.13 million tonnes of chemical fertilizers was used which in 1980-81 increased to 5.52 million tonnes and further to 12.54 million tonnes in 1990-91. In

1995-96, the use of chemical fertilizers was recorded to the tune of 15.7 million tonnes.

8. **Development of Agricultural Land:** Efforts have been made to develop agricultural land during the five year plans. Major success has been achieved in the levelling of land, terracing of fields and contour building. Land surveys are also being conducted.
9. **Animal Husbandry:** Animal husbandry has assumed a much broader role in the overall agricultural development. Presently, this sector accounts for 25% of gross value of agricultural output. India's vast livestock population offers tremendous potential for meeting domestic demand for milk, egg, meat, wool, etc.
10. **Land Reforms:** In a bid to increase agricultural productivity, land reforms are of immense use. Since the dawn of independence, Govt, of India has undertaken several land reform measures. For instance, Abolition of zamidari system, Fixation of ceilings on Land Holdings, Consolidation of Land Holdings, co-operative farming etc.

References

Dwivedi, S. and Bhat, A. 2014. Crisis in Indian Agriculture: Present Scenario and its Remedial Measures in the *Edited Book entitled "Indian Agriculture: Issues & Challenges"* published by Kunal Books, Daryaganj, New Delhi.

Dwivedi, S. and and Sharma, M. 2011. Agricultural Marketing System and Status of Indian Agriculture published in the *Edited Book entitled, "Technological Innovations in Dryland Farming Systems"* published by IBDC publisher, Lucknow.

Rao, C.H. Hanumantha. 2000. Declining Demand for Foodgrains in Rural India: Causes and Implications. *Economic and Political Weekly,* 22 (January): 201-206.

Tripathi, A. and Prasad, A. R. 2009. Agricultural Development in India since Independence: A Study on Progress, Performance, and Determinants. *Journal of Emerging Knowledge on Emerging Markets.* 1(1): 1-33.

INTEGRATED PEST MANAGEMENT: CHALLENGES AND FUTURE STRATEGIES

24

D. P. Abrol and Uma Shankar

24.0 INTRODUCTION

A spouse indicated the marriage partner, he or she called "our wife or our husband", not "my wife or my husband," Closing gaps in crop yield while enhancing sustainability is among the greatest challenges for achieving food security (Garibaldi, 2015).

It is beyond doubt that Integrated Pest Management is the powerful tool in the hands of farming community engaged in sustainable agriculture. In one hand, it has played a crucial role in enhancing the yield potential in the sustainable way to meet out the burgeoning population demand and on the other to keep the natural resources (land and water) and environment safe for the future farming generations to come.

But, the scenario has been different in many developing countries having small area to feed or support the rapidly growing population, where IPM has started some 20 years late rather than developed countries and could not keep the essence of IPM intact in achieving the mandate of an IPM programme.

Chemical pesticides have been set as the last resort in IPM practices, but farmers still applying the pesticides as the first resort for getting quick results. At the same time, pesticides proved to be more dangerous due to their indiscriminate and excessive use, contaminating food (milk, honey, cereals, vegetables, and fruits) and the environment (ground water, soil, etc.), resulting in pest resistance, pest resurgence, loss of beneficial fauna, pesticide poisoning and frequent localized pest outbreaks (Abrol and Shankar, 2014).

We have travelled a long way of five decades since the inception of economic threshold level by Stern *et al.* in 1959 and at this juncture we have to redefine and analyze the challenges and strategies of IPM particularly in developing countries to boost the safe and sustainable food production.

Growing population pressure has hastened environmental degradation and depletion of essential natural resources (UNFPA, 2008). Nearly 1 billion people in the world are undernourished or suffer from chronic diseases as a result of food insecurity due to population growth, climate change, and urban development (FAO, 2010). In the next 50 years, the global population is expected to reach 9 billion, doubling the demand for food, feed, and crop (FAO, 2009). According to World Health Organization (WHO), more than 20,000 unintentional deaths and 3 million poisonings occur due to misuse of pesticides in the Third World every year (Lowel, 1998).

Available estimates reveal that global pesticide use has increased 50-fold since 1950, and 2.5 million tonnes of agricultural pesticides are now used each year worldwide (Tyler, 2002). No doubt pesticides are credited to have saved millions of lives by controlling diseases, such as malaria and yellow fever which are insect borne; however, their use causes a variety of adverse health effects and environmental pollution as described above.

Alternate pest control methods and the restricted use of pesticides can minimize the risk of pesticide usage (Soundararajan, 2012). Pesticides can prove to be the most effective instruments in crop protection and if correctly used, their effect is fast and complete, which makes them applicable against nearly every pest (Oomen and Bouma, 2003).

However, the introduction of high inputs of agro-chemicals during the Green Revolution era have proved to be more dangerous due to their overuse, which resulted in the deterioration of soil and plant health, pesticide contamination of food and the environment, and pesticide resistance and pest outbreaks.

Consequently, the need arose for IPM strategies to produce safe food and reduce the negative externalities caused by pesticides (Kogan, 1998; Wilson and Tisdell, 2001; Abrol and Shankar, 2012; Shankar and Abrol, 2012). In the near future, these challenges are expected to worsen further if measures are not taken to address them.

24.1 Scenario of IPM in India

In India, Integrated Pest management has mostly remained as an academic exercise, and could not be able to leave the long term impressions in terms of its implementation and adoption among the farming community. It is the matter of grave concern that farmers are not willingly coming forward to adopt the IPM technology due to lack of proper knowledge.

World-wide integrated pest management (IPM) is the accepted policy decision for pest management. However, in reality this often becomes "integrated pesticide management". The strategy of IPM and its implementation and evaluation has always struggled with interpretation and true progress (Peshin and Pimentel, 2014). There are several challenges in successful implementation and adoption of Integrated Pest management programme which include-

24.2 Scenario of Pesticide Pressure or Integrated Pesticide Management

Worldwide, about 4 million tons of pesticides per year are used although their distribution is uneven in different countries (FAOSTAT, 2010). Despite the fact that the consumption of pesticides in India is still very low, about 0.5 kg/ha of pesticides against 6.60 and 12.0 kg/ha in Korea and Japan, respectively, there has been a widespread contamination of food commodities with pesticide residues, basically due to non-judicious use of pesticides. In India, 51% of food commodities are contaminated with pesticide residues and out of these, 20% have pesticides residues above the maximum residue level values on a worldwide basis (Gupta, 2004).

Among the various pesticides used in India, 40% of all the pesticides used belong to organochlroine class of chemical pesticides (FAO, 2005; Gupta, 2004). The other major category is organophosphate pesticides. Regarding the usage of technical pesticides, insecticides account for 80% of total pesticide used in the country, followed by herbicides and fungicides. Globally herbicides are the leading category followed by insecticides and fungicides. In India, the share of herbicides is insignificant (MAFF, 1999-2000). Lack of training in pesticide use, ignorance about potential dangers to health and environment, poor literacy, inappropriate mixing and application methods, repeated and excessive dosage of pesticides application are some related factors relevant to pesticide usage and safety measures in India (Abhilash and Singh, 2009).

Although pesticides are generally profitable in agriculture, their use does not always decrease crop losses. For example, despite the more than 10-fold increase in insecticide (organochlorines, organophosphates, and carbamates) use in the United States from 1945 to 2000, total crop losses from insect damage have nearly doubled from 7 to 13% (Pimentel *et al.* 1991). Still farming community is using the excessive and indiscriminate use of pesticides. Pesticides continue to play the key role in arthropod pest management. However, some key species (*Heliothis armigera* Hubner, *Plutella xylostella* L.) are developing high levels of insecticide resistance and with the human and environmental health concerns related to pesticide use, cultural techniques are increasingly being used. The combination of genetic resistance, hygiene, and monitoring of crops for threshold levels of infestation, allows the most economic and effective use of chemical controls with the result that economic yields can be maximized.

24.3 Environmental degradation/losses

Extensive and indiscriminate usage of chemical pesticides has resulted in environmental degradation, adverse effects on human health and other organisms, eradication of beneficial insects and the development of resurgence and resistant to pesticides in insects pests. Reliance on single control tactics have resulted in environmental degradation, contamination of food products, problems of residues and resistance in target species, thereby seriously impairing the sustainability (Vega *et al.*, 2009; Miller *et al.*, 2010).

It is therefore essential to devise a sound management system that is based on ecological principles resulting in sustainable agricultural production without disturbing the balance of nature (Overton, 1996; Lewis *et al.*, 1997; Kennedy and Sutton, 2000).

24.4 Lack of proper taxonomy of Pests and Biocontrol agents

Identification or proper diagnosis is the pre-requisite in the IPM programme. We often talk of biodiversity loss, but unless we have a detailed account of the existing species, the loss cannot be pinpointed.

Numerous examples can be cited where wrong classification has led to misinterpretations. Taxonomy plays an important role in the management of pests and weeds. To illustrate this point, *Salvinia molesta* (kariba weed), native of Brazil, is an aquatic fern and one of the world's worst weeds. The environmental damage caused by it has been enormous. It chokes lakes, reservoirs, slow-moving rivers, irrigation systems, rice paddies, fish ponds, etc. with continuous metre-thick mats of dense vegetation. In addition to rendering the water useless for normal use, its presence can lead to the breeding of mosquitoes. Initially, the weed was identified as *Salvinia auriculata.* A weevil, *Cyrtobagous singularis*, from Trinidad was used in Africa to control it, but the effort failed. Later, this weed was identified as *S. molesta*, whose growth in Queensland was controlled by *Cyrtobagous* from Brazil. It is evident from these examples as to how effective control or mitigation measures could be implemented2.

Similarly, identification of an effective biological control agent for Azolla depended on expert taxonomic work. Floating water fern/fairy fern (*Azolla filiculoides*) has for years been a highly effective invasive species in South Africa, creating problems in inland waterways. The weevil, *Stenopelmusrufinasus* was found effective in cleaning up sites heavily infested with *Azolla* within months. Proactive taxonomy of biotypes of whiteflies, *Siphoniusphillyreae* and *Bemisiatabaci* causing viral epidemic in crops in Argentina allowed effective implementation of biological control programmes via natural enemies such as *Encarsiahispidia*, *E. protransvena* and *E. transvena* (Viscarret *et al.*, 2003).

In spite of its importance as an inevitable field for all types of research, taxonomy faces many challenges and remains a neglected subject.

There is no national repository centre, museum, or maintenance of taxonomic collections. Many museums have no curators and several universities have no faculty positions for taxonomists. A demand for taxonomy is on the rise in the wake of the global biodiversity crisis. There is a need to promote scientific identification and documentation requires immediate attention.

Due to lack of infrastructure at academic institutions and a well-structured research programme, seldom are students willing to pursue taxonomy. There are a handful of job openings for taxonomists and pay scale is low. Further, inadequate funding in taxonomy largely diverts students to studying phylogeny. It is also difficult to publish in the field as there are only a few journals dedicated to taxonomy research. The miniscule journals however do not accept large revisionary work or monographs.

This calls for building a network among institutions and organizations involved in taxonomic collections. It is high time that the national institutes and funding agencies encourage taxonomic work and provide financial assistance to strengthen the knowledge base in taxonomy (Abrol, 2013).

24.5 Problems in Pest Management in context of Climate Change

Climate change no longer is a matter of opinion or speculation. Of concern now is the assessment of the extent of the changes and their potential impacts. Climate change, food insecurity, and energy demand are major concerns for modern agriculture and their impact is increasing rapidly. The last decade has seen new developments in food production: the genetic engineering of organisms and the organic chemical-free agriculture. Biotechnology and release of genetically modified organisms (GMOs), such as engineered soybean, colza, maize, and tomatoes, did promise a solution to food security needs and nutritional problems (Khush, 2002). According to the main private biotechnology companies (Aventis, Monsanto, Novartis, Zeneca, etc.), these GMOs may be resistant to insect pests, molds, frost, dry conditions, etc., and could revolutionize agriculture (Pingali and Traxler, 2002). For example, soybean and other plants were modified to be tolerant to glyphosate, a common herbicide used to fight weeds allowing for much higher crop yields. However, because the weeds become increasingly resistant to this herbicide, the use of these genetically modified (GM) plants renders the farmers dependent on the use of more and more glyphosate.

24.6 Lack of Information and Communication Technology (ICT) in IPM

Practically, IPM is concerned with most constitutional levels of agroecosystems, from populations and communities down to individual viruses or genotypes, and genomes and genes, as well as up to the levels of landscape and global ecosystems. In fact, IPM practices are involved in a complex course responding to climatic change, soil dynamics, vegetation evolution and human activities.

ICT is any technology that enables communication and the electronic capture, processing and transmission of information. Radio, television and print media are vital in many developing countries. Over the last decade, 'new' ICTs, such as mobile phones and Internet-associated applications such as Voice over Internet Protocol (VoIP), have become available to growing numbers worldwide. Developing countries face challenges when harnessing the potential of ICT for economic development (Michelle and Fong, 2009). The potential of ICT for the speedy dissemination of information to farmers needs to be realized (Meera *et al.*, 2004). Integrated pest management (IPM) practices have to solve many problems of sensitivity and intractability in the sustainable development of agriculture. ICT has proved to be a powerful tool in pest forecasting as a prop to giving priority to prevention, as pest forecasting involves data acquisition, processing and information dissemination. ICT can also be very helpful in terms of enforcing IPM (Shen *et al.*, 2012).

24.7 IPM Implementation and adoption

In the 1970s and 1980s, the first IPM program under the Operational Reseach Project (ORP) focused on pilot programs using a prescriptive approach to demonstrate IPM practices in cotton and rice crops in a cluster of villages in seven states. The government of India adopted IPM as the main strategy for plant protection in 1985. In the early 1990s, the farmer field school (FFS) model was adopted to implement IPM by educating farmers and extension workers. Between 1990 (before many adhoc IPM programs began) and 2002 (when *Bt* cotton was introduced) pesticide use active ingredient (a.i.) by weight decreased by35 %, mainly because hexa-chloro-cyclohexane, accounting for 30% of the total pesticides, was banned in 1997 and low-dosage pesticides were introduced. Only about 2–4% of the total cultivated area, including only 5% of the farmers, however, is covered under IPM programs, so whether IPM has reduced overall pesticide use in Indian agriculture is debatable. Although the introduction of *Bt* cotton has reduced insecticide use in cotton by almost 50%, mass pesticide use in Indian agriculture overall has increased by 9% since 2002 (Peshin *et al.*, 2014).

Bt crops are compatible with IPM strategies but *Bt* crops alone are not sustainable. Over reliance on transgenic crops has already led to the weed and insect resistance which may lead farmers into a transgenic-cum-pesticide treadmill. Experiences with implementation of pesticide action plans and IPM programs around the world confirm that reduction in pesticide use by mass is not the robust indicator to measure success of IPM.

Low volume pesticides propelled the pesticide use reduction in many countries. The pesticide treatment frequency index and the environmental impact quotient are better evaluation indicators tomeasure the impact of IPM programs.What experiences with IPM technology and IPM extension that are documented in this book can be bracketed successful and viable? In many instances IPM technologies developed at the research level have not been effectively scaled upto industry-wide practice because of the lack of a well conceived and evaluated extension process. Different extension approaches are needed in different situations for greater adoption of IPM by the farmers. IPM practices in most cases are tested for success at pilot scale but fail to factor in the constraints, mainly the IPM attributes, for replication in large scale (Peshin and Pimentel, 2014).

Some of the strategies which need to be adopted to make the IPM programme successful include-

24.8 Ecological Pest Management Programme

The aim of this new approach is to shift management strategies so that they have less reliance on chemicals and more on the biology of pest sand their interactions with crops. Thus, ecologically-based IPM combining all approaches – physical, cultural, chemical and biological – is the only option for sustaining productivity and maintaining the health of ecosystems (Kennedy and Sutton, 2000). Most of the IPM inventories like cultural, mechanical, biological, plant resistance and biotechnological approaches are basically compatible and supportive tactics in the IPM strategy. Hence, strengthening of IPM infrastructure, especially for surveillance and forecasting the outbreak of localized insect pests and diseases and mass multiplication of bio-control agents for filed use, should be given adequate attention and priority. IPM programmes need to be designed in a way to manage pests on the one hand and ensure the build-up of beneficial organisms on the other (Kennedy and Sutton, 2000).

24.9 Bio-Intensive Integrated Pest Management (BIPM) Programme

BIPM is the recent trend to reduce the pesticide pressure and to enhance the farmers' interest to fetch higher remuneration for their produce. Consumers Union of the United States Department of Agriculture defines biointensive IPM as, "A systems approach to pest management based on understanding of pest ecology. It begins with steps to accurately diagnose the nature and source ofpest problems, and then relies on a range of preventive tactics and biological controls to keep pest populations within acceptable limits. Reduced-risk pesticides are usedif other tactics have not been adequately effective, as a last resort, and with care tominimize risks."

BIPM also opens new horizons to opt for a better choice, as biological control and use of bio rational products, which are less toxic and only affect the target pest. It also includes biopesticides derived from microbials, parasitoids, predators, botanicals and all conventional non-chemical methods of pest management. It refers to the more dynamic and ecologically informed approach to IPM that considers the farm as a vital part of an agro-ecosystem. BIPM approach is a biological form of agriculture in which as mall area is intensively cultivated, using natural ingredients to rebuild and revitalize the soil health. Initially, it is more labor intensive than conventional approaches, therefore, it is better suited to the small farm areas. It lays emphasis on the use of predominantly indigenous or region specific cultivars, crop diversity, good soilhealth and water conditions which results in little or no pest problem (Kaul *et al.*, 2009).

24.10 Ensuring the availability of Biological control agents and IPM devices

Efforts shall also be made to ensure the timely availability of Biological control agents on demand to farmers to help them adopt IPM in the true spirit by encouraging the private sector, Government organizations, central and state agricultural universities in providing such support services. Government must have to strengthen the programme through some new policy to establish more number of regional biocontrol laboratories, encourage the use of bio-pesticides and bio-control agents and safer and efficacious quality IPM devices or products for the farming community.

24.11 Habitat Manipulation or Ecological Engineering

It is evident that the crop diversification tends to increase natural enemy abundance and diversity, providing a system more resilient to pest population increase. Overall farming diversity within the agroecosystem may also affect biological control by natural enemies, due in part to a wider range of flowering plants that provide nectar (carbohydrate) and pollen (protein) resources to insects during more times of the growing season. Thus, pest outbreaks tend to be less common in polycultures than in monocultures (Andow, 1991). The response of beneficial insect populations to habitat manipulation depends upon their ability to use or exploit one or more of the plant components of the agroecosystem (Altieri and Nicholls, 2004). Flowering plant strips adjacent to fields help support beneficial insect biodiversity in agricultural landscapes (Baggen and Gurr, 1998; Carreck and Williams, 2002; Fiedler and Landis, 2007a, 2007b; Tuell *et al.*, 2008). Much of the testing of flowering plants has been done with non-native annual or biennial flowering species, although these often blooming one growing season requiring annual sowing. A well-designed flowering border adjacent to a crop field will provide necessary resources and alternative food source for natural enemies during periods when crop pest and crop flower numbers are low, thus maintaining high populations of natural enemies supported by the provision of nutrients throughout the season (Landis *et al.*, 2000).

24.12 Farmers demonstration through Community participation for enhancing the adoption of IPM

The dangers of acute pesticide poisoning are well known to Indian farmers, but the need is to highlight the chronic exposure which can cause many neurological diseases. The current need is to address the awareness for long term moderate exposure. Further, Indian farmers are lagging behind in the use of new application technologies for pesticides spraying. The crux of the problem is that though, the technology is available in the country and is being used by few progressive farmers, majority of them are not in a position to implement these new technologies either due to lack of awareness or means. The farmers need to be educated by the crop protection industry since they are the main end-users of the products. This realization leads to development of indigenous technology as required by the farmers (Pina and Forcada 2004; Falconer and Hodge, 2000). Farmers in their quests to controlling pests and increasing yields ignore the problems of health and environment hazard.

24.13 Avoidance and excessive use of pesticides

Currently, the pesticides are being used on 25% of the cultivated area and, the consumption of pesticide is showing a slight declining trend, probably due to shift of farmers toward biopesticides, natural plant sources and other alternative methods. Despite such a large consumption of pesticides, it is estimated that crop losses vary between 10–30% due to pests alone. In monetary terms, these losses amount to Rs. 290,000 million per year (Agnihotri, 1999). It has been observed that their long-term, low-dose exposure are increasingly linked to human health effects such as immune-suppression, hormone disruption, diminished intelligence, reproductive abnormalities, and cancer. In this light, problems of pesticide safety, regulation of pesticide use, use of biotechnology, and biopesticides, and use of pesticides obtained from natural plant sources such as neem extracts are some of the future strategies for minimizing human exposure to pesticides.

24.14 Rational use of new interventions and Biotechnological Approaches

New technological interventions such as transgenics and the offing of biorational compounds may become the more suitable choice for the farming community for maintaining the balance in an agro-ecosystem for sustainable agriculture. These new interventions may prove to be environmentally safe, economically sound and socially viable in a long run.

Major technological advances in biochemistry, molecular biology, genetics and biotechnology have facilitated in designing plants with improved resistance to insect pests. The potential of genetic transformation technology has been widely recognized over the last two decades. Recent advances in molecular biology, plant tissue culture and genetic engineering have clearly demonstrated the possibility of incorporating foreign genes for desired characters while preserving theexisting traits of improved genotypes. *Bt* is a Gram positive, aerobic, sporulating bacterium that synthesizes crystalline proteins during sporulation. These crystalline (Cry) proteins are highly insecticidal at very low concentrations to lepidopterans, coleopterans, dipterans, lice, mites and nematodes, and are non-toxic to mammals and other organisms.

By using biotechnological approaches, it has become possible to use *Bt* more effectively and rationally by introducing the insecticidal crystal proteins (ICPs) of *Bt* into crop plants. The crystal toxins are classified on the basis of amino acid sequence homology. The first transgenic tobacco plants using cry genes were developed against tobacco hornworm in 1987. The commercialization of *Bt* transgenic crops started in 1996 with the introduction of bollworm-resistant cotton (Bollgard) carrying the *Bt* toxin gene in the USA, while *Bt* cotton was commercialized in India in 2002. Plants possess a wide array of defense proteins including the PIs and lectins induced in response to insect attack.

24.15 Organic Way to get safer and Remunerative foods/Organic farming

The principal essence of organic food production aims to encourage and enhance biological cycles within farming systems to maintain and increase long term fertility of soil, to minimize different forms of pollution caused by synthetic chemicals (fertilizers and pesticides) and to produce food of high quality in sufficient quantity. It has been reported that more than 75% of consumers are very concerned regarding pesticides residues in food. All these issues have a direct bearing on the marketing of the products (Hallman *et al.*, 2003). However, this process is costly, quite labor intensive, and in some cases ineffective (Abhilash and Singh, 2009).

Sikkim became the 1st state in India to gone organic which portrays the sound picture of state to promote the pesticide free Agri-Horticulture produce and simultaneously Agri-Tourism to provide job opportunities to the rural masses. Organic scenario will definitely assist in maintaining the balance of an ecosystem, promotion for safe and residue free produce and thereby fetching the enhanced remuneration from their organic certified produce.

24.16 Conclusion

Keeping the above facts in mind, it is imperative to properly educate farmers for judicious use of pesticides. Other alternatives such as use of biotechnology, use of biopesticides, and use of pesticides obtained from natural products such as neem, pongamia oil etc. should be encouraged.

Secondly, Government should have to frame policy to propagate and strengthen the adoption of IPM to the farmers through demonstration of IPM technologies which will not only ensure the production of safe food but also reduce economic losses resulting through contamination of food, degradation of environment, ecosystem, and indirect losses to the human health.

References

Abhilash, P. C. and Singh, N. 2009. Pesticide use and application: An Indian scenario. *Journal of Hazardous Materials*, 165: 1–12.

Abrol, D. P. 2013. Future of taxonomy in the 21st century – whither or wither. *Current Science*, 104(12): 1594.

Abrol, D. P. and Shankar, U. 2012. History, Overview and Principles of Ecologically-based Pest Management, pp. 1–26. In: Integrated Pest Management: Principles and Practices (Abrol and Shankar eds.) CAB International 2012, UK. ISBN : 9781845938088.

Abrol, D. P. and Shankar, U. 2014. Pesticides, Food Safety and Integrated Pest Management, pp. 167–199, *In*: Pimental, D., Peshin, R., (eds.) Integrated Pest management- Pesticide problems, Springer Science+Business Media.

Agnihotri, N. P. 1999. Pesticide safety Evaluation and Monitoring, Division of Agriculture chemicals, Indian Agricultural Research Institute, New Delhi p. 14.

Altieri, M. A. and Nicholls, C. I. 2004. Biodiversity and Pest Management in Agroecosystems (2nd ed.), Food Products Press, ISBN 1560229255, New York.

Andow, D. A. 1991. Vegetational diversity and arthropod population response. *Annual Review of Entomology*, 36: 561–586.

Baggen, L. R. and Gurr, G. M. 1998. The influence of food on *Copidosoma koehleri* (Hymenoptera: Encyrtidae), and the use of flowering plants as a habitat management tool to enhance biological control of potato moth, *Phthorimaea operculella* (Lepidoptera: Gelechiidae). *Biol. Control*, 11:9–17.

Carreck, N. L. and Williams, I. H. 2002. Food for insect pollinators on farmland: insect visits to flowers of annual seed mixtures. *Journal of Insect Conservation*, 6: 13–23.

Falconer, K. and Hodge, I. 2000. Using economic incentives for pesticide usage reductions: responsiveness to input taxation and agricultural systems. *Agric. Syst.*, 63: 175–194.

FAO. 2005. The state of food insecurity in the world 2005. Economic and social department, Food and Agriculture Organization of the United Nations, Rome, Italy. http://www.fao.org/ docrep/008/a0200e/a0200e00.htm. Accessed on 30 May 2013.

FAO. 2009. High level expert forum—How to feed the world in 2050. Rome: Food and Agriculture Organization of the United Nations. http://www.fao.org/fileadmin/templates/wsfs/docs/expert_paper/How_to_Feed_the_World_in_2050.pdf.

FAO. 2010. The State of Food Insecurity in the World-Addressing food insecurity in protracted crises. Food and Agriculture Organization of the United Nations, Rome.

Fiedler, A. K. and Landis, D. A. 2007a. Attractiveness of Michigan native plants to arthropod natural enemies and herbivores. *Environ Entomol.,* 36:751–765.

Fiedler, A. K. and Landis, D. A. 2007b. Plant characteristics associated with natural enemy abundance at Michigan native plants. *Environ Entomol.*, 36:878–886.

Food and Agriculture Organisation of the United Nations. FAOSTAT. (2010). http://faostat.fao. org/site/424/default.aspx#%23ancor. Accssesed on 15 May 2013.

Garibaldi, L. A. 2015. Pollination and food security: win-win scenarios between pollinator diversity and crop yield for smallholdings worldwide. 44th APIMONDIA International Apicultural Congress, 16-20 September, 2015, Daejeon, South Korea.

Gupta, P. K. 2004. Pesticide exposure — Indian scene. *Toxicology*, 198: 83–90.

Hallman, W. K., Hebden, W. C., Aquino, H. L., Cuite, C. L. and Lang, J. T. 2003. Public perceptions of genetically modified foods: a national study of American knowledge and opinion. Publication no. 1003- 004. http://ideas.repec.org/p/ags/rutfwp/18174.html

Kaul, V., Shankar, U. and Khushu, M. 2009. Bio-intensive Integrated Pest Management in fruit Crop Ecosystem. *In:* Integrated Pest Management: Innovation Development Process Vol-1, 2009. Peshin, R. and Dhawan, A. K. (eds.), Springer Verlag. pp. 631–666. DOI 10.1007/978-1-4020-8992-3 21.

Kennedy, G. G. and Shutton, T. B. 2000. Emerging Technologies for Integrated Pest Management: Concept Research and Implementation. APS Press, St Paul, Minnesota.

Khush, G. S. 2002. The promise of biotechnology in addressing current nutritional problems in developing countries. *Food Nutr. Bull.*, 23(4):354–357.

Kogan, M. 1998. Integrated Pest Management: Historical Perspectives and Contemporary Developments. *Annual Review of Entomology*, 43:243–270.

Landis, D. A., Wratten, S. D. and Gurr, G. M. 2000. Habitat management to conserve natural enemies of arthropod pests in agriculture. *Annual Review of Entomology,* 45: 175–201.

Lewis, W. J., van Lenteren, J. C., Phatak, S. C. and Tumlinson, J. H. II. 1997. A total system approach to sustainable pest management. *Proceedings of the National Academy of Sciences USA,* 94: 12243–12248.

Lowel, J. F. 1998. Producing food without pesticides. C.W.S. New York, pp 1–5.

MAFF. 1999-2000. Annual Report of Working Party on Pesticide Residues 1998/1999, Health and Safety Executive, MAFF, London, 1999–2000.

Meera, S. N., Jhamtani, A. and Rao, D.U.M. 2004. Information and communication technology in agricultural development: a comparative analysis of three projects from India. AgREN Network Paper No.135, ODI, January 2004. 20p. Available at: www.odi.org.uk/agren/papers/agrenpaper_135.pdf

Michelle, W. and Fong, L. 2009. Digital divide: the case of developing countries. *Issues in Informing Science and Information Technology*, 471–478.

Miller, J. R., McGhee, P. S., Siegert, P. Y., Adams, C. G., Huang, J., Grieshop, M. J. and Gut, L. J. 2010. General principles of attraction and competitive attraction as revealed by large-cage studies of moths responding to sex pheromone. *Proceedings of the National Academy of Sciences USA,* 107: 22–27.

Oomen, P. A. and Bouma, E. 2003. Role of Pesticides in IPM. Wageningen, Netherlands: Plant Protection Service.

Overton, J. 1996. Ecologically Based Pest Management: New Solutions for a New Century. National Academy Press, Washington, DC.

Peshin, R. and Pimentel, D. (eds.) 2014. Integrated Pest Management Experiences with Implementation, Global Overview, Vol. 4. Springer Science+Business Media. p. 574.

Peshin, R., Kranthi, K. R. and Sharma, R. 2014. Pesticide use and experiences with Integrated Pest Management programs and *Bt* Cotton in India. pp. 269–306, *In*: Peshin, R. and Pimentel, D. (eds.), Integrated Pest Management-Experiences with Implementation, Global Overview, Vol. 4, Springer Science+Business Media Dordrecht, 2014.

Pimentel, D., McLaughlin, L., Zepp, A., Lakitan, B., Kraus, T., Kleinman, P., Vancini, F., Roach, W. J., Graap, E., Keeton, W. S. and Selig, G. 1991. Environmental and economic impacts of reducing U.S. agricultural pesticide use. *In*: Pimentel, D. (Ed.), Handbook of pest management in agriculture (Vol. I, 2nd ed.). Boca Raton, Florida: CRC Press.

Pina, C. M. and Forcada, S. A., 2004. Effects of an environmental tax on pesticides in Mexico. *Industry and Environment*, 27: 34 –38.

Pingali, P.L. and Traxler, G. 2002. "Changing locus of agricultural research: will the poor benefit from biotechnology and privatization trends," *Food Policy*, 27:223–238.

Shankar, U. and Abrol, D. P. 2012. Role of Integrated Pest Management in Food and Nutritional Security. pp. 408-432. *In*: Integrated Pest Management: Principles and Practices (eds. Abrol and Shankar) CAB International 2012, UK. ISBN : 9781845938088.

Shen, Zuo-Rui., Cheng, Deng-Fa., Zhang, Yun-Hui. and Gao, Ling-Wang. 2012. Role of Information and Communication Technology in Integrated Pest Management. *In*: Integrated Pest Management (eds Abrol, D.P. and Shankar, U.) CABI.

Soundararajan, R. P. (ed.) 2012. Pesticides - Advances in Chemical and Botanical Pesticides. In: Tech Janeza Trdine, Rijeka, Croatia. http://dx.doi.org/10.5772/2609. Accessed 27 March 2013.

Tuell, J. K., Fiedler, A. K., Landis, D. A. and Isaacs, R. 2008. Visitation by wild and managed bees (Hymenoptera: Apoidea) to Eastern U.S. native plants for use in conservation programs. *Environ Entomol.*, 37:707–718.

Tyler, M. G. Jr. 2002. *Living in the Environment: Principles, Connections and Solutions, 12th Edition.* Wadsworth/Thomson Learning, Belmont, California, USA.

UNFPA. 2008. Population and Climate Change- Framework of UNFPA's Agenda. United Nations Population Fund (UNFPA). http://www.unfpa.org/pds/climate/docs/climate_change_unfpa.pdf.

Vega, F. E., Infante, F., Castillo, A. and Jaramillo, J. 2009. The coffee berry borer, *Hypothenemus hampei* (Ferrari) (Coleoptera: Curculionidae): a short review, with recent findings and future research directions. *Terr. Arthropod Rev.*, 2: 129 –147.

Viscarret, M. M., Torres-Jerez, I., Agostini de Manero, E., López, S. N., Botto, E. E. and Brown, Y. J. K. 2003. Mitochondrial DNA evidence for a distinct new world group of *Bemicia tabaci* (Gennadius) (Hemiptera: Aleyrodidae) indigenous to Argentina and Bolivia and presence of the old world B biotype in Argentina. *Ann. Entomol. Soc. Am.*, 96: 65–72.

Wilson, C. and Tisdell, C. 2001. Why farmers continue to use pesticides despite environmental, health and sustainability costs. *Ecological Economics*, 39:449-462.

CITATION INDEX